Lecture Notes in Control and Information Sciences

Volume 462

AF789576

About this Series

This series aims to report new developments in the fields of control and information sciences—quickly, informally and at a high level. The type of material considered for publication includes:

1. Preliminary drafts of monographs and advanced textbooks
2. Lectures on a new field, or presenting a new angle on a classical field
3. Research reports
4. Reports of meetings, provided they are
    (a) of exceptional interest and
    (b) devoted to a specific topic. The timeliness of subject material is very important.

More information about this series at http://www.springer.com/series/642

Madhu N. Belur · M. Kanat Camlibel
Paolo Rapisarda · Jacquelien M.A. Scherpen
Editors

Mathematical Control Theory II

Behavioral Systems and Robust Control

Springer

Editors
Madhu N. Belur
Department of Electrical Engineering
Indian Institute of Technology
Mumbai
India

M. Kanat Camlibel
Johann Bernoulli Institute for Mathematics and Computer Science
University of Groningen
Groningen
The Netherlands

Paolo Rapisarda
Department of Electronics and Computer Science
University of Southampton
Southampton
UK

Jacquelien M.A. Scherpen
Engineering and Technology institute Groningen
University of Groningen
Groningen
The Netherlands

ISSN 0170-8643 ISSN 1610-7411 (electronic)
Lecture Notes in Control and Information Sciences
ISBN 978-3-319-21002-5 ISBN 978-3-319-21003-2 (eBook)
DOI 10.1007/978-3-319-21003-2

Library of Congress Control Number: 2015942816

Mathematics Subject Classification: 34H05, 34H15, 47N70, 70Q05, 93B05, 93B52, 93C05, 93C10, 93C15, 93C35

Springer Cham Heidelberg New York Dordrecht London

Printed on acid-free paper

Springer International Publishing AG Switzerland is part of Springer Science+Business Media
(www.springer.com)

Foreword

This volume is offered to Prof. Dr. Harry L. Trentelman in celebration of his birthday. It contains papers by his collaborators, including a number of former Ph.D. students and postdoctoral fellows.

This is the second of a series of two books, appearing in connection with the workshop "Mathematical systems theory: from behaviors to nonlinear control" dedicated to the 60th birthdays of Arjan van der Schaft and Harry Trentelman, both at the Johann Bernoulli Institute for Mathematics and Computer Science and at the Jan C. Willems Center for Systems and Control at the University of Groningen.

Preface

I mean, what is an un-birthday present?
A present given when it isn't your birthday, of course.

(L. Carroll, Through the Looking-Glass,
and What Alice Found There.
Chapter VI: Humpty Dumpty)

To those who are familiar with him since many decades, the scientist and human being Harry L. Trentelman displays a remarkable *almost invariance property*. For as long as we have known him, the adjectives that best continue to describe Harry's attitude to science and academic work are: resilient, creative, energetic, hard-working. When one is almost giving up trying to prove some result and is ready to add yet another assumption to make life a bit easier, Harry always comes to the rescue (typically the next morning, after a night of hard thinking) with a simple and effective solution. In science, Harry is also rigorous, original, open-minded, and skeptical: we fondly recall his ability to shoot down our most cherished pet theories with one sharp question, and his tireless efforts in correcting our sloppiness in thinking or writing. One should add that Harry has always been the epitome of scientific integrity, that he has consistently displayed an excellent taste in his choice of problems and in finding elegant solutions to them, and that he is an ambitious, kind, helpful, conscientious, and truly educational supervisor to his Master and Ph.D. students. All these qualities have brought Harry to make pioneering and lasting scientific contributions to areas as diverse as geometric control theory, the behavioral approach, and most recently systems over networks. The excellency of his work has been officially recognized in his recent nomination as IEEE Fellow and Senior Editor for the IEEE Transactions on Automatic Control.

The adjectives that come to mind when trying to describe his personality are: charming, sociable, enthusiastic, optimistic, witty. His particular brand of humor would require far better writers than us to be described adequately; suffice it to say that nobody who has sat next to him at conference dinners can easily forget his cheeky, brilliant one-liners. All of those who know him even tangentially remember his outspoken passion for sports and outdoor activities, his appreciation of good music, his love for and involvement in his family, and his taste for dancing.

Here is a man who loves life so unconditionally as to be an example to all who meet him. He does not philosophise about how to live life fully; he shows one how to do it in practice. What then, one could ask, does such a man deserve as a gift for the anniversary of almost threescores years of a life? What could be the best token of our appreciation for him as a man and a scientist?

As his friends and former students we pondered long over the obvious choices. A luxurious penthouse in Manhattan? A shiny Lamborghini? A stately Tuscan villa? After much deliberation we discarded all these options: none was good enough to really translate our affection for and esteem of Harry (and given his unswerving devotion to Volvo family cars, he would not have liked the Lamborghini anyway).

Almost driven to desperation by our inability to make our feelings concrete, we finally found the perfect gift: a collection of scientific papers written specifically for this occasion by friends, colleagues, and former students! The confirmation that this was the right idea was the enthusiasm with which it was accepted by the contributors to this volume, who quickly produced the high-quality works gathered in this volume.

What you are holding in your hands then, Harry, is the best "sixtieth un-birthday" gift we could think of. We present it to you with affection, admiration, and our best wishes for several decades more of top-level scientific productivity.

Mumbai, India — Madhu N. Belur
Groningen, The Netherlands — M. Kanat Camlibel
Southampton, UK — Paolo Rapisarda
Groningen, The Netherlands — Jacquelien M.A. Scherpen
May 2015

Contents

Contributors

A.C. Antoulas Department of Electrical and Computer Engineering, Rice University, Houston, TX, USA; School of Engineering and Science, Jacobs University Bremen, Bremen, Germany

Madhu N. Belur Department of Electrical Engineering, Indian Institute of Technology Bombay, Mumbai, India

Chayan Bhawal Department of Electrical Engineering, Indian Institute of Technology Bombay, Mumbai, India

M.K. Camlibel Johann Bernoulli Institute for Mathematics and Computer Science, University of Groningen, Groningen, The Netherlands

A.R.F. Everts Johann Bernoulli Institute for Mathematics and Computer Science, University of Groningen, Groningen, The Netherlands

Paul A. Fuhrmann Department of Mathematics, Ben Gurion University of the Negev, Beer Sheva, Israel

Uwe Helmke Institute of Mathematics, University of Würzburg, Würzburg, Germany

Osamu Kaneko Institute of Science and Engineering, Kanazawa University, Kanazawa, Ishikawa, Japan

Sandeep Kumar Department of Electrical Engineering, Indian Institute of Technology Bombay, Mumbai, India

Nima Monshizadeh Faculty of Mathematics and Natural Sciences, Engineering and Technology Institute Groningen, University of Groningen, Ag Groningen, The Netherlands

Diego Napp Department of Mathematics, CIDMA—Center for Research and Development in Mathematics and Applications, University of Aveiro, Campus Universitario de Santiago, Aveiro, Portugal

Debasattam Pal Department of Electrical Engineering, Indian Institute of Technology Bombay, Mumbai, India

Harish K. Pillai Department of Electrical Engineering, IIT Bombay, Mumbai, India

P. Rapisarda Vision, Learning and Control Group, School of Electronics and Computer Science, University of Southampton, Southampton, UK

Paula Rocha Faculty of Engineering, Department of Electrical and Computer Engineering, SYSTEC—Research Center for Systems and Technologies, University of Porto, Porto, Portugal

Ali Saberi School of Electrical Engineering and Computer Science, Washington State University, Pullman, WA, USA

A. Sanand Amita Dilip Department of Electrical Engineering, IIT Bombay, Mumbai, India

Shiva Shankar Chennai Mathematical Institute, Chennai (Madras), India

Anton A. Stoorvogel Department of Electrical Engineering, Mathematics and Computer Science, University of Twente, Enschede, The Netherlands

Kiyotsugu Takaba Department of Electrical and Electronic Engineering, Ritsumeikan University, Kusatsu, Shiga, Japan

Jochen Trumpf Research School of Engineering, Australian National University, Canberra, Australia

Arjan van der Schaft Jan C. Willems Center for Systems and Control, Johann Bernoulli Institute for Mathematics and Computer Science, University of Groningen, Groningen, The Netherlands

Jacob van der Woude Delft Institute of Applied Mathematics, Delft University of Technology, Delft, The Netherland

Meirong Zhang School of Electrical Engineering and Computer Science, Washington State University, Pullman, WA, USA

Chapter 1
Open Loop Control of Higher Order Systems

Paul A. Fuhrmann and Uwe Helmke

Abstract In this paper we solve the problem of computing open loop controls that steer a higher order system from the equilibrium to a prescribed state, or partial state. Using unimodular embeddings of coprime polynomial factorizations, we derive an explicit formula for the inverse of the reachability map. Proceeding along the same lines we outline the connection to flat outputs and develop an independent approach for open loop control.

1.1 Introduction

In the pioneering work of Kalman [6], the basic concepts of reachability and observability for linear state space systems

$$\begin{aligned} x_{t+1} &= Ax_t + Bu_t \\ y_t &= Cx_t + Du_t \end{aligned} \tag{1.1}$$

were introduced in a module theoretic framework. The modeling of electrical or mechanical networks led in a natural way to higher order polynomial system representations

$$\begin{aligned} T(\sigma)x &= U(\sigma)u \\ y &= V(\sigma)x + W(\sigma)u, \end{aligned} \tag{1.2}$$

a class of systems, for which no a priori state space has been available; see Rosenbrock [9]. Fuhrmann [3] succeeded to associate with any system (1.2) a canonical state

P.A. Fuhrmann
Department of Mathematics, Ben Gurion University of the Negev, Beer Sheva, Israel
e-mail: fuhrmannbgu@gmail.com

U. Helmke (✉)
Institute of Mathematics, University of Würzburg, 97074 Würzburg, Germany
e-mail: helmke@mathematik.uni-wuerzburg.de

M.N. Belur et al. (eds.), *Mathematical Control Theory II*,
Lecture Notes in Control and Information Sciences 462,
DOI 10.1007/978-3-319-21003-2_1

space, the polynomial (or rational) model of the nonsingular polynomial matrix $T(z)$. This led to representation free definitions of reachability and observability for higher order systems in terms of the associated shift realization. Proceeding to a higher level of abstraction, Jan C. Willems, see [8], introduced the theory of behaviors to analyze higher order systems of the form

$$R(\sigma)w = 0 \tag{1.3}$$

where $R(z)$ denotes a rectangular polynomial system matrix. In this context, reachability and observability are defined in terms of spaces of trajectories, i.e., the behavior of (1.3). For all underlying representations of a linear system the fundamental question of reachability is: Can every state be reached from the zero state by a suitable control sequence? A more important question is that of constructive reachability: Given a state, or a partial state, how can one compute control sequences that steer the system from rest to the desired state? This problem is very closely related to the motion planning problem, i.e., the goal of designing desired system trajectories that join two given states.

In this paper we will take a middle ground and focus our analysis on higher order Rosenbrock type systems (1.2). As the **reachability map** of the state space system (1.1) maps controls to states, the natural approach for constructive reachability is to invert it. However, even if reachability of the given system is assumed, the reachability map is surjective but not injective, so it has no regular inverse. In order to obtain a regular inverse, we need to factor out its kernel and restrict the map to the associated quotient space. Moreover, the reachability map is a homomorphism over the ring of polynomials $\mathbb{F}[z]$ and the **reduced reachability map** can be inverted by embedding an intertwining relation in a doubly coprime factorization. This leads to an explicit formula for the inverse of the reduced reachability map that, strangely, seems to be new. Each embedding in a unimodular matrix is tantamount to the addition of a special output, which following [7], we refer to as a **flat output**. The concept of flat outputs was first introduced by [1] in connection with state feedback linearizations as a tool for control of higher order nonlinear systems. Flat outputs turn out to be a useful tool in the solution of the terminal state problem and we proceed to a rather complete analysis of flat systems using doubly coprime factorizations. Of course, the use of coprime factorizations for analyzing flatness of higher order systems is not new. While Levine and Nguyen [7] have studied a restricted class of systems (1.2) where $U(z)$ is a constant full column rank matrix, Trentelman [10] characterized flat outputs for arbitrary behaviors. Our approach, that links flatness to the study of the reachability map, though seems to be new.

This paper is dedicated to Harry Trentelman, a wonderful colleague and friend, whose elegant work [10] has been a major source of inspiration for us. Of course it is not the only paper by him that we admire! While all three of us had for long been involved in developing linear systems theory, it may be interesting to note that our current interests shifted almost simultaneously to networks; [4, 11]. We are convinced that methods from algebraic systems theory, such as, e.g., flatness and open loop control, are bound to play an increasingly important role for observing and controlling networks of systems. The future thus looks bright and we look forward to further exciting work by Harry and to continue learning from him.

1.2 Functional Models and the Shift Operator

In this section, following [2, 3], we summarize the basic results of the theory of polynomial models and introduce the shift realization. Proofs can be found in the recent monograph [5]. Throughout this paper, $\mathbb{F}$ denotes an arbitrary field.

1.2.1 Polynomial and Rational Models

Polynomial models are concrete representations of quotient modules $\mathbb{F}[z]^m/D(z)\mathbb{F}[z]^m$, defined for nonsingular polynomial matrices $D(z) \in \mathbb{F}[z]^{m\times m}$. Explicitly, the polynomial model associated with $D(z)$ is defined as the $\mathbb{F}$—vector space

$$X_D = \{f \in \mathbb{F}[z]^m \mid D(z)^{-1}f(z) \quad \text{is strictly proper}\}.$$

In contrast, the rational model of D is the $\mathbb{F}$—vector space defined as

$$X^D = \{h \in z^{-1}\mathbb{F}[[z^{-1}]]^m \mid D(z)h(z) \quad \text{is a polynomial}\}.$$

It is easily seen that X_D and X^D are both finite dimensional $\mathbb{F}$—vector spaces of dimension $\dim X_D = \dim X^D = \deg \det D$. We note that $f(z) \in X_D$ holds if and only if $D(z)^{-1}f(z) \in X^D$. To introduce a module structure on these spaces we proceed as follows. Let $\mathbb{F}((z^{-1}))^m$ denote the vector space of truncated Laurent series, i.e., $f(z) = \sum_{j=-\infty}^{n_f} f_j z^j$, $f_j \in \mathbb{F}^m$. Thus, f_{-1} denotes the residue of $f(z)$. We denote the canonical projections onto the strictly proper and polynomial parts, respectively, by $\pi_- : \mathbb{F}((z))^m \longrightarrow z^{-1}\mathbb{F}[[z^{-1}]]^m$ and $\pi_+ : \mathbb{F}((z))^m \longrightarrow \mathbb{F}[z]^m$. The backward shift $\sigma : z^{-1}\mathbb{F}[[z^{-1}]]^m \longrightarrow z^{-1}\mathbb{F}[[z^{-1}]]^m$ acts on vectors of strictly proper functions via

$$\sigma(h) = \pi_-(zh(z)), \tag{1.4}$$

i.e., as $\sigma(\sum_{j=1}^{\infty} h_j z^{-j}) = \sum_{j=1}^{\infty} h_{j+1}z^{-j}$. Defining a projection map $\pi_D : \mathbb{F}[z]^m \longrightarrow \mathbb{F}[z]^m$ by

$$\pi_D f = D\pi_-(D^{-1}f), \qquad f \in \mathbb{F}[z]^m,$$

we obtain the isomorphism

$$X_D = \operatorname{Im} \pi_D \simeq \mathbb{F}[z]^m / D(z)\mathbb{F}[z]^m,$$

which gives a concrete, but noncanonical, representation for the quotient module. The **shift operator** $S_D : X_D \longrightarrow X_D$ is defined by

$$S_D f = \pi_D(zf) = zf - D(z)\xi_f, \qquad f \in X_D,$$

where $\xi_f = (D^{-1}f)_{-1}$. A special case of interest is provided by the matrix pencil of an $n \times n$—matrix A. It is an elementary observation that the shift operator on X_{zI-A} is conjugate to the matrix A. This is the starting point for the polynomial model state-space realization theory.

The polynomial model X_D becomes an $\mathbb{F}[z]$-module by using the S_D-induced module structure, i.e.,

$$p \cdot f = \pi_D(pf), \qquad p \in \mathbb{F}[z], f \in X_D.$$

Similarly, the rational model X^D is endowed with the $\mathbb{F}[z]$-module structure

$$p \cdot h = p(\sigma)h = \pi_-(ph), \qquad p(z) \in \mathbb{F}[z],\ h(z) \in X^D.$$

It is easily seen that the linear map

$$X_D \longrightarrow X^D, \quad f \mapsto D^{-1}f$$

is an isomorphism of $\mathbb{F}[z]$—-modules. We next relate coprimeness to the very important concept of doubly coprime factorizations. Let

$$G(z) = P(z)Q(z)^{-1} = T(z)^{-1}U(z)$$

be a right and left coprime factorization of $G(z) \in \mathbb{F}(z)^{p\times m}$, respectively. This implies the **intertwining relation**

$$U(z)Q(z) = T(z)P(z). \tag{1.5}$$

The next result, see [5], characterizes intertwining relations in terms of unimodular embeddings.

Theorem 1.1 (Doubly Coprime Factorization) *Let $U(z) \in \mathbb{F}[z]^{p\times m}$, $T(z) \in \mathbb{F}[z]^{p\times p}$ be right coprime and $P(z) \in \mathbb{F}[z]^{p\times m}$, $Q(z) \in \mathbb{F}[z]^{m\times m}$ be left coprime with*

$$T(z)P(z) = U(z)Q(z).$$

Then, there exist unique polynomial matrices $V(z) \in \mathbb{F}[z]^{m\times p}$, $\overline{V}(z) \in \mathbb{F}[z]^{m\times p}$, $W(z) \in \mathbb{F}[z]^{m\times m}$, $\overline{W}(z) \in \mathbb{F}[z]^{p\times p}$ *with*

$$\begin{pmatrix} T(z) & -U(z) \\ V(z) & W(z) \end{pmatrix} \begin{pmatrix} \overline{W}(z) & P(z) \\ -\overline{V}(z) & Q(z) \end{pmatrix} = \begin{pmatrix} I_p & 0 \\ 0 & I_m \end{pmatrix},$$

such that $V(z)T(z)^{-1}$ *and* $Q(z)^{-1}\overline{V}(z)$ *are strictly proper.*

1.2.2 The Shift Realizations

The following result from Fuhrmann [2, 3] is central as it allows us to write down a canonical state space realization, the so-called **shift realization** that is associated with any coprime polynomial matrix factorization of a proper transfer function. This establishes a canonical link between state space methods and polynomial system matrices. It implies particularly that any state space realization (A, B, C) can be regarded as the shift realization on the polynomial model space X_{zI-A}.

Theorem 1.2 (Shift Realization) *Consider any polynomial system matrix*

$$\mathscr{P} = \begin{pmatrix} T(z) & -U(z) \\ V(z) & W(z) \end{pmatrix} \in \mathbb{F}[z]^{(r+p)\times(r+m)}$$

with $T(z) \in \mathbb{F}[z]^{r\times r}$ *nonsingular. Let* $G(z)$ *be the associated* $p \times m$ *rational transfer function defined as*

$$G(z) = V(z)T(z)^{-1}U(z) + W(z). \tag{1.6}$$

Assume that G *is proper with expansion* $G(z) = G_0 + \sum_{i=1}^{\infty} G_i z^{-i}$. *Then*

1. *The system defined, in the state space* X_T, *by the quadruple of maps* A, B, C, D, *with* $A : X_T \longrightarrow X_T$, $B : \mathbb{F}^m \longrightarrow X_T$, $C : X_T \longrightarrow \mathbb{F}^p$ *and* $D : \mathbb{F}^m \longrightarrow \mathbb{F}^p$, *by*

$$\Sigma_{VT^{-1}U+W} := \begin{cases} Af = S_T f & f \in X_T \\ B\xi = \pi_T(U\xi), & \xi \in \mathbb{F}^m \\ Cf = (VT^{-1}f)_{-1} & f \in X_T \\ D = G_0 \end{cases} \tag{1.7}$$

   is a realization of $G(z)$. *We will refer to (1.7) as the* **shift realization**.
2. *The realization is observable if and only if* $V(z)$ *and* $T(z)$ *are right coprime and reachable if and only if* $T(z)$ *and* $U(z)$ *are left coprime.*

3. *The reachability and observability maps of the realization (1.7) are given by*

$$\mathscr{R}u = \pi_T(Uu), \qquad u \in \mathbb{F}[z]^m$$

and

$$\mathscr{O}f = \pi_-(VT^{-1}f), \qquad f \in X_T.$$

1.3 Open Loop Control for Shift Realizations

We present a simple approach to the terminal state problem for the shift realization of higher order systems from the point of view of inverting the reachability map. This leads directly to the problem of unimodular embedding and hence, indirectly, to the study of flat outputs. Our explicit formula for the inverse of the reachability map seems to be new and is of independent interest.

1.3.1 State Space Representations

We begin our investigation with a system given in state space form as

$$x_{t+1} = Ax_t + Bu_t. \tag{1.8}$$

We assume that $(A, B) \in \mathbb{F}^{n\times n} \times \mathbb{F}^{n\times m}$ is a reachable pair and that the system has been at rest till time $t = -\tau$. Given a prescribed state $x_* \in \mathbb{F}^n$, our aim is to compute control sequences $u_{-\tau}, u_{-\tau+1}, \dots, u_{-1}$ that steer the system from the origin to the state x_* at time $t = 0$. This leads to the equation

$$x_* = \sum_{i=1}^{\tau} A^{i-1}Bu_{-i}.$$

By our assumption on the reachability of the pair (A, B), this equation is solvable for all $\tau \geq n$. The associated polynomial $u(z) = \sum_{j=0}^{\tau-1} u_{-j-1}z^j$ is called the input polynomial.

To arrive at an algebraic formulation of the problem, we identify $\mathbb{F}^n$, endowed with the $\mathbb{F}[z]$-module structure induced by A, with the polynomial model X_{zI-A}. Next, we recall the definition of the **reachability map** $\mathscr{R}_{(A,B)} : \mathbb{F}[z]^m \longrightarrow X_{zI-A}$, by

$$\mathscr{R}_{(A,B)} \sum_{i=0}^{s} u_i z^i = \pi_{zI-A} B \sum_{i=0}^{s} u_i z^i = \sum_{i=0}^{s} A^i Bu_i, \qquad \sum_{i=0}^{s} u_i z^i \in \mathbb{F}[z]^m, \tag{1.9}$$

or, equivalently, by

$$\mathscr{R}_{(A,B)}u = \pi_{zI-A}Bu, \qquad u(z) \in \mathbb{F}[z]^m. \tag{1.10}$$

The reachability map is an $\mathbb{F}[z]$-module homomorphism. Clearly, to compute controls that steer to a state x_*, one has to invert the reachability map $\mathscr{R}_{(A,B)}$. We note that $\mathscr{R}_{(A,B)}$ has a large kernel, which is a full submodule of $\mathbb{F}[z]^m$, hence is representable as $Q\mathbb{F}[z]^m$ for a nonsingular $Q(z) \in \mathbb{F}[z]^{m\times m}$. To get an invertible map, we factor out the kernel. Denote by $\mathscr{R}$ the **reduced reachability map**, namely the map induced by $\mathscr{R}_{(A,B)}$ on $\mathbb{F}[z]^m/Q\mathbb{F}[z]^m$, which we identify with the polynomial model X_Q. Thus $\mathscr{R} : X_Q \longrightarrow X_{zI-A}$ is given by

$$\mathscr{R}u = \pi_{zI-A}Bu, \qquad u(z) \in X_Q. \tag{1.11}$$

To determine the polynomial matrix Q that occurs in the reduced reachability map, let

$$P(z)Q(z)^{-1} = (zI - A)^{-1}B \tag{1.12}$$

be a right coprime factorization of the transfer function $(zI - A)^{-1}B$, with $Q(z) \in \mathbb{F}[z]^{m\times m}$ nonsingular and $P(z) \in \mathbb{F}[z]^{n\times m}$. By Theorem 1.1, the intertwining relation

$$BQ(z) = (zI - A)P(z) \tag{1.13}$$

can be embedded in the doubly coprime factorization

$$\begin{pmatrix} zI - A & -B \\ V(z) & W(z) \end{pmatrix} \begin{pmatrix} \overline{W}(z) & P(z) \\ -\overline{V}(z) & Q(z) \end{pmatrix} = \begin{pmatrix} I_n & 0 \\ 0 & I_m \end{pmatrix} \tag{1.14}$$

with $V(z)(zI - A)^{-1}$ and $Q(z)^{-1}\overline{V}(z)$ strictly proper. In particular, $C := V(z)$ is a constant matrix. We will see later that the output equation

$$y_t = Cx_t$$

defines a flat output of (1.1). The next result lists basic properties of the reduced reachability map and gives an explicit formula for the inverse.

Theorem 1.3 *Let $P(z)Q(z)^{-1}$ be a right coprime factorization of $(zI - A)^{-1}B$ and let (A, B) be reachable.*

1. The reachability map $\mathscr{R}_{(A,B)}$ is an $\mathbb{F}[z]$-homomorphism with kernel

$$\operatorname{Ker} \mathscr{R}_{(A,B)} = Q(z)\mathbb{F}[z]^m. \tag{1.15}$$

The reachability map induces the isomorphism

$$\mathscr{R} : X_Q \longrightarrow X_{zI-A}, \quad \mathscr{R}(u) = \pi_{zI-A} Bu. \tag{1.16}$$

2. *Given a state $x_* \in \mathbb{F}^n$, there exists a unique input polynomial $u_{min}(z) \in X_Q$ that steers the system from the zero state to x_* at time $t = 0$. The associated input sequence is given by the reverse coefficients of $u_{min}(z) = \mathscr{R}^{-1} x_*$. Specifically, in terms of the doubly coprime factorization (1.14),*

$$u_{min}(z) = \mathscr{R}^{-1} x_* = \pi_Q \overline{V} x_*. \tag{1.17}$$

3. *An arbitrary solution $u_*(z)$ to the steering problem is given by*

$$u_*(z) = u_{min}(z) + Q(z) g(z), \tag{1.18}$$

with $g(z) \in \mathbb{F}[z]^m$.

Proof Note that $\pi_{zI-A}(Bu) = 0$ if and only $(zI - A)^{-1} Bu(z) = P(z)Q(z)^{-1}u(z)$ is a polynomial. Using the coprimeness of both sides of (1.12), it is easily seen that the polynomial matrix $P(z)$ is right prime, i.e., there exists a polynomial matrix $M(z)$ with $M(z)P(z) = I_m$. Thus $P(z)Q(z)^{-1}u(z)$ is a polynomial if and only if $Q(z)^{-1}u(z)$ is a polynomial, i.e., if and only if $u \in Q(z)\mathbb{F}[z]^m$. This proves the first part.

From the doubly coprime factorization (1.14), one obtains the Bezout equation

$$(zI - A)\overline{W}(z) + B\overline{V}(z) = I_n$$

and therefore

$$(zI - A)^{-1} B\overline{V}(z) = (zI - A)^{-1} - \overline{W}(z).$$

By strict properness of $Q(z)^{-1}\overline{V}(z)$, we compute

$$\mathscr{R}\pi_Q(\overline{V}x_*) = \pi_{zI-A}\left(B\pi_Q(\overline{V}x_*)\right) = (zI - A)\pi_-\left((zI - A)^{-1}B\overline{V}x_*\right).$$

This proves

$$\mathscr{R}\pi_Q(\overline{V}x_*) = (zI - A)\pi_-\left((zI - A)^{-1}x_*\right) = x_*$$

and verifies the second claim. The last assertion follows from (1.15). ■

1.3.2 High Order System Representations

In many situations, the system under consideration is given not in state space terms, but rather by difference equations of higher order. Thus, assuming $\xi(z), u(z)$ to be strictly proper Laurent series, the system is given by an equation of the form

$$T(\sigma)\xi(z) = U(\sigma)u(z), \tag{1.19}$$

with σ the backward shift, $T(z) \in \mathbb{F}[z]^{r\times r}$ nonsingular and $U(z) \in \mathbb{F}[z]^{r\times m}$. We assume that $T(z)^{-1}U(z)$ is strictly proper, i.e., that (1.19) represents a strictly causal system. Using the shift realization (1.7), we can associate with the system equation (1.19) a natural state space realization

$$x_{t+1} = Ax_t + Bu_t, \tag{1.20}$$

defined on the polynomial model X_T as the state space. Here

$$A = S_T, \quad B = \pi_T(U\cdot).$$

Our goal is to extend Theorem 1.3 to the present situation, i.e., to compute controls that steer the system (1.20) from the zero state to an arbitrary state $f(z)$ in the state space X_T. By the Shift Realization Theorem, the pair (A, B) is reachable if and only if $T(z), U(z)$ are left coprime. Moreover, the **reachability map** of the shift realization (1.20) is given as

$$\mathscr{R}_{A,B} : \mathbb{F}[z]^m \longrightarrow X_T, \quad \mathscr{R}_{A,B}u = \pi_T(Uu).$$

As in Theorem 1.3 one shows that the kernel of the reachability map is a full submodule $Q(z)\mathbb{F}[z]^m$. Therefore $\mathscr{R}_{A,B}$ induces the **reduced reachability map** $\mathscr{R} : X_Q \longrightarrow X_T$ by $\mathscr{R}u = \pi_T Uu$. It defines an isomorphism of $\mathbb{F}[z]$ modules.

Proceeding as before we note that there exist right coprime polynomial matrices $Q(z) \in \mathbb{F}[z]^{r\times r}$, $P(z) \in \mathbb{F}[z]^{r\times m}$ that satisfy the intertwining relation:

$$T(z)P(z) = U(z)Q(z). \tag{1.21}$$

Without loss of generality, $Q(z)$ can be chosen column proper with column degrees $\mu_1 \geq \cdots \geq \mu_m \geq 0$. Note further that the intertwining relation (1.21) can be embedded in the following doubly coprime factorization:

$$\begin{pmatrix} T(z) & -U(z) \\ V(z) & W(z) \end{pmatrix} \begin{pmatrix} \overline{W}(z) & P(z) \\ -\overline{V}(z) & Q(z) \end{pmatrix} = \begin{pmatrix} I_r & 0 \\ 0 & I_m \end{pmatrix}, \tag{1.22}$$

such that $V(z)T(z)^{-1}$ and $Q(z)^{-1}\overline{V}(z)$ are strictly proper. In that case, the pair $(V(z), T(z))$ induces, by way of the shift realization, an observable pair in the state space X_T. The following theorem, the counterpart of Theorem 1.3, summarizes the main results.

Theorem 1.4 *Let* $(T(z), U(z)) \in \mathbb{F}[z]^{r\times r} \times \mathbb{F}[z]^{r\times m}$ *be left coprime polynomial matrices with* $T(z)^{-1}U(z)$ *strictly proper. Then*

1. *The kernel of the reachability map* $\mathscr{R}_{A,B} : \mathbb{F}[z]^m \longrightarrow X_T$ *is* $Q(z)\mathbb{F}[z]^m$. *The reachability map induces the* $\mathbb{F}[z]$ *module isomorphism*

$$\mathscr{R} : X_Q \longrightarrow X_T, \quad \mathscr{R}(u) = \pi_T(Uu).$$

2. *There exists a unique control sequence whose input polynomial is in* X_Q *and that steers the shift realization (1.20) from the zero state to* $f(z) \in X_T$. *It is given by the (reversed) coefficients of* $u_{min}(z) = \mathscr{R}^{-1}f$. *Specifically, in terms of the doubly coprime factorization (1.22),*

$$u_{min}(z) = \mathscr{R}^{-1}f = \pi_Q \overline{V} f. \tag{1.23}$$

3. *An arbitrary control* $u_*(z)$ *steers the shift realization (1.20) from the zero state to* $f(z)$ *at time* $t = 0$ *if and only if, for some* $g(z) \in \mathbb{F}[z]^m$,

$$u_*(z) = u_{min}(z) + Q(z)g(z). \tag{1.24}$$

Proof The proof is analogous to that of Theorem 1.3 and is therefore omitted. ■

1.4 Flat Outputs and the Control of Partial States

In the preceding sections we solved the open loop control task for higher order systems by steering to an arbitrary state x_τ of the associated shift realization. It is of course also of interest to compute controls that steer the origin to a given partial state ξ_τ. This makes contact with flat outputs, using image representations of the behavior defined by the higher order system. It seems that the paper [10] by Trentelman was the first where the equivalence between flatness and the existence of image representations of behaviors was observed. We begin with a brief analysis of flat outputs in the context of higher order systems and then derive explicit formula for controls that assign partial states.

1.4.1 Flat Systems

We introduce the notion of flatness for higher order linear systems and show how one can compute explicit steering controllers via coprime factorizations. In the literature on flatness we did not find any systematic account discussing flat outputs for higher order linear input/output systems. The only exceptions we are aware of are the papers by [7], that characterize flat outputs for a restricted class of higher order systems, and [10], that treats flatness in the larger context of behaviors.

Consider a linear higher order system

$$T(\sigma)\xi(z) = U(\sigma)u(z) \tag{1.25}$$

where $T(z) \in \mathbb{F}[z]^{r\times r}$ is a nonsingular polynomial matrix, $U(z) \in \mathbb{F}[z]^{r\times m}$ and strictly proper transfer function $T(z)^{-1}U(z)$. Here σ denotes the backwards shift operator on $z^{-1}\mathbb{F}[[z^{-1}]]^r$, which acts on strictly proper series $\xi(z) \in z^{-1}\mathbb{F}[[z^{-1}]]^r$ in the usual way as $\sigma\xi(z) = \pi_-(z\xi(z))$. Let $\mathscr{B}_{(T,U)}$ denote the **behavior** associated with $(T(z), U(z))$, i.e., the solution set of (1.25) in the space of strictly proper power series as

$$\mathscr{B}_{(T,U)} := \{\mathrm{col}\,(\xi(z), u(z)) \in z^{-1}\mathbb{F}[[z^{-1}]]^{r+m} \mid T(\sigma)\xi(z) = U(\sigma)u(z)\}.$$

Note, that $\mathscr{B}_{(T,U)}$ is a behavior in the terminological sense, i.e., a closed, backward shift invariant linear subspace of $z^{-1}\mathbb{F}[[z^{-1}]]^{r+m}$.

Definition 1.5 Consider any pair of polynomial matrices $V(z) \in \mathbb{F}[z]^{m\times r}$, $W(z) \in \mathbb{F}[z]^{m\times m}$. A linear output

$$y(z) = V(\sigma)\xi(z) + W(\sigma)u(z) \in z^{-1}\mathbb{F}[[z^{-1}]]^m \tag{1.26}$$

is called a **flat output** of system (1.25) if there exist polynomial matrices $P(z) \in \mathbb{F}[z]^{r\times m}$, $Q(z) \in \mathbb{F}[z]^{m\times m}$ with the following two properties:

1. The behavior has the **image representation**

$$\mathscr{B}_{(T,U)} = \{\mathrm{col}\,(P(\sigma)y(z), Q(\sigma)y(z)) \in z^{-1}\mathbb{F}[[z^{-1}]]^{r+m} \mid y(z) \in z^{-1}\mathbb{F}[[z^{-1}]]^m\}. \tag{1.27}$$

2. The **output condition** holds

$$y(z) = V(\sigma)P(\sigma)y(z) + W(\sigma)Q(\sigma)y(z), \quad \forall y(z) \in z^{-1}\mathbb{F}[[z^{-1}]]^m, \tag{1.28}$$

i.e., $V(\sigma)P(\sigma) + W(\sigma)Q(\sigma) = id$.

A system (1.25) is called **flat** if it possesses a flat output. The polynomial matrices $P(z)$, $Q(z)$ are called the **representing parameters** of (1.25), while $V(z)$, $W(z)$ are called the **flat output parameters**. The set

$$\Pi^{(P,Q)} = \{y(z) \in z^{-1}\mathbb{F}[[z^{-1}]]^m \mid (P(z)y(z), Q(z)y(z)) \in z^{-1}\mathbb{F}[[z^{-1}]]^{r+m}\} \tag{1.29}$$

is called the **set of flat parameters**.

The right coprimeness of $P(z)$, $Q(z)$ is equivalent to (1.27) being an **observable image representation**. Similarly, the assumption on left coprimeness of $T(z)$, $U(z)$ is a reachability condition. Note further that the output condition (1.28) is equivalent to the Bezout identity

$$V(z)P(z) + W(z)Q(z) = I_m. \tag{1.30}$$

This implies that the pairs of polynomial matrices $(V(z), W(z))$ and $(P(z), Q(z))$ are left coprime and right coprime, respectively. Note further that the inclusion

$$\{\mathrm{col}\,(P(\sigma)y(z), Q(\sigma)y(z)) \in z^{-1}\mathbb{F}[[z^{-1}]]^{r+m} \mid y(z) \in z^{-1}\mathbb{F}[[z^{-1}]]^m\} \subset \mathcal{B}_{(T,U)} \tag{1.31}$$

holds if and only if the polynomial matrices $P(z)$, $Q(z)$ satisfy

$$T(z)P(z) = U(z)Q(z). \tag{1.32}$$

We note in passing that the reverse inclusion

$$\mathcal{B}_{(T,U)} \subset \{\mathrm{col}\,(P(\sigma)y(z), Q(\sigma)y(z)) \in z^{-1}\mathbb{F}[[z^{-1}]]^{r+m} \mid y(z) \in z^{-1}\mathbb{F}[[z^{-1}]]^m\}$$

implies that the polynomial matrix $Q(z)$ is nonsingular. In fact, the reverse inclusion implies $z^{-1}\mathbb{F}[[z^{-1}]]^m \subset Q(\sigma)z^{-1}\mathbb{F}[[z^{-1}]]^m$. This implies $z^{-1}\mathbb{F}[[z^{-1}]]^m \subset \Gamma(\sigma)Q(\sigma)P(\sigma)z^{-1}\mathbb{F}[[z^{-1}]]^m$, for any biproper rational matrix $\Gamma(z)$ and any unimodular polynomial matrix $P(z)$. Thus we can assume that $Q(z)$ is in left Wiener-Hopf canonical form, from which the claim easily follows.

Theorem 1.6 *The following assertions are equivalent:*

1. *The system (1.25) has a flat output.*
2. *$T(z), U(z)$ are left coprime.*
3. *The shift realization (1.7) for (1.25) is reachable.*

Proof It suffices to prove the equivalence of the first two parts. The equivalence of items two and three follows from Theorem 1.2.

Any unimodular polynomial matrix $R(z) \in \mathbb{F}[z]^{(r+m)\times(r+m)}$ acts as a module isomorphism $R(\sigma)$ on $z^{-1}\mathbb{F}[[z^{-1}]]^{r+m}$ that maps a behavior $\mathcal{B}_{(T,U)}$ onto the behavior $\mathcal{B}_{(T,U)R}$. For any unimodular matrix $L(z) \in \mathbb{F}[z]^{r\times r}$, we obtain the equality of modules $\mathcal{B}_{(LT,LU)} = \mathcal{B}_{(T,U)}$. Thus, without loss of generality, we can assume that the $r \times (m+r)$ matrix $(T(z)\ \ U(z)) = (D(z)\ \ 0)$ is in Smith normal form with $T(z) = D(z) = \mathrm{diag}\,(d_1(z), \dots, d_r(z))$ and $U(z) = 0$. Thus $(\xi_1, \dots, \xi_r, u) \in \mathcal{B}_{(T,U)}$ holds if and only if $d_i(\sigma)\xi_i = 0$ for $i = 1, \dots, r$ and $U(\sigma)u = 0$.

Conversely, assume that (1.26) is a flat output of (1.25). Let $P(z)$, $Q(z)$ denote the associated polynomial matrices. Then $D(z)P(z) = T(z)P(z) = U(z)Q(z) = 0$ and therefore the first row vector $P_1(z)$ of $P(z)$ vanishes. Choose any element $\mathrm{col}\,(\xi_1, \dots, \xi_r, u) \in \mathcal{B}_{(T,U)}$ with $d_1(\sigma)\xi = 0$ and $\xi_1 \neq 0$. we obtain $d_1(z) \neq 0$, as $T(z)$ is nonsingular. Thus, such an element clearly exists if $d_1(z) \neq 0$ is not a constant polynomial. On the other hand, by (1.27) we have $\xi_1 = P_1(\sigma)y = 0$, which is a contradiction. Therefore, $d_1(z)$ is a constant polynomial. Similarly, all the other nonzero invariant factors $d_i(z)$, $i = 1, \ldots, r$ are constant polynomials. Suppose, that for some $1 \leq i \leq r$, $d_i(z) = 0$ is the zero polynomial. Then the $i - th$ row vector $P_i(z)$ of $P(z)$ satisfies $P_i(z) = 0$. Thus, (1.27) implies that any element $(\xi_1, \ldots, \xi_r, u) \in \mathscr{B}_{(T,U)}$ satisfies $\xi_i = 0$. Since the elements of $\mathscr{B}_{(T,U)}$ are characterized by $d_i(\sigma)\xi_i = 0, i = 1, \ldots, r$ and $U(\sigma)u = 0$, we can always find $(\xi, u) \in \mathscr{B}_{(T,U)}$ with $\xi_i \neq 0$. This is a contradiction. Therefore, flatness implies that all invariant factors of $T(z), U(z)$ are nonzero constants. Thus $T(z), U(z)$ are left coprime. ∎

We next turn to the task of characterizing flat outputs. This can be done using coprime factorizations. In fact, assume that we have a left coprime pair of polynomial matrices $T(z), U(z)$ with $T(z)$ nonsingular. Then there exists a right coprime factorization

$$P(z)Q(z)^{-1}$$

of $T(z)^{-1}U(z)$. Thus, in particular, we obtain $T(z)P(z) = U(z)Q(z)$. Choose any polynomial solution $V(z), W(z)$ of the Bezout equation

$$V(z)P(z) + W(z)Q(z) = I_m. \tag{1.33}$$

Then $y = V(\sigma)\xi + W(\sigma)u$ is a flat output and the solution set $\mathscr{B}_{(T,U)}$ of (1.25) is given as (1.27). Thus, the representing parameters $P(z), Q(z)$ are obtained by any right coprime factorization of the transfer function $T(z)^{-1}U(z)$, while the flat output parameters are obtained by solving the Bezout Eq. (1.33).

The next result characterizes the set of all flat outputs for a given system (1.25). For an extension of the result to behaviors we refer to [10].

Theorem 1.7 *Assume that $T(z), U(z)$ are left coprime. The following conditions are equivalent:*

(i) (1.26) is a flat output of system (1.25).
(ii) There exist right coprime polynomial matrices $P(z) \in \mathbb{F}[z]^{r\times m}$, $Q(z) \in \mathbb{F}[z]^{m\times m}$ with $Q(z)$ nonsingular and

$$\begin{aligned} T(z)P(z) - U(z)Q(z) &= 0 \\ V(z)P(z) + W(z)Q(z) &= I_m. \end{aligned} \tag{1.34}$$

(iii) The polynomial system matrix

$$\begin{pmatrix} T(z) & -U(z) \\ V(z) & W(z) \end{pmatrix} \tag{1.35}$$

is unimodular.

Proof The direction $(i) \implies (ii)$ is already shown. For a proof of $(ii) \implies (i)$ assume that polynomial matrices $P(z)$, $Q(z)$ exist that satisfy (1.34). Then P, Q are right coprime and there exists an extension to a unimodular matrix

$$\begin{pmatrix} X(z) & P(z) \\ Y(z) & Q(z) \end{pmatrix}$$

using suitable polynomial matrices $X(z)$, $Y(z)$. By (1.34) this implies

$$\begin{pmatrix} T(z) & -U(z) \\ V(z) & W(z) \end{pmatrix} \begin{pmatrix} X(z) & P(z) \\ Y(z) & Q(z) \end{pmatrix} = \begin{pmatrix} A(z) & 0 \\ B(z) & I_m \end{pmatrix} = \begin{pmatrix} A(z) & 0 \\ 0 & I \end{pmatrix} \begin{pmatrix} I & 0 \\ B(z) & I \end{pmatrix},$$

with polynomial matrices $A(z) \in \mathbb{F}[z]^{r\times r}$, $B(z) \in \mathbb{F}[z]^{m\times r}$. But this implies that $A(z)$ is a common left factor of $T(z)$, $U(z)$. Since T, U are assumed to be left coprime we conclude that $A(z)$ is unimodular. Thus, the system matrix

$$\begin{pmatrix} T(z) & -U(z) \\ V(z) & W(z) \end{pmatrix}$$

is unimodular, too, and we obtain a representation

$$\begin{pmatrix} T(z) & -U(z) \\ V(z) & W(z) \end{pmatrix} \begin{pmatrix} \overline{X}(z) & P(z) \\ \overline{Y}(z) & Q(z) \end{pmatrix} = \begin{pmatrix} I & 0 \\ 0 & I \end{pmatrix}.$$

Therefore,

$$\begin{pmatrix} \overline{X}(z) & P(z) \\ \overline{Y}(z) & Q(z) \end{pmatrix} \begin{pmatrix} T(z) & -U(z) \\ V(z) & W(z) \end{pmatrix} = \begin{pmatrix} I & 0 \\ 0 & I \end{pmatrix} \tag{1.36}$$

holds. Assume that $\xi(z), u(z)$ is any element of $\mathscr{B}_{(T,U)}$. Then (1.36) implies that $(\xi(z), u(z)) = (P(z)y(z), Q(z)y(z))$ for $y(z) = V(z)\xi(z) + W(z)u(z)$. This shows that (1.27), (1.28) hold. Thus (1.28) is a flat output and we are done.

By unimodularity of (1.35) there exist polynomial matrices $P(z)$, $Q(z)$ such that

$$\begin{pmatrix} T(z) & -U(z) \\ V(z) & W(z) \end{pmatrix} \begin{pmatrix} P(z) \\ Q(z) \end{pmatrix} = \begin{pmatrix} 0 \\ I_p \end{pmatrix}.$$

Thus condition (ii) is satisfied which implies (i). This proves $(iii) \implies (i)$. Conversely, assume that (1.26) is a flat output of system (1.25). Then the preceding proof of $(i) \implies (ii)$ shows that (1.35) is unimodular. This completes the proof. ■

By the uniqueness part in the unimodular embedding, the preceding result has the following consequence:

Corollary 1.8 *Let $T(z)^{-1}U(z) = P(z)Q(z)^{-1}$ be left and right coprime factorizations of a strictly proper transfer function. Then there exists a unique flat output of (1.25)*

$$y = V(\sigma)\xi(z) + W(\sigma)u(z) \tag{1.37}$$

with $V(z)T(z)^{-1}$ strictly proper. The output (1.37) is called the ***canonical flat output*** *of (1.25).*

The set of flat parameters enables one to solve terminal state control problems in a constructive way. Thus, the description of all flat parameters becomes important. In general, for any flat output, the strictly proper series $y(z) \in \Pi^{(P,Q)}$ yield admissible input-state trajectories $\mathrm{col}\,(x(z), u(z)) = \mathrm{col}\,(P(\sigma)y(z), Q(\sigma)y(z)) \in \mathscr{B}_{(T,U)}$. Therefore, the inclusion

$$\{\mathrm{col}\,(P(\sigma)y(z), Q(\sigma)y(z)) \mid y(z) \in \Pi^{(P,Q)}\} \subset \mathscr{B}_{(T,U)} \tag{1.38}$$

holds, which however, is only a proper inclusion. Nevertheless, $\Pi^{(P,Q)}$ is very useful to solve control problems with fixed initial and terminal states.

Proposition 1.9 *Assume that $G(z) := T(z)^{-1}U(z)$ is strictly proper and let $G(z) = P(z)Q(z)^{-1}$ be a right coprime factorization. Let $Q(z)$ be column proper with column degrees $\mu_1 \geq \cdots \geq \mu_m \geq 0$.*

1. *For the space of flat parameters $\Pi^{(P,Q)}$, the equality*

$$\Pi^{(P,Q)} = \Pi^{Q} := \{y(z) \in z^{-1}\mathbb{F}[[z^{-1}]]^m \mid Q(z)y(z) \in z^{-1}\mathbb{F}[[z^{-1}]]^m\} \tag{1.39}$$

   is satisfied. Moreover, $\mathrm{Ker}\,\pi^{Q} = \Pi^{Q}$.
2. *For any nonsingular polynomial matrix $Q'(z) \in \mathbb{F}[z]^{m\times m}$ with $Q'(z)Q(z)^{-1}$ biproper there is the direct sum decomposition of vector spaces*

$$\Pi^{(P,Q)} \oplus X^{Q'} = z^{-1}\mathbb{F}[[z^{-1}]]^m. \tag{1.40}$$

3. *If $y \in \Pi^{Q}$, then y is a flat output of the system (1.25).*
4. *The space of flat parameters has the representation*

$$\Pi^{(P,Q)} = \{y(z) \in z^{-1}\mathbb{F}[[z^{-1}]]^r \mid y_{i,j} = 0, i = 1, \ldots, r, j = 0, \ldots, \mu_i - 1\}, \tag{1.41}$$

   where $y(z) = \mathrm{col}\,(y_1(z), \ldots, y_r(z))$ and $y_i(z) = \sum_{j=0}^{\infty} y_{i,j}z^{-j-1}$.

Proof Clearly, the inclusion $\Pi^{(P,Q)} \subset \Pi^{Q}$ holds. In view of the strict properness of $P(z)Q(z)^{-1}$, it follows that, for any $y(z) \in \Pi^{Q}$, the condition $Q(z)y(z) \in z^{-1}\mathbb{F}[[z^{-1}]]^m$ implies $P(z)y(z) = (P(z)Q(z)^{-1})(Q(z)y(z)) \in z^{-1}\mathbb{F}[[z^{-1}]]^r$. Thus $\Pi^{Q} \subset \Pi^{(P,Q)}$ and equality (1.39) follows. Our assumption that $Q(z)$ is column proper implies that $Q(z)$ is properly invertible. This implies the equality $\Pi^Q = \operatorname{Ker}\pi^Q$.

Note that, for the rational model X^Q, there are two equivalent representations, namely

$$X^Q = \operatorname{Ker} Q(\sigma) = \operatorname{Im}\pi^Q, \tag{1.42}$$

where π^Q is the projection map for the rational models. Note further that $X^Q = X^{Q'}$ holds for any nonsingular polynomial matrices Q, Q' with $Q(z)Q'(z)^{-1}$ biproper. Since $z^{-1}\mathbb{F}[[z^{-1}]]^m = \operatorname{Im}\pi^Q \oplus \operatorname{Ker}\pi^Q$, the direct sum (1.40) follows.

The left coprimeness of $T(z), U(z)$ implies the reachability of the behavior $\mathscr{B}_{(T,U)}$ and hence the existence of an image representation of it, namely $\mathscr{B}_{(T,U)} = \operatorname{Im}\begin{pmatrix} P(\sigma) \\ Q(\sigma) \end{pmatrix}$. Thus, if $y \in \Pi^Q$, then $\begin{pmatrix} \xi \\ u \end{pmatrix} = \begin{pmatrix} P(\sigma) \\ Q(\sigma) \end{pmatrix} y = \begin{pmatrix} P(z) \\ Q(z) \end{pmatrix} y$. Using the doubly coprime factorization (1.43), we compute

$$y = \begin{pmatrix} V(z) & W(z) \end{pmatrix} \begin{pmatrix} P(z) \\ Q(z) \end{pmatrix} y = \begin{pmatrix} V(z) & W(z) \end{pmatrix} \begin{pmatrix} \xi \\ u \end{pmatrix},$$

which shows that y is a flat output.

By our assumption, we can write $Q(z) = \Gamma(z)\Delta(z)$, where $\Gamma(z)$ is biproper, i.e., proper and properly invertible, and $\Delta(z) = \operatorname{diag}(z^{\mu_1}, \ldots, z^{\mu_r})$. This implies that $\Pi^Q = \Pi^\Delta$ and, using (1.39), the representation (1.41) follows. ∎

1.4.2 Partial State Assignment for Higher Order Systems

We now turn to the task of generalizing our preceding approach to construct inputs for controlling partial states. Thus, for the higher order system

$$T(\sigma)\xi = U(\sigma)u,$$

we search for an input sequence that steers $\xi = 0$ in finite time $\tau > 0$ to a desired partial state ξ_*. Here we assume that $T(z) \in \mathbb{F}[z]^{r\times r}$ is nonsingular and $T(z), U(z)$ are left coprime. Thus, the setting will be the spaces of formal power series in z^{-1} with zero constant term, that is, the input functions $u(z)$ belonging to $z^{-1}\mathbb{F}[[z^{-1}]]^m$, and similarly for $\xi(z)$ and $y(z)$. Since we want to include the present time $t = 0$, we will write the trajectories as $\xi(z) = \sum_{j=0}^{\infty} \xi_j z^{-j-1}$, and similarly for all other variables.

To explain the flatness approach to open loop control of (1.25) we make use of the unimodular embedding (1.22). To the system given by (1.25), we associate a flat output

$$y(z) = V(\sigma)\xi + W(\sigma)u, \tag{1.43}$$

obtaining

$$\begin{pmatrix} 0 \\ I \end{pmatrix} y = \begin{pmatrix} T(\sigma) & -U(\sigma) \\ V(\sigma) & W(\sigma) \end{pmatrix} \begin{pmatrix} \xi \\ u \end{pmatrix}. \tag{1.44}$$

As a consequence of the reachability assumption, the behavior Ker $\begin{pmatrix} T(\sigma) & -U(\sigma) \end{pmatrix}$ has an image representation, given by

$$\text{Ker} \begin{pmatrix} T(\sigma) & -U(\sigma) \end{pmatrix} = \text{Im} \begin{pmatrix} P(\sigma) \\ Q(\sigma) \end{pmatrix}. \tag{1.45}$$

Using the doubly coprime factorization (1.22), we compute

$$\begin{aligned} \begin{pmatrix} P(\sigma) \\ Q(\sigma) \end{pmatrix} y &= \begin{pmatrix} \overline{W}(\sigma) & P(\sigma) \\ -\overline{V}(\sigma) & Q(\sigma) \end{pmatrix} \begin{pmatrix} 0 \\ I \end{pmatrix} y = \begin{pmatrix} \overline{W}(\sigma) & P(\sigma) \\ -\overline{V}(\sigma) & Q(\sigma) \end{pmatrix} \begin{pmatrix} T(\sigma) & -U(\sigma) \\ V(\sigma) & W(\sigma) \end{pmatrix} \begin{pmatrix} \xi \\ u \end{pmatrix} \\ &= \begin{pmatrix} \xi \\ u \end{pmatrix}, \end{aligned}$$

that is,

$$\begin{aligned} \xi &= P(\sigma)y \\ u &= Q(\sigma)y. \end{aligned} \tag{1.46}$$

Equations (1.46), which are at the heart of the flatness approach, are very suggestive. They indicate that the top one is an underdetermined system of equations, to be solved for the flat output y. The second equation then gives the required input.

We illustrate this process for the task of computing a controller that steers the system (1.25) from the zero state at time $t = 0$ to a prescribed partial state $\xi_* \in \mathbb{F}^r$ at time $t = \tau$. We note that, in general, the variable ξ is not a state variable. Thus, in the absence of a map from the state space X_T to the space $\mathbb{F}^r$ of partial states, one cannot apply the preceding approach.

The next result gives an explicit approach to solve this partial state assignment problem.

Theorem 1.10 *Let $(T(z), U(z)) \in \mathbb{F}[z]^{r \times r} \times \mathbb{F}[z]^{r \times m}$ be left coprime polynomial matrices with $T(z)$ nonsingular and $T(z)^{-1}U(z)$ strictly proper. Let $T(z)^{-1}U(z) = P(z)Q(z)^{-1}$ be a right coprime factorization with $Q(z) \in \mathbb{F}[z]^{m \times m}$, $P(z) \in \mathbb{F}[z]^{r \times m}$ and $Q(z)$ column proper with column degrees $\mu_1 \geq \cdots \geq \mu_m \geq 0$.*

A controller $u^*(z)$ steers the system (1.25) from the zero state at time 0 to a prescribed partial state $\xi_\tau = \xi^*$ at time $\tau \geq \mu_1$ if and only if $u^*(z) = Q(z)y(z)$ for $y(z) \in \Pi^Q$ that satisfies

$$\xi^* = \sum_{i=0}^{\mu_1 - 1} P_i y_{\tau+i}. \tag{1.47}$$

Proof Let $y \in \Pi^Q$ satisfy (1.47). Then $u^* = Q(\sigma)y$ is strictly proper and $\xi = P(\sigma)y$ satisfies $\xi_\tau = \xi_*$. This proves the sufficiency of the condition. For the necessity part assume that (ξ, u^*) be a solution trajectory of (1.25) with $\xi_\tau = \xi_*$. By the flatness property of the solution behavior, it follows that $\xi = P(\sigma)y$ and $u^* = Q(\sigma)y$ for some strictly proper y. Since u^* is strictly proper this shows that $y \in \Pi^Q$. This completes the proof. ■

1.4.3 Linear State Space Systems

As an illustration of the preceding results, we analyze flatness for linear state space systems

$$x_{t+1} = Ax_t + Bu_t, \quad x_0 = 0, \quad t \in \mathbb{N}_0. \tag{1.48}$$

Here $A \in \mathbb{F}^{n\times n}$ and $B \in \mathbb{F}^{n\times m}$. Using the strictly proper formal power series

$$x(z) = \sum_{j=1}^{\infty} \frac{x_j}{z^{j+1}}, \quad u(z) = \sum_{j=0}^{\infty} \frac{u_j}{z^{j+1}}, \tag{1.49}$$

then (1.48) becomes equivalent to the equation among formal power series as

$$(zI - A)x(z) = Bu(z). \tag{1.50}$$

By the Hautus criterion, the polynomial matrices $T(z) = zI - A$ and $U(z) = B$ are left coprime if and only if (A, B) is reachable. Choose any factorization

$$(zI - A)^{-1}B = P(z)Q(z)^{-1}$$

with $P(z), Q(z)$ right coprime. Then

$$y(z) = V(z)x(z) + W(z)u(z)$$

is a flat output of (1.50) if and only if $V(z), W(z)$ are a solution of the Bezout identity

$$V(z)P(z) + W(z)Q(z) = I_m.$$

We aim to describe all finite input sequences $u_0, \dots, u_T$ that control the initial state $x_0 = 0$ into a desired terminal state $x_* = x_T$. For any reachable pair (A, B), the reachability map defines a isomorphism of modules

$$\mathscr{R}_{(A,B)} : X_Q \longrightarrow X_{zI-A}, \quad \mathscr{R}_{(A,B)}f = \pi_{zI-A}(Bf(z)).$$

For any flat parameter $y(z) \in \Pi^{(P,Q)}$, the state and input strictly proper formal power series are given as

$$x(z) = P(z)y(z) \tag{1.51}$$
$$u(z) = Q(z)y(z). \tag{1.52}$$

Thus, for any $T > 0$ and $u(z) = \sum_{j=0}^{\infty} u_j z^{-j-1}$ the coefficients of $\pi_+(z^T u(z)) = \sum_{j=0}^{T-1} u_j z^{T-j-1}$ yield the finite input sequence $u_{T-1}, \ldots, u_0$ (in reverse order) that steers $x_0 = 0$ into x_T.

By reachability of (A, B), the polynomial matrix $P(z)$ is right proper, i.e.,

$$P(z) = \begin{pmatrix} P_1(z) \\ \vdots \\ P_n(z) \end{pmatrix},$$

with n linearly independent row vectors of polynomials $P_i(z) \in \mathbb{F}[z]^{1\times m}$. Assume further that $Q(z)$ is column proper with column indices $\kappa_1 \geq \cdots \geq \kappa_m$. Note that the column indices κ_i coincide with the reachability indices of (A, B). Since $P(z)Q(z)^{-1} = (zI - A)^{-1}B$ is strictly proper, this implies that the jth column of $P(z)$ has maximal degree κ_j. Thus we obtain the expansion

$$P(z) = \sum_{i=0}^{\kappa_1 - 1} P^{(i)} z^i$$

with a rectangular $n \times m\kappa_1$—matrix of coefficients

$$[P] := \left(P^{(0)} \ \ldots \ P^{(\kappa_1 - 1)}\right)$$

of full row rank. By the degree constraint on the columns of $P(z)$, the jth column of $P^{(i)}$ satisfies $P_j^{(i)} = 0$ for $i \geq \kappa_j$. Let $M(P)$ denote the $n \times n$ submatrix of $[P]$ with columns $P_j^{(i)}$, $i < \kappa_j$, ordered lexicographically. Since $[P]$ has full column rank, thus $M(P)$ is invertible.

The next result characterizes the canonical flat output of (A, B).

Theorem 1.11 *Let (A, B) have reachability indices $\kappa_1 \geq \ldots \geq \kappa_m$ and right coprime factorization $P(z)Q(z)^{-1} = (zI - A)^{-1}B$. Let $Q(z)$ be column proper.*

1. *(A, B) is reachable if and only if the polynomial row vectors $P_1(z), \ldots, P_n(z)$ are $\mathbb{F}-$ linearly independent with* deg $p_{ij}(z) < \kappa_j, i = 1, \ldots, n, j = 1, \ldots, m$. *Equivalently, $M(P)$ is invertible.*
2. *System (1.48) has a flat output if and only if (A, B) is reachable. In either case, there exists a unique $C \in \mathbb{F}^{m\times n}$ such that*

$$y = Cx$$

is a flat output.

Proof The first part is already shown, as well as that reachability of (A, B) is equivalent to the existence of a flat output. Since $P(z)Q(z)^{-1}$ is strictly proper, the row vectors $P_i(z)$ are a basis for polynomial model X_Q, where the elements of X_Q are viewed as row vectors. Thus there exists a matrix $C \in \mathbb{F}^{m\times n}$ such that $CP(z) = I_m$. Next, consider the polynomial system matrix

$$\mathscr{P}(z) := \begin{pmatrix} zI - A & -B \\ C & 0 \end{pmatrix}.$$

Embedding $P(z)$, $Q(z)$ into the unimodular matrix

$$\begin{pmatrix} P(z) & X(z) \\ Q(z) & Y(z) \end{pmatrix},$$

then the product is

$$\begin{pmatrix} zI - A & -B \\ C & 0 \end{pmatrix} \begin{pmatrix} P(z) & X(z) \\ Q(z) & Y(z) \end{pmatrix} = \begin{pmatrix} 0 & \overline{X}(z) \\ CP(z) & \overline{Y}(z) \end{pmatrix} \tag{1.53}$$

for suitable polynomial matrices $\overline{X}, \overline{Y}$. By (1.53), the matrix $(0\ \overline{X})$ is the product of a left prime matrix and a unimodular one. Thus $(0\ \overline{X})$ is left prime and therefore $\overline{X}$ is unimodular. Since $CP(z) = I$ this shows that the product is unimodular and thus $\mathscr{P}$ is unimodular. This shows that $y = Cx$ is a flat output for (1.48). ■

From the preceding argument, the condition for a flat output is

$$x_T = \sum_{j=0}^{\kappa_1 - 1} P_j y_{T+j}. \tag{1.54}$$

Decompose each vector $y_k \in \mathbb{F}^m$ as

$$y_k = \begin{pmatrix} y_{k,1} \\ \vdots \\ y_{k,m} \end{pmatrix}.$$

Thus (1.54) is equivalent to

$$x_T = \sum_{j=1}^{m} \sum_{i=0}^{\kappa_j - 1} P_j^{(i)} y_{T+i,j},$$

or, equivalently, to

$$M(P)\overline{y} = x_T$$

where $\overline{y} \in \mathbb{F}^n$ contains all components $y_{T+i,j}$, $i < \kappa_j$, ordered lexicographically. Thus we have shown

Theorem 1.12 *Let $T \geq \kappa_1$, (A, B) be reachable and let $x_T \in \mathbb{F}^n$ be an arbitrary terminal state vector. Let $\overline{y} = \mathrm{col}\,(y_{T+i,j} | i < \kappa_j, 1 \leq j \leq m) \in \mathbb{F}^n$ be the unique solution of*

$$M(P)\overline{y} = x_T. \tag{1.55}$$

Define a sequence of vectors $y_T, \ldots, y_{T+\kappa_1-1}$ by assigning arbitrary values to the other components of y_{T+i}. Select arbitrary vectors $y_{\kappa_1}, \ldots, y_{T-1}$ and set $y_0 = \cdots = y_{\kappa_1-1} := 0$. Then the set of all input sequences $u = (u_0, \ldots, u_{T-1})^\top$ that steers $x_0 = 0$ in time T to x_T is the set of all vectors

$$u = \begin{pmatrix} Q_{\kappa_1} & & & & \\ \vdots & \ddots & & & \\ Q_0 & \cdots & Q_{\kappa_1} & & \\ & \ddots & & \ddots & \\ & & Q_0 & \cdots & Q_{\kappa_1} \end{pmatrix} \begin{pmatrix} y_{\kappa_1} \\ \vdots \\ y_{T+n-1} \end{pmatrix} \tag{1.56}$$

1.5 Conclusions

The open loop control problems, discussed in this paper, can be solved for some classes of exactly controllable, infinite dimensional linear systems. For that case, one can single out minimal norm solutions. We leave these problems for the future research.

Acknowledgments The research of U. Helmke is partially supported by grant HE 1858/13-1 from the Deutsche Forschungsgemeinschaft DFG.

References

1. Fliess, M., Levine, J., Martin, Ph, Rouchon, P.: Flatness and defect of nonlinear systems: introductory theory and applications. Int. J. Control **61**, 1327–1361 (1995)
2. Fuhrmann, P.A.: Algebraic system theory: an analyst's point of view. J. Franklin Inst. **301**, 521–540 (1976)
3. Fuhrmann, P.A.: On strict system equivalence and similarity. Int. J. Control **25**, 5–10 (1977)

4. Fuhrmann, P.A., Helmke, U.: Strict equivalence, controllability and observability of networks of linear systems. Math. Control Signals Syst. **25**, 437–471 (2013)
5. Fuhrmann, P.A., Helmke, U.: The Mathematics of Networks of Linear Systems. Springer, New York (2015)
6. Kalman, R.E., Falb, P., Arbib, M.: Topics in Mathematical System Theory. McGraw-Hill, New York (1969)
7. Levine, J., Nguyen, D.V.: Flat output characterization for linear systems using polynomial matrices. Syst. Control Lett. **48**, 69–75 (2003)
8. Polderman, J.W., Willems, J.C.: Introduction to Mathematical System Theory. Springer, New York (1997)
9. Rosenbrock, H.H.: State-Space and Multivariable Theory. Wiley, New York (1970)
10. Trentelman, H.L.: On flat behaviors and observable image representations. Syst. Control Lett. **51**, 51–55 (2004)
11. Trentelman, H.L., Takaba, K., Monshizadeh, N.: Robust synchronization of uncertain linear multi-agent systems. IEEE Trans. Autom. Control **58**, 1511–1523 (2013)

Chapter 2
Bilinear Differential Forms and the Loewner Framework for Rational Interpolation

P. Rapisarda and A.C. Antoulas

Abstract The Loewner approach, based on the factorization of a special-structure matrix derived from data generated by a dynamical system, has been applied successfully to realization theory, generalized interpolation, and model reduction. We examine some connections between such approach and that based on bilinear- and quadratic differential forms arising in the behavioral framework.

2.1 Introduction

The Loewner framework was initiated in [17, 18] in the context of tangential interpolation and partial realization problems (see also [1, 4]). Its relevance for the problem of modeling from frequency response measurements and for model order reduction has been reported in a series of publications (see [2, 3]), resulting also in important applications in the (reduced-order) modeling of physical systems from data (see [15, 16]). Time series modeling from a behavioral perspective has been introduced in [30, 31], specialized to the vector exponential case in [32], and applied to metric interpolation problems in [13, 14, 27].

The purpose of this paper is to illustrate some connections between these two approaches. The relation between rational interpolation and partial realization

---

Dedicated to Prof. Harry L. Trentelman- friend, colleague, and for the first author also co-supervisor- on the occasion of his "sixtieth birthday"

P. Rapisarda (✉)
Vision, Learning and Control Group, School of Electronics and Computer Science,
University of Southampton, Southampton SO17 1BJ, UK
e-mail: pr3@ecs.soton.ac.uk

A.C. Antoulas
Department of Electrical and Computer Engineering, Rice University,
Houston, TX 77005, USA
e-mail: aca@rice.edu

A.C. Antoulas
School of Engineering and Science, Jacobs University Bremen, Bremen, Germany

M.N. Belur et al. (eds.), *Mathematical Control Theory II*,
Lecture Notes in Control and Information Sciences 462,
DOI 10.1007/978-3-319-21003-2_2

problems and the behavioral framework for data modeling is well known, see [7]; we will concentrate here on the analogies and insights coming from a more recently introduced approach (see [21, 25]) that while essentially behavioral (i.e., trajectory-based) also uses Gramian-based ideas to derive models from data. An important tool in such approach is the calculus of bilinear- and quadratic differential forms (B/QDFs in the following), introduced in [33] and applied successfully in many areas of systems and control (see [22, 28]). In this paper we show that several results derived in the Loewner approach can be formulated also in terms of the two-variable polynomial matrix representations of B/QDFs derived from the system parameters. Of particular relevance is that the factorization of the Loewner matrix—an important step of the Loewner approach in obtaining state models from data—can be given a trajectory-based interpretation based on B/QDFs.

The paper is organized as follows. In Sect. 2.2 we illustrate the essential concepts of the Loewner approach, of bilinear- and quadratic differential forms, and of behavioral systems theory. In Sect. 2.3 we show how the Loewner matrix and some of its properties can be formulated in the polynomial language of the representations of B/QDFs. In Sect. 2.4 we show how the computations of state equations based on Loewner matrix factorizations have a straightforward interpretation in terms of bilinear differential forms. Finally, Sect. 2.5 contains an exposition of directions of current and future research.

2.1.1 Notation

The space of n-dimensional real (complex) vectors is denoted by $\mathbb{R}^n$ (respectively, $\mathbb{C}^n$), and that of $m \times n$ real matrices by $\mathbb{R}^{m\times n}$. $\mathbb{R}^{\bullet\times m}$ denotes the space of real matrices with m columns and an unspecified finite number of rows. Given matrices $A, B \in \mathbb{R}^{\bullet\times m}$, $\mathrm{col}(A, B)$ denotes the matrix obtained by stacking A over B.

The ring of polynomials with real coefficients in the indeterminate ξ is denoted by $\mathbb{R}[\xi]$; the ring of two-variable polynomials with real coefficients in the indeterminates ζ and η is denoted by $\mathbb{R}[\zeta, \eta]$. $\mathbb{R}^{r\times q}[\xi]$ denotes the set of all $r \times q$ matrices with entries in ξ, and $\mathbb{R}^{n\times m}[\zeta, \eta]$ that of $n \times m$ polynomial matrices in ζ and η. The set of rational $m \times n$ matrices is denoted by $\mathbb{R}^{m\times n}(\xi)$.

The set of infinitely differentiable functions from $\mathbb{R}$ to $\mathbb{R}^q$ is denoted by $\mathfrak{C}^\infty(\mathbb{R}, \mathbb{R}^q)$. $\mathfrak{D}(\mathbb{R}, \mathbb{R}^q)$ is the subset of $\mathfrak{C}^\infty(\mathbb{R}, \mathbb{R}^q)$ consisting of compact support functions. Given $\lambda \in \mathbb{C}$, we denote by $e^{\lambda\cdot}$ the exponential function whose value at t is $e^{\lambda t}$.

2.2 Background Material

We restrict ourselves to the minimum amount of information necessary to understand the rest of the paper. For more details and a thorough introduction to behavioral system theory, bilinear/quadratic differential forms, and the Loewner framework we refer to [17, 19, 33], respectively.

2.2.1 Behavioral System Theory

The basic object of study in the behavioral framework is the set of trajectories, the *behavior* of a system. In this paper we consider *linear differential behaviors*, i.e., subsets of $\mathfrak{C}^\infty(\mathbb{R}, \mathbb{R}^q)$ that consist of solutions $w : \mathbb{R} \to \mathbb{R}^q$ to systems of linear, constant coefficient differential equations:

$$R\left(\frac{d}{dt}\right) w = 0. \tag{2.1}$$

where $R \in \mathbb{R}^{\bullet \times q}[\xi]$. A representation (2.1) is called a kernel representation of the *behavior*

$$\mathfrak{B} := \left\{ w \in \mathfrak{C}^\infty(\mathbb{R}, \mathbb{R}^q) \mid R\left(\frac{d}{dt}\right) w = 0 \right\},$$

and we associate to it in a natural way the polynomial matrix $R \in \mathbb{R}^{\bullet \times q}[\xi]$. Note that $\mathfrak{B}$ admits different kernel representations; such a representation is *minimal* if the number of rows of R is minimal among all possible representations of $\mathfrak{B}$. We denote with $\mathfrak{L}^q$ the set of all linear time-invariant differential behaviors with q variables.

If a behavior is controllable (see Chap. 5 of [19] for a definition), then it also admits an *image representation*. Let

$$w = M\left(\frac{d}{dt}\right) \ell, \tag{2.2}$$

where $M \in \mathbb{R}^{q \times l}[\xi]$ and ℓ is an auxiliary variable also called a *latent variable*; i.e.,

$$\mathfrak{B} := \{ w \in \mathfrak{C}^\infty(\mathbb{R}, \mathbb{R}^q) \mid \exists\, \ell \in \mathfrak{C}^\infty(\mathbb{R}, \mathbb{R}^l) \text{ such that (2.2) holds} \} =: \operatorname{im} M\left(\frac{d}{dt}\right).$$

We call (2.2) an image representation of $\mathfrak{B}$.

The latent variable ℓ in (2.2) is called *observable* from w if $[w = M(\frac{d}{dt})\ell = 0] \Longrightarrow [\ell = 0]$. A controllable behavior always admits an observable image representation. The set of linear differential controllable behaviors whose trajectories take their values in $\mathbb{R}^q$ is denoted by $\mathfrak{L}^q_{\text{cont}}$.

A latent variable ℓ is a *state variable* for $\mathfrak{B}$ if there exist $E, F \in \mathbb{R}^{\bullet \times \bullet}$, $G \in \mathbb{R}^{\bullet \times q}$ such that

$$\mathfrak{B} = \left\{ w \mid \exists\, \ell \text{ s.t. } E\frac{d\ell}{dt} + F\ell + Gw = 0 \right\}, \tag{2.3}$$

i.e., if $\mathfrak{B}$ has a representation of first order in ℓ and zeroth order in w. The minimal number of state variables needed to represent $\mathfrak{B}$ in this way is called the *McMillan degree* of $\mathfrak{B}$, denoted by $n(\mathfrak{B})$.

A state variable for $\mathfrak{B}$ can be computed as the image of a polynomial differential operator called a *state map* (see [9, 20, 26, 29]); such polynomial can act either on the external variable w, or on the latent variable ℓ of an image representation of $\mathfrak{B}$.

Finally, we introduce the notion of *dual* (or *adjoint*, see [29]) behavior. Let $\mathfrak{B} \in \mathfrak{L}^q$ and let $J = J^\top \in \mathbb{R}^{q\times q}$ be an involution, i.e., $J^2 = I_q$. We call

$$\mathfrak{B}^{\perp_J} := \left\{ w' \in \mathfrak{c}^\infty(\mathbb{R}, \mathbb{R}^q) \mid \int_{-\infty}^{+\infty} w'^\top J w \, dt = 0 \text{ forall } w \in \mathfrak{B} \cap \mathfrak{c}^\infty(\mathbb{R}, \mathbb{R}^q) \right\} \tag{2.4}$$

the *J-dual behavior* of $\mathfrak{B}$; if $J = I_l$, we denote it simply by $\mathfrak{B}^\perp$. It can be shown that if $\mathfrak{B} = \text{im } M\left(\frac{d}{dt}\right) = \ker\ R\left(\frac{d}{dt}\right)$, then $\mathfrak{B}^{\perp_J} = \text{im } JR^\top\left(-\frac{d}{dt}\right) = \ker\ M^\top\left(-\frac{d}{dt}\right) J$. Note that if R induces a minimal kernel representation and M an observable image representation of $\mathfrak{B}$, then $M^\top(-\xi)J$ induces a minimal kernel representation and $JR^\top(-\xi)$ an observable image representation of $\mathfrak{B}^{\perp_J}$.

2.2.2 Bilinear- and Quadratic Differential Forms

Let $\Phi \in \mathbb{R}^{q_1\times q_2}[\zeta, \eta]$; then $\Phi(\zeta, \eta) = \sum_{h,k} \Phi_{h,k}\zeta^h\eta^k$, where $\Phi_{h,k} \in \mathbb{R}^{q_1\times q_2}$ and the sum extends over a finite set of nonnegative indices. $\Phi(\zeta, \eta)$ induces the *bilinear differential form* (abbreviated with BDF in the following) L_Φ acting on $\mathfrak{C}^\infty$-trajectories defined by

$$\begin{aligned} L_\Phi &: \mathfrak{C}^\infty(\mathbb{R}, \mathbb{R}^{q_1}) \times \mathfrak{C}^\infty(\mathbb{R}, \mathbb{R}^{q_2}) \to \mathfrak{C}^\infty(\mathbb{R}, \mathbb{R}) \\ L_\Phi(w_1, w_2) &:= \sum_{h,k} (\frac{d^h w_1}{dt^h})^\top \Phi_{h,k} \frac{d^k w_2}{dt^k} \end{aligned}$$

If $q_1 = q_2 = q$, then $\Phi(\zeta, \eta)$ also induces the *quadratic differential form* (abbreviated QDF in the following) Q_Φ acting on $\mathfrak{C}^\infty$-trajectories defined by

$$\begin{aligned} Q_\Phi : &\quad \mathfrak{C}^\infty(\mathbb{R}, \mathbb{R}^q) \to \mathfrak{C}^\infty(\mathbb{R}, \mathbb{R}) \\ Q_\Phi(w) &:= \sum_{h,k} (\frac{d^h w}{dt^h})^\top \Phi_{h,k} \frac{d^k w}{dt^k}. \end{aligned}$$

Without loss of generality we can assume that a QDF is induced by a *symmetric* two-variable polynomial matrix $\Phi(\zeta, \eta)$, i.e., one such that $\Phi(\zeta, \eta) = \Phi(\eta, \zeta)^\top$; we denote the set of such matrices by $\mathbb{R}_s^{q\times q}[\zeta, \eta]$.

$\Phi(\zeta, \eta) \in \mathbb{R}^{q_1\times q_2}[\zeta, \eta]$ (and consequently also the BDF L_Φ) can be identified with its *coefficient matrix*

$$\widetilde{\Phi} := \left[\Phi_{h,k}\right]_{h,k=0,\ldots,\infty},$$

in the sense that

$$\Phi(\zeta,\eta) = \begin{bmatrix} I_{q_1} & \zeta I_{q_1} & \cdots \end{bmatrix} \tilde{\Phi} \begin{bmatrix} I_{q_2} \\ \eta I_{q_2} \\ \vdots \end{bmatrix}.$$

Although $\tilde{\Phi}$ is infinite, only a finite number of its entries are nonzero, since the highest power of ζ and η in $\Phi(\zeta,\eta)$ is finite. Note that $\Phi(\zeta,\eta)$ is symmetric if and only if $\tilde{\Phi}^\top = \tilde{\Phi}$.

Factorizations of the coefficient matrix of a B/QDF and factorizations of the two-variable polynomial matrix corresponding to it are related as follows:

Preposition 2.1 *Let $\Phi \in \mathbb{R}^{q_1\times q_2}[\zeta,\eta]$, and let $\tilde{\Phi}$ be its coefficient matrix. Then the following two statements are equivalent:*

1. *There exist real matrices $\tilde{F}$, $\tilde{G}$ with n rows such that*

$$\tilde{\Phi} = \tilde{F}^\top \tilde{G};$$

2. *There exist polynomial matrices $F \in \mathbb{R}^{n\times q_1}[\xi]$, $G \in \mathbb{R}^{n\times q_2}[\xi]$ with coefficient matrices $\tilde{F}$, $\tilde{G}$, i.e., $F(\xi) = \tilde{F}\begin{bmatrix} I_{q_1} \\ \xi I_{q_1} \\ \vdots \end{bmatrix}$ and $G(\xi) = \tilde{G}\begin{bmatrix} I_{q_2} \\ \xi I_{q_2} \\ \vdots \end{bmatrix}$, such that*

$$\Phi(\zeta,\eta) = F(\zeta)^\top G(\eta).$$

Proof This follows from the discussion on p. 1709 of [33]. □

Factorizations as those of Proposition 2.1, which moreover correspond to the minimal value $n = \operatorname{rank}(\tilde{\Phi})$, are called *minimal* (or *canonical* as in [33]). Note that the matrices $\tilde{F}$ and $\tilde{G}$ involved in a minimal factorization of $\tilde{\Phi}$ are of *full row rank*. Minimal factorizations are not unique; using standard linear algebra arguments the following proposition can be proved in a straightforward way.

Preposition 2.2 *Given a minimal factorization $\tilde{\Phi} = \tilde{F}^\top \tilde{G}$, every other minimal factorization $\tilde{\Phi} = \tilde{F}'^\top \tilde{G}'$ can be obtained from it by premultiplication of $\tilde{F}$ and $\tilde{G}$ by a nonsingular $n\times n$ matrix S, respectively, $S^{-\top}$. In view of Proposition 2.1 this implies that $\Phi(\zeta,\eta) = F(\zeta)^\top G(\eta) = F'(\zeta)^\top G'(\eta)$ with $F'(\xi) := SF(\xi)$, $G'(\xi) := S^{-\top}G(\xi)$.*

Given L_Ψ, its *derivative* is the BDF L_Φ defined by

$$L_\Phi(w_1,w_2) := \frac{d}{dt}(L_\Psi(w_1,w_2)),$$

for all $w_i \in \mathfrak{C}^\infty(\mathbb{R}, \mathbb{R}^{q_i})$, $i = 1, 2$; this holds if and only if

$$\Phi(\zeta, \eta) = (\zeta + \eta)\Psi(\zeta, \eta) \tag{2.5}$$

(see [33], p. 1710). An analogous result holds for QDFs. From this two-variable characterization it follows that if $L_\Phi = \frac{d}{dt}L_\Psi$, then $\Phi(-\xi, \xi) = 0_{q_1 \times q_2}$; it can be shown (see Theorem 3.1, p. 1711 of [33]) that also the converse implication holds true.

Finally, we introduce a standard result in B/QDF theory of great importance for the rest of this paper. The first part of the result is a straightforward consequence of the relation (2.5) between the two-variable representation of a B/QDF and its derivative; the second part follows from Proposition 10.1, p. 1730 of [33].

Preposition 2.3 *Let $R \in \mathbb{R}^{g \times q}[\xi]$ and $M \in \mathbb{R}^{q \times l}[\xi]$ induce a minimal kernel, respectively, observable image representation of $\mathfrak{B} \in \mathfrak{L}^q$. There exists $\Psi \in \mathbb{R}^{g \times l}[\zeta, \eta]$ such that*

$$R(-\zeta)M(\eta) = (\zeta + \eta)\Psi(\zeta, \eta)\ . \tag{2.6}$$

Moreover, there exist polynomial matrices $Z \in \mathbb{R}^{\bullet \times g}[\xi]$ and $X \in \mathbb{R}^{\bullet \times l}[\xi]$ such that

$$\Psi(\zeta, \eta) = Z(\zeta)^\top X(\eta), \tag{2.7}$$

and $Z\left(\frac{d}{dt}\right)$ is a minimal state map for $\mathfrak{B}^\perp$ and $X\left(\frac{d}{dt}\right)$ is a minimal state map for $\mathfrak{B}$.

State maps such as Z and X in (2.7) are called *matched*. Factorizations such as (2.7) can be computed factorizing canonically the coefficient matrix $\widetilde{\Psi}$ as illustrated in Proposition 2.1, see also Proposition 2.2.

2.2.3 Rational Interpolation and Modeling of Vector Exponential Time Series

Define the *left* and *right interpolation data* as the triples in $\mathbb{C} \times \mathbb{C}^p \times \mathbb{C}^m$ and $\mathbb{C} \times \mathbb{C}^m \times \mathbb{C}^p$, respectively:

$$\begin{aligned} &\{(\mu_i, \ell_i^*, v_i^*)\}_{i=1,\ldots,k_1}\ , \qquad \mu_i \in \mathbb{C}, \ell_i^* \in \mathbb{C}^{1\times p}, v_i^* \in \mathbb{C}^{1\times m} \\ &\{(\lambda_i, r_i, w_i)\}_{i=1,\ldots,k_2}\ , \qquad \lambda_i \in \mathbb{C}, r_i \in \mathbb{C}^{m\times 1}, w_i \in \mathbb{C}^{p\times 1}. \end{aligned} \tag{2.8}$$

In the rest of this paper, we will assume for simplicity of exposition that the μ_is *and λ_is are distinct*; the general case follows with straightforward modifications of the statements and the arguments. We will also assume that $\{\mu_i\}_{i=1,\ldots,k_1} \cap \{\lambda_j\}_{j=1,\ldots,k_2} = \emptyset$.

Let $H \in \mathbb{R}^{p\times m}(\xi)$ be a proper rational matrix. H *satisfies the interpolation constraints* if

$$\begin{aligned} \ell_i^* H(\mu_i) &= v_i^* \ , \ i = 1, \dots, k_1 \\ H(\lambda_i) r_i &= w_i \ , \ i = 1, \dots, k_2 \ . \end{aligned} \tag{2.9}$$

Rational interpolation can be stated as behavioral modeling of vector exponential functions (see [7]). Assume that $H \in \mathbb{R}^{p\times m}(\xi)$ satisfies the interpolation constraints, and let $H(\xi) = N(\xi)D(\xi)^{-1} = P(\xi)^{-1}Q(\xi)$ be right, respectively, left coprime factorizations of $H(\xi)$, with $N \in \mathbb{R}^{p\times m}[\xi]$, $D \in \mathbb{R}^{m\times m}[\xi]$, $P \in \mathbb{R}^{p\times p}[\xi]$, $Q \in \mathbb{R}^{p\times m}[\xi]$. We associate to the right coprime factorization of $H(\xi)$ the observable image representation

$$M(\xi) := \begin{bmatrix} D(\xi) \\ N(\xi) \end{bmatrix} \tag{2.10}$$

and to the left coprime factorization the minimal controllable kernel representation

$$R(\xi) := \begin{bmatrix} Q(\xi) & -P(\xi) \end{bmatrix} . \tag{2.11}$$

It follows from standard results in behavioral system theory (see Ch. 5 of [19]) that

$$\ker \ \left[Q\left(\tfrac{d}{dt}\right) \ -P\left(\tfrac{d}{dt}\right) \right] = \mathrm{im} \begin{bmatrix} D\left(\frac{d}{dt}\right) \\ N\left(\frac{d}{dt}\right) \end{bmatrix} =: \mathfrak{B}. \tag{2.12}$$

Under the standing assumption that *$D(\mu_i)$ and $P(\lambda_i)$ are nonsingular at μ_i, respectively λ_i*, we rewrite (2.9) equivalently as

$$\begin{aligned} \begin{bmatrix} v_i^* & -\ell_i^* \end{bmatrix} \begin{bmatrix} D(\mu_i) \\ N(\mu_i) \end{bmatrix} &= 0 \ , \ i = 1, \dots, k_1 \\ \begin{bmatrix} Q(\lambda_i) & -P(\lambda_i) \end{bmatrix} \begin{bmatrix} r_i \\ w_i \end{bmatrix} &= 0 \ , \ i = 1, \dots, k_2. \end{aligned} \tag{2.13}$$

From the equalities (2.13) it follows that

$$\begin{aligned} \begin{bmatrix} v_j^* & -\ell_j^* \end{bmatrix} &\in \text{row span} \begin{bmatrix} Q(\mu_j) & -P(\mu_j) \end{bmatrix} \\ \begin{bmatrix} r_i \\ w_i \end{bmatrix} &\in \mathrm{im} \begin{bmatrix} D(\lambda_i) \\ N(\lambda_i) \end{bmatrix}, \end{aligned}$$

$j = 1, \dots, k_1$, $i = 1, \dots, k_2$. We conclude that the interpolation constraints (2.9) (and the Eqs. (2.13)) are equivalent with

$$
\begin{aligned}
w_i(\cdot) &:= \begin{bmatrix} r_i \\ w_i \end{bmatrix} e^{\lambda_i \cdot} \in \mathfrak{B}\ ,\ i = 1, \ldots, k_2 \\
w'_j(\cdot) &:= \begin{bmatrix} v_j \\ -\ell_j \end{bmatrix} e^{-\mu_j \cdot} \in \mathfrak{B}^{\perp}\ ,\ j = 1, \ldots, k_1,
\end{aligned} \tag{2.14}
$$

where $\mathfrak{B}^{\perp}$ is the dual behavior $\mathfrak{B}^{\perp} = \text{im} \begin{bmatrix} Q^{\top}\left(-\frac{d}{dt}\right) \\ -P^{\top}\left(-\frac{d}{dt}\right) \end{bmatrix} = \ker \left[D^{\top}\left(-\frac{d}{dt}\right)\right]$ $\left[N^{\top}\left(-\frac{d}{dt}\right)\right]$. In the language of [31], $\mathfrak{B}$ and $\mathfrak{B}^{\perp}$, respectively, are *unfalsified models* for the trajectories (2.14). Thus every solution of the interpolation problem yields an unfalsified model for the exponential trajectories associated with the data; and conversely, every minimal kernel or observable image representation of such an unfalsified model for such trajectories yields a solution of the interpolation problem.

From (2.13) it follows that there exist vectors $s_j \in \mathbb{C}^{1\times p}$, $j = 1, \ldots, k_1$ and p_i, $i = 1, \ldots, k_2$, uniquely defined because of observability and of minimality and controllability, such that

$$
\begin{aligned}
\left[v_j^* \ -\ell_j^*\right] &= s_j^* \left[Q(\mu_j) \ -P(\mu_j)\right] \\
\begin{bmatrix} r_i \\ w_i \end{bmatrix} &= \begin{bmatrix} D(\lambda_i) \\ N(\lambda_i) \end{bmatrix} p_i .
\end{aligned} \tag{2.15}
$$

It is straightforward to check that such vectors define (unique) latent variable trajectories $p_i e^{\lambda_i \cdot}$ and $s_j e^{-\mu_j \cdot}$ for the image representations $\mathfrak{B} = \text{im}\ M\left(\frac{d}{dt}\right)$, $\mathfrak{B}^{\perp} = \text{im}\ R^{\top}\left(-\frac{d}{dt}\right)$, respectively.

2.3 The Loewner Matrix and Its Properties

The *Loewner matrix* associated with the interpolation data (2.8) is defined by

$$
\mathbb{L} := \left[\frac{v_i^* r_j - \ell_i^* w_j}{\mu_i - \lambda_j}\right]_{i=1,\ldots,k_1;\, j=1,\ldots,k_2} . \tag{2.16}
$$

The *shifted Loewner matrix* is defined by

$$
\mathbb{L}_{\sigma} := \left[\frac{\mu_i v_i^* r_j - \lambda_j \ell_i^* w_j}{\mu_i - \lambda_j}\right]_{i=1,\ldots,k_1;\, j=1,\ldots,k_2} . \tag{2.17}
$$

The first result of this paper connects the Loewner matrix and the two-variable polynomial matrix $\Psi(\zeta, \eta)$ in (2.6), and is the fundamental connection between the two approaches.

Preposition 2.4 *Let $\Psi(\zeta,\eta) \in \mathbb{R}^{p\times m}[\zeta,\eta]$ be defined by (2.6), with M and R defined by (2.10) and (2.11), and s_i and p_j defined as in (2.15). Then*

$$\mathbb{L} = -\left[s_i^* \Psi(-\mu_i, \lambda_j) p_j\right]_{i=1,\ldots,k_1;\, j=1,\ldots,k_2}. \tag{2.18}$$

Proof It follows from the equations (2.15) that if $H \in \mathbb{R}^{p\times m}(\xi)$ satisfies the interpolation constraints, then the Loewner matrix (2.16) can also be written as

$$\mathbb{L} = \left[\frac{s_i^*\left[Q(\mu_i) \; -P(\mu_i)\right]\begin{bmatrix} D(\lambda_j) \\ N(\lambda_j)\end{bmatrix} p_j}{\mu_i - \lambda_j}\right]_{i=1,\ldots,k_1,\, j=1,\ldots,k_2}, \tag{2.19}$$

where s_i and p_j are defined by (2.15). The claim follows easily from this equation and the definition of $\Psi(\zeta,\eta)$. □

If all $-\mu_i$ and λ_i are all on one and the same side of the imaginary axis (e.g., the left-hand side) then the two-variable polynomial (2.6) is associated with a BDF, and the Loewner matrix has the interpretation of a Gramian, as illustrated in the following result.

Preposition 2.5 *Partition the variables in $\mathfrak{B}$, respectively, $\mathfrak{B}^\perp$ by $w' := \begin{bmatrix} y' \\ u' \end{bmatrix} \in \mathfrak{C}^\infty(\mathbb{R}, \mathbb{C}^{m+p})$, respectively, $w := \begin{bmatrix} u \\ y \end{bmatrix} \in \mathfrak{C}^\infty(\mathbb{R}, \mathbb{C}^{m+p})$. Assume that $\lambda_i, -\mu_j \in \mathbb{C}_-$, $i = 1,\ldots,k_1$, $j = 1,\ldots,k_2$.*

Define the bilinear form $\langle , \rangle$ on $\mathfrak{B}' \cap \mathfrak{D}(\mathbb{R}, \mathbb{R}^q) \times \mathfrak{B} \cap \mathfrak{D}(\mathbb{R}, \mathbb{R}^q)$ by

$$\langle w', w\rangle := \int_0^{+\infty} y'^* u + u'^* y \, dt.$$

Then

$$\mathbb{L}_{i,j} = \langle w_i', w_j\rangle,$$

where w_i', w_j are defined by (2.14).

Proof The claim follows integrating $w_i'^\top w_j$ on the half line. □

The equality (2.18) is instrumental in obtaining the following result, analogous to Lemma 2.1 in [17].

Preposition 2.6 *Denote by n the McMillan degree of $\mathfrak{B}$. If $k_1, k_2 \geq n$, then* rank $\mathbb{L} = n$.

Proof Using the factorization (2.7) of $\Psi(\zeta,\eta)$, conclude that $\mathbb{L} = -S^* P$, where S and P are defined by

$$S := \left[Z(-\mu_1^*)s_1 \ldots Z(-\mu_{k_1}^*)s_{k_1}\right] \in \mathbb{C}^{n\times k_1}$$
$$P := \left[X(\lambda_1)p_1 \ldots X(\lambda_{k_2})p_{k_2}\right] \in \mathbb{C}^{n\times k_2}.$$

We now prove that under the assumption that the λ_is are distinct, the matrix P has full row rank n; a similar argument yields the same property for S.

Assume by contradiction that $\text{rank}(P) = r < n$; then there exist $\alpha_i \in \mathbb{C}$, $i = 1, \ldots, k_2$, not all zero, such that $P\text{col}(\alpha_i)_{i=1,\ldots,k_2} = 0$. Let $F \in \mathbb{R}^{m\times m}[\xi]$ be such that ker $\left(F\left(\frac{d}{dt}\right)\right)$ equals the subspace of $\mathfrak{C}^\infty(\mathbb{R},\mathbb{R}^m)$ spanned by $v_i e^{\lambda_i \cdot}$, $i = 1, \ldots, k_2$; such F always exists (see section XV of [32]). Now consider the following equations:

$$\begin{aligned} w &= M\left(\frac{d}{dt}\right)\ell \\ x &= X\left(\frac{d}{dt}\right)\ell \\ 0 &= F\left(\frac{d}{dt}\right)\ell. \end{aligned} \tag{2.20}$$

The external behavior $\mathfrak{B}' \subset \mathfrak{B}$ described by these equations is autonomous (see [19]), of dimension k_2. Moreover $X\left(\frac{d}{dt}\right)$ is a state map for $\mathfrak{B}'$, since it is a state map for $\mathfrak{B}$. Consider the trajectory $\hat{\ell}$ defined by $\hat{\ell}(t) := \sum_{i=1}^N \alpha_i p_i e^{\lambda_i t}$, and let $\ell = \hat{\ell}$ in (2.20); then the value of $\hat{x} := X\left(\frac{d}{dt}\right)\hat{\ell}$ at $t = 0$ is zero. Since $\mathfrak{B}'$ is autonomous, it follows that $\hat{w} := M\left(\frac{d}{dt}\right)\hat{\ell}$ is also zero. From the observability of M it follows then that $\hat{\ell} = 0$, which is in contradiction with the assumption that not all α_i's are equal to zero. Consequently, P has rank n. □

Another result well known in the Loewner framework (see the first formula in (12) p. 640 of [17]) follows in a straightforward way from (2.18) and Proposition 2.3.

Preposition 2.7 *Define the matrices*

$$\begin{aligned} M &:= \text{diag}(-\mu_i)_{i=1,\ldots,k_1} \\ \Lambda &:= \text{diag}\,(\lambda_j)_{j=1,\ldots,k_2} \\ S &:= \left[s_i^*\left[Q(\mu_i) - P(\mu_i)\right]\right]_{i=1,\ldots,k_1} \in \mathbb{C}^{k_1\times q} \\ W &:= \left[\begin{bmatrix} D(\lambda_j) \\ N(\lambda_j)\end{bmatrix} p_j\right]_{j=1,\ldots,k_2} \in \mathbb{C}^{q\times k_2}. \end{aligned}$$

$\mathbb{L}$ *satisfies the Sylvester equation*

$$M\mathbb{L} + \mathbb{L}\Lambda = -S^* W. \tag{2.21}$$

Proof Observe that

$$Q(-\zeta)^\top D(\eta) - P(-\zeta)^\top N(\eta) = \zeta \frac{Q(-\zeta)^\top D(\eta) - P(-\zeta)^\top N(\eta)}{\zeta+\eta} + \eta \frac{Q(-\zeta)^\top D(\eta) - P(-\zeta)^\top N(\eta)}{\zeta+\eta}.$$

The claim follows in a straightforward way substituting ζ with $-\mu_i^*$, η with λ_j, and multiplying on the left by s_i^* and on the right by p_j. □

Remark 2.8 In the special case of lossless- and self-adjoint port-Hamiltonian systems, the results of Propositions 2.6 and 2.7 coincide with results obtained in the B/QDF approach in [25]. Note that Proposition 2.4, on which the Loewner approach is fundamentally based, is valid for any linear differential system, while the results illustrated in [25] are valid only under the assumption of conservativeness or self-adjointness.

The transfer function $H(s) \in \mathbb{R}^{m\times m}[s]$ of a *lossless port-Hamiltonian system* (see [22, 25] for the definition) satisfies the equality $-H(-s)^\top = H(s)$. From such property, using the right and left coprime factorizations already introduced we conclude that given the image representation M, a kernel representation is

$$R(s) = M(-s)^\top \begin{bmatrix} 0 & I_m \\ I_m & 0 \end{bmatrix} = \begin{bmatrix} N(-s)^\top & D(-s)^\top \end{bmatrix}.$$

Thus for this class of systems the two-variable polynomial matrix $\Psi(\zeta, \eta)$ defined in Proposition 2.3 is

$$\Psi(\zeta,\eta) = \frac{\begin{bmatrix} N(\zeta)^\top & D(\zeta)^\top \end{bmatrix} \begin{bmatrix} D(\eta) \\ N(\eta) \end{bmatrix}}{\zeta+\eta}.$$

If we consider *symmetric* data, i.e., $k_1 = k_2$, $\mu_i = \lambda_i$ and $s_i = p_i$, $i = 1, \dots, k_1$, then it is a matter of straightforward verification to check that the Loewner matrix (2.16) coincides with the *Pick* matrix defined in formula (1) in [25]. Moreover, if the frequencies μ_i and λ_j lie all on one and the same side of the complex plane, the Pick (i.e., Loewner) matrix has a straightforward interpretation as a Gramian for the trajectories in the indefinite inner product on the half real line induced by

$$J := \begin{bmatrix} 0 & I_m \\ I_m & 0 \end{bmatrix},$$

see formulas (2.8) and (2.11) of [25].

Under the assumptions mentioned above, the rank result of Proposition 2.6 of this paper coincides with the result of Proposition 2.1 of [25], and the Sylvester equation result of Proposition 2.7 coincides with that of Proposition 2.2 of [25].

The transfer function $H(s) \in \mathbb{R}^{m\times m}[s]$ of a *self-adjoint port-Hamiltonian system* (see [25] for the definition) satisfies the equality $H(s)^\top = H(s)$, from which using the right and left coprime factorizations already introduced we conclude that given an image representation M, a kernel representation is

$$R(s) = M(s)^\top \begin{bmatrix} 0 & I_m \\ -I_m & 0 \end{bmatrix} = \begin{bmatrix} N(s)^\top & -D(s)^\top \end{bmatrix}.$$

Thus for this class of systems the two-variable polynomial matrix $\Psi(\zeta, \eta)$ defined in Proposition 2.3 is

$$\Psi(\zeta, \eta) = \frac{\begin{bmatrix} N(-\zeta)^\top & -D(-\zeta)^\top \end{bmatrix} \begin{bmatrix} D(\eta) \\ N(\eta) \end{bmatrix}}{\zeta + \eta}.$$

If we consider *symmetric* data, i.e., $k_1 = k_2$, $\mu_i = \lambda_i$ and $s_i = p_i$, $i = 1, \dots, k_1$, and if the frequencies λ_i lie all on the right or left half-plane, then the Loewner matrix (2.16) coincides with the *Pick* matrix of formula (34) in [25]. In this case, the Loewner matrix has an interpretation as Gramian for the indefinite inner product on the half real line induced by

$$J' := \begin{bmatrix} 0 & I_m \\ -I_m & 0 \end{bmatrix}.$$

Results analogous to Proposition 2.6 and Proposition 2.7 of this paper appear as Proposition 2.6 and Proposition 2.7, respectively, in [25]. □

Remark 2.9 In this chapter we restrict ourselves to the problem of modeling continuous-time trajectories. Gramian-based ideas for the identification of state-space systems in the discrete-time case under the assumption of losslessness have been illustrated in [23]. □

The shifted Loewner matrix (2.17) can be associated with a two-variable polynomial matrix in the following way. From the right and left coprime factorizations of H define

$$\Psi'(\zeta, \eta) := \frac{\zeta Q(-\zeta)^\top D(\eta) + P(-\zeta)^\top N(\eta)\eta}{\zeta + \eta}; \tag{2.22}$$

note that $\Psi'(\zeta, \eta)$ is a polynomial matrix, since substituting $-\xi$ in place of ζ and ξ in place of η in $\zeta Q(-\zeta)^\top D(\eta) + P(-\zeta)^\top N(\eta)\eta$ yields the zero matrix. The following result follows in a straightforward way from (2.22).

Preposition 2.10 *Let $\Psi' \in \mathbb{R}^{k_1\times k_2}[\zeta, \eta]$ be defined by (2.22). Then*

$$\mathbb{L}_\sigma = -\begin{bmatrix} s_i^* \Psi'(-\mu_i, \lambda_j) p_j \end{bmatrix}_{i=1,\dots,k_1;\, j=1,\dots,k_2}.$$

If the frequencies λ_i, $-\mu_i$ are all on one and the same side of the imaginary axis (e.g., the left-hand side) then the two-variable polynomial (2.22) is associated with the following BDF, and the Loewner matrix has the interpretation of a Gramian, as illustrated in the following result.

Preposition 2.11 *Assume that $\lambda_i, -\mu_i \in \mathbb{C}_-$ and partition w' and w as in Proposition 2.5. Define the following BDF on $\mathfrak{B}' \times \mathfrak{B}$:*

$$\langle\langle w', w\rangle\rangle := \int_0^{+\infty} \left(\frac{d}{dt}y'\right)^\top u + u'^\top \left(\frac{d}{dt}y\right) \, dt \; ;$$

then

$$\sigma\mathbb{L}_{i,j} = \langle\langle w_i', w_j\rangle\rangle,$$

where w_i', w_j are defined in (2.14).

Proof The claim follows integrating $\left(\frac{d}{dt}v_i\right)^\top r_j + \ell_i'^\top \left(\frac{d}{dt}w_j\right)$ on the half line. □

Another dynamical interpretation of the shifted Loewner matrix can be given as follows: associate to the behavior $\mathfrak{B}$ defined in (2.12) the behavior

$$\mathfrak{B}' := \left\{ \mathrm{col}(y', u') \mid \exists\, \mathrm{col}(y, u) \in \mathfrak{B} \text{ s.t. } y' := \frac{d}{dt}y,\, u' = u \right\} . \tag{2.23}$$

To each trajectory (2.14) in $\mathfrak{B}$, $\mathfrak{B}^\perp$ one can associate a corresponding trajectory in $\mathfrak{B}'$ by "differentiating the output variable". It is straightforward to see that the shifted Loewner matrix is the Loewner matrix of such new set of interpolation data, or equivalently, the Loewner matrix associated with the transfer function $sH(s)$. Now following an argument analogous to that used in proving Proposition 2.7, one can prove that $\mathbb{L}_\sigma$ satisfies the following Sylvester equation:

$$M\mathbb{L}_\sigma + \mathbb{L}_\sigma \Lambda = -S'P',$$

where M, L are as in Proposition 2.7 and

$$S' := \left[s_i^* \left[Q(\mu_i)\mu_i \; - P(\mu_i)\right]\right]_{i=1,\dots,k_1} \in \mathbb{C}^{k\times(l+g)}$$

$$P' := \left[\begin{bmatrix} D(\lambda_j) \\ \lambda_j N(\lambda_j)\end{bmatrix} p_j\right]_{j=1,\dots,k_2} \in \mathbb{C}^{(l+g)\times q}.$$

This is the counterpart of the second formula in (12) p. 640 of [17].

2.4 Computation of Interpolants

Generalized state-space formulas of interpolants based on the Loewner matrix and the shifted Loewner matrix are given in Lemma 5.1 p. 643 of [17]. The dimension of the generalized state variable equals the number of right interpolation data, and thus in general this procedure does not produce a minimal order interpolant; on the other hand, the interpolant is constructed directly from the Loewner and shifted Loewner matrices, without need of further computations. In Sect. 5.2 of [17] formulas for a minimal order interpolant are obtained in terms of the *short singular value decomposition* of the matrix $\nu\mathbb{L} - \mathbb{L}_\sigma$, where $\nu \in \{\mu_j\} \cup \{\lambda_i\}$, under the assumption (20) on p. 645 *ibid.* In this section we show how analogous results can be derived in the B/QDF approach; we examine separately the mono-directional interpolation problem (where only the right or left interpolation constraints need to be satisfied) and the bidirectional one.

Given a matrix $S \in \mathbb{R}^{k_1 \times k_2}$, a *rank-revealing factorization* of S is any factorization $S = U_1 U_2$ with $U \in \mathbb{R}^{k_1 \times n}$, $U_2 \in \mathbb{R}^{n \times k_2}$ of full rank $n = \text{rank } S$; such a factorization can be computed in a straightforward way from a singular value decomposition of S. The results presented in this section are based on the following fundamental result connecting rank-revealing factorizations of the Loewner matrix and state trajectories corresponding to the vector exponential ones (2.14) in the external variables of the primal- and the dual system.

Preposition 2.12 *Let $\mathbb{L} = Z^* V$ be any rank-revealing factorization of the Loewner matrix associated with the data (2.8); denote by V_i, respectively Z_i, the ith column of V, respectively, Z.*

There exists a minimal state representation (2.3) of $\mathfrak{B}$, respectively $\mathfrak{B}^\perp$, such that $V_i e^{\lambda_i \cdot}$, respectively, $Z_i e^{-\mu_i \cdot}$, are minimal state trajectories of $\mathfrak{B}$, respectively, $\mathfrak{B}^\perp$.

Proof The claim follows straightforwardly from Propositions 2.3 and 2.4. □

Different rank-revealing factorizations of $\mathbb{L}$ yield different state trajectories and thus different realizations; see [24] for an application to the computation of canonical realizations.

2.4.1 Mono-directional Interpolants and Factorizations of the Loewner Matrix

We first show that under suitable assumptions on the number of interpolation data, a minimal state representation (2.3) of an interpolant of the *right* interpolation data can be computed from a rank-revealing factorization of $\mathbb{L}$.

Preposition 2.13 *Assume* $k_1, k_2 \geq n = \text{rank}(\mathbb{L})$*, and let* $\mathbb{L} = Z^* V$ *be a rank-revealing factorization with* $Z \in \mathbb{C}^{n \times k_1}$ *and* $V \in \mathbb{C}^{n \times k_2}$*. Define*

$$M := \text{diag}(-\mu_i)_{i=1,\ldots,k_1} \in \mathbb{C}^{k_1 \times k_1}$$
$$S := \left[s_i^* \left[Q(\mu_i) \; -P(\mu_i)\right]\right]_{i=1,\ldots,k_1} \in \mathbb{C}^{k_1 \times q}$$

Then a minimal state representation (2.3) of a right interpolant for the data $\left(\lambda_i, \begin{bmatrix} r_i \\ w_i \end{bmatrix}\right)$*,* $i = 1, \ldots, k_2$ *is*

$$Z^* \frac{d}{dt} x + \left(M Z^*\right) x + S w = 0. \tag{2.24}$$

Proof We prove that the external behavior of (2.24) contains the trajectories $\begin{bmatrix} r_i \\ w_i \end{bmatrix} e^{\lambda_i \cdot}$, $i = 1, \ldots, k_2$, i.e., that there exist trajectories x_i, $i = 1, \ldots, k_2$ such that (2.24) is satisfied. Denote by v_i the ith column of the matrix V of the rank-revealing factorization of $\mathbb{L}$, and define $x_i(\cdot) := v_i e^{\lambda_i \cdot}$, $i = 1, \ldots, k_2$. It follows from Proposition 2.12 and the Sylvester Eq. (2.21) that with such positions (2.24) is satisfied. □

Remark 2.14 Formula (2.24) is similar to formula (15) p. 642 of [17], which gives a input-state-output representation of an interpolant of McMillan degree k_1. Note however that the McMillan degree of (2.24) equals $\text{rank}(\mathbb{L})$.

Remark 2.15 Proposition 2.13 implies that the rational matrix $-(sZ^* + MZ^*)^{-1} S$ satisfies the equations

$$(\lambda_i Z^* + M Z^*)^{-1} S \begin{bmatrix} r_i \\ w_i \end{bmatrix} = v_i \;, \; i = 1, \ldots, k_2,$$

where v_i is the ith column of the matrix V associated with the rank-revealing factorization of $\mathbb{L}$. Thus the matrix V plays a role analogous to that of the generalized tangential controllability matrix of p. 639 of [17]. □

Remark 2.16 When minimal, respectively, observable, kernel, and image representations of $\mathfrak{B}$ are known, a state representation (2.3) of $\mathfrak{B}$ can be obtained directly from the coefficient matrices of $Z(\xi)$ and $X(\xi)$ in (2.7), see sect. 2.5 of [29]. □

In order to find an input-state-output (iso) representation

$$\begin{aligned} E \frac{d}{dt} x &= A x + B u \\ y &= C x + D u \end{aligned} \tag{2.25}$$

of an interpolant, assume $k_1, k_2 \geq n = \text{rank}(\mathbb{L})$, and compute a rank-revealing factorization $\mathbb{L} = Z^* V$. Define

$$U := \begin{bmatrix} r_1 & \dots & r_{k_1} \end{bmatrix} \in \mathbb{C}^{m \times k_1}$$
$$Y := \begin{bmatrix} w_1 & \dots & w_{k_1} \end{bmatrix} \in \mathbb{C}^{p \times k_1}.$$

The following result, whose proof is straightforward and hence omitted, characterizes ISO representations of right interpolants.

Preposition 2.17 *A quintuple* $(E, A, B, C, D) \in \mathbb{R}^{n \times n} \times \mathbb{R}^{n \times n} \times \mathbb{R}^{n \times m} \times \mathbb{R}^{n \times p} \times \mathbb{R}^{m \times m}$ *defines an ISO representation of a right interpolant if and only if*

$$\begin{bmatrix} E & -A & -B & 0_{n \times p} \\ 0 & C & D & -I_p \end{bmatrix} \begin{bmatrix} V\Lambda \\ V \\ U \\ Y \end{bmatrix} = 0. \tag{2.26}$$

It follows from Proposition 2.17 that in order to find an ISO representation of a right interpolant it suffices to find a matrix whose rows form a basis for the space orthogonal to im $\begin{bmatrix} V\Lambda \\ V \\ U \\ Y \end{bmatrix}$, and with the special structure

$$\begin{bmatrix} E & -A & -B & 0_{n \times p} \\ 0 & C & D & -I_p \end{bmatrix}.$$

This can be achieved with standard linear algebra computations; we will not deal with such details here.

Remark 2.18 In Proposition 2.4 and section VI of [25] explicit formulas in terms of the matrices arising from a rank-revealing factorization of $\mathbb{L}$ are given for computing A, B, C, D of an input-state-output representation

$$\frac{d}{dt}x = Ax + Bu$$
$$y = Cx + Du$$

of a right interpolant for data generated by conservative- and adjoint port-Hamiltonian systems (see Remark 2.8 of this paper). Moreover, a parametrization for all such interpolants is also given. □

Remark 2.19 Following an argument analogous to that used in proving Proposition 2.13 it can be shown that a state representation (2.3) of an interpolant for the *left* interpolation data can be computed defining $E := V^*$, $F := V^*\mathrm{diag}(\lambda_i)$, $G := W^*$. Moreover, a result analogous to that of Proposition 2.17 holds true also for left interpolants; we will not state it explicitly. □

2.4.2 Bidirectional Interpolation and BDFs

In Theorem 5.1 of [17] formulas are given for the matrices E, A, B, and C of an ISO representation (2.25) of a left and right interpolant. In the following we show that these can be given an interpretation in terms of BDFs, and in case the interpolation points are all on the same side of the imaginary axis, in terms of factorization of the Loewner and shifted Loewner matrix.

In the following, besides the ISO representation (2.25) we consider its *dual* (note that the terminology "dual" is not uniform in the literature; on this issue see also [8, 10, 11]), defined by

$$\begin{aligned} E^\top \frac{d}{dt} z &= -A^\top z - C^\top u' \\ y' &= -B^\top z, \end{aligned} \tag{2.27}$$

where $z \in \mathfrak{C}^\infty(\mathbb{R}, \mathbb{R}^n)$, $u' \in \mathfrak{C}^\infty(\mathbb{R}, \mathbb{R}^p)$, $y' \in \mathfrak{C}^\infty(\mathbb{R}, \mathbb{R}^m)$.

The following two results are crucial for computing E and A from factorizations of the Loewner matrices.

Proposition 2.20 *Let* $\mathrm{col}(x, u, y)$ *and* $\mathrm{col}(z, u', y')$ *be full trajectories of the behaviors described by (2.25) and (2.27), respectively. Then*

$$\frac{d}{dt}\left(z^\top E x\right) = -u'^\top y - y'^\top u = -\begin{bmatrix} u'^\top & y'^\top \end{bmatrix} \begin{bmatrix} 0 & I_p \\ I_m & 0 \end{bmatrix} \begin{bmatrix} u \\ y \end{bmatrix}. \tag{2.28}$$

Proof The claim follows from the following chain of equalities:

$$\begin{aligned} \frac{d}{dt}\left(z^\top E x\right) &= \left(\frac{d}{dt} z^\top\right) E x + z^\top E \left(\frac{d}{dt} x\right) = \left(-z^\top A - u'^\top C\right) x + z^\top (Ax + Bu) \\ &= -u'^\top y - y'^\top u. \end{aligned}$$

We now state another important result.

Proposition 2.21 *Let* $\mathrm{col}(x, u, y)$ *and* $\mathrm{col}(z, u', y')$ *be full trajectories of the behaviors described by (2.25) and (2.27), respectively. Then*

$$\frac{d}{dt}\left(z^\top A x\right) = -u'^\top \left(\frac{d}{dt} y\right) + \left(\frac{d}{dt} y'^\top\right) u. \tag{2.29}$$

Proof The claim follows from the following chain of equalities:

$$
\begin{aligned}
u'^{\top}\left(\frac{d}{dt}y\right)-\left(\frac{d}{dt}y'^{\top}\right)u &= u'^{\top}\left(C\frac{d}{dt}x\right)-\left(-\frac{d}{dt}z^{\top}B\right)u\\
&=\left(u'^{\top}C\right)\frac{d}{dt}x+\frac{d}{dt}z^{\top}(Bu)\\
&=-\left(\frac{d}{dt}z^{\top}E+z^{\top}A\right)\frac{d}{dt}x+\frac{d}{dt}z^{\top}\left(E\frac{d}{dt}x-Ax\right)\\
&=-\frac{d}{dt}\left(z^{\top}Ax\right)
\end{aligned}
$$

The next result follows in a straightforward way from Propositions 2.20 and 2.21 and reformulates (2.28) and (2.29) in two-variable polynomial terms.

Preposition 2.22 *Let $R \in \mathbb{R}^{p\times(p+m)}[\xi]$, respectively, $M \in \mathbb{R}^{(m+p)\times m}[\xi]$ be a minimal kernel, respectively, observable image representation of the external behavior $\mathfrak{B}$ of (2.25). Define*

$$
\begin{aligned}
\Psi(\zeta,\eta) &:= R(-\zeta)M(\eta)\\
\Psi'(\zeta,\eta) &:= R(-\zeta)\begin{bmatrix}0 & -I_p\eta\\ \zeta I_m & 0\end{bmatrix}M(\eta).
\end{aligned}
$$

There exist state maps $X, Z \in \mathbb{R}^{\bullet\times m}[\xi]$ for $\mathfrak{B}$ and $\mathfrak{B}^{\perp}$, respectively, such that

$$
\begin{aligned}
\Psi(\zeta,\eta) &= (\zeta+\eta)Z(\zeta)^{\top}EX(\eta)\\
\Psi'(\zeta,\eta) &= (\zeta+\eta)Z(\zeta)^{\top}AX(\eta).
\end{aligned}
\tag{2.30}
$$

The following is an important consequence of Propositions 2.20, 2.21 and 2.22.

Preposition 2.23 *Let (2.25) be an ISO representation of a bidirectional interpolant. There exist $X', X \in \mathbb{C}^{n\times k}$ such that*

$$
\begin{aligned}
\mathbb{L} &= X'^{*}EX\\
\mathbb{L}_s &= X'^{*}AX.
\end{aligned}
\tag{2.31}
$$

Moreover, the columns of X', respectively, X correspond to the directions of (exponential) state trajectories of the dual, respectively, primal system, corresponding to the external trajectories (2.14).

Proof The claim follows by substituting μ_i in place of ζ and λ_i in place of η in (2.30), and multiplying on the left by s_i^* and on the right by p_j. □

Remark 2.24 If $-\mu_i$ and λ_j lie on the same half plane, the result of Proposition 2.23 can be proved integrating by parts (2.28) and (2.29) along the trajectories (2.14). □

To compute E and A from $\mathbb{L}$ and $\mathbb{L}_s$, respectively, observe that from (2.31) it follows that

$$\begin{aligned}\begin{bmatrix}\mathbb{L} \ \mathbb{L}_s\end{bmatrix} &= X'^* \begin{bmatrix}EX \ AX\end{bmatrix}\\ \begin{bmatrix}\mathbb{L}\\ \mathbb{L}_s\end{bmatrix} &= \begin{bmatrix}X'^* E\\ X'^* A\end{bmatrix} X.\end{aligned} \tag{2.32}$$

These factorizations are the counterpart of those in formula (2.25) of [5], with $Y = X'^*$, $\Sigma_\ell \widetilde{X}^* = \begin{bmatrix}EX \ AX\end{bmatrix}$ and $\widetilde{Y}\Sigma_r = \begin{bmatrix}X'^* E\\ X'^* A\end{bmatrix}$. A "short" SVD of the two matrices on the left-hand side of (2.32) yields matrices X'^* and X with orthonormal rows; under such assumption we recover E and A by projection of $\mathbb{L}$ and $\mathbb{L}_s$ as

$$\begin{aligned}E &= X'\mathbb{L}X^*\\ A &= X'\mathbb{L}_s X^*,\end{aligned}$$

respectively, see the first two formulas (22) p. 646 of [17].

The matrices B, C of a representation (2.25) can be obtained as follows. From the output equation $y' = -B^\top z$ of the dual system (2.27) it follows that $V = -B^\top X'$, where

$$V := \begin{bmatrix}\ell_1 \dots \ell_{k_1}\end{bmatrix} \in \mathbb{C}^{m\times k_1}.$$

Assuming that X' has been obtained via a short SVD, it follows that

$$B = -X'V^*.$$

This is the third equation in (2.28) p. 17 of [6]. Analogously, from the output equation $y = Cx$ of the primal system (2.25) it follows that $W = CX$, where

$$W := \begin{bmatrix}w_1 \dots w_{k_2}\end{bmatrix} \in \mathbb{C}^{m\times k_2}.$$

Consequently

$$C = WX^*,$$

the fourth equation in (2.28) p. 17 of [6].

Remark 2.25 The BDFs used to compute E and A in Propositions 2.20 and 2.21 are not the same; such difference goes against the interpretation of the shifted Loewner matrix as the Loewner matrix associated with the transfer function $sH(s)$. It is currently investigated whether such asymmetry depends on our possibly nonstandard definition of the dual system (2.27), or whether there is an intrinsic motivation to it. □

2.5 Conclusions

We have shown that several results in the Loewner framework for interpolation can be given a direct interpretation in the language of bilinear differential forms and their two-variable polynomial matrix representations. We have shed new light on known results in the Loewner framework (e.g., the rank result of Proposition 2.6, the Sylvester equation in Proposition 2.7), and we have also given insights of a more fundamental nature (e.g., the correspondence between state trajectories and factorizations in Proposition 2.12, the interpretation of the Loewner matrices as Gramians, see Propositions 2.5 and 2.11).

For reasons of space we have refrained from illustrating the correspondences between the Loewner approach to model order reduction and that based on BDFs (see Sect. 3 of [6], section V of [25]); this will be pursued elsewhere. Current research questions include the formulation of recursive interpolation in the BDF framework, and the extension to parametric interpolation and parametric model order reduction (see [12]).

Acknowledgments The authors would like to thank Prof. Dr. A.J. van der Schaft for stimulating discussions.

The results presented here were obtained during the second author's visit (supported by a travel grant of the UK Engineering and Physical Sciences Research Council) to the Jan C. Willems Center for Systems and Control, Johann Bernoulli Institute for Mathematics and Computer Science, University of Groningen, The Netherlands.

References

1. Anderson, B.D.O., Antoulas, A.C.: Rational interpolation and state variable realizations. Linear Algebr. Appl. **137**(138), 479–509 (1995)
2. Antoulas, A.C.: Advances in Design and Control. Approximation of large-scale dynamical systems. SIAM, Philadelphia (2008)
3. Antoulas, A.C.: On the construction of passive models from frequency response data. Automatisierungstechnik **AT–56**, 447–452 (2008)
4. Antoulas, A.C., Anderson, B.D.O.: On the scalar rational interpolation problem. IMA J. Math. Control. Inf. **3**, 61–88 (1986)
5. Antoulas, A.C., Ionita, A.C., Lefteriu, S.: On two-variable rational interpolation. Linear Algebr. Appl. **436**, 2889–2915 (2012)
6. Antoulas, A.C., Lefteriu, S., Ionita, A.C.: A tutorial introduction to the Loewner framework for model reduction. In: Benner, P., Cohen, A., Ohlberger, M., Willcox, K. (eds.) Model Reduction and Approximation for Complex Systems, Birkhäuser, ISNM Series (2015)
7. Antoulas, A.C., Willems, J.C.: A behavioral approach to linear exact modeling. IEEE Trans. Autom. Control. **38**, 1776–1802 (1993)
8. Cobb., D.: Controllability, observability, and duality in singular systems. IEEE Trans. Autom. Control. **29**(12), 1076–1082 (1984)
9. Fuhrmann, P.A., Rapisarda, P., Yamamoto, Y.: On the state of behaviors. Linear Algebr. Appl. **424**(2–3), 570–614 (2007)
10. Ilchmann, A., Mehrmann, V.: A behavioral approach to time-varying linear systems. Part 1 general theory. SIAM J. Control. Optim. **44**(5), 1725–1747 (2005)

11. Ilchmann, A., Mehrmann, V.: A behavioral approach to time-varying linear systems. Part 2: descriptor systems. SIAM J. Control. Optim. **44**(5), 1748–1765 (2005)
12. Ionita, A.C., Antoulas, A.C.: Parametrized model order reduction from transfer function measurements, in reduced order methods for modeling and computational reduction. In: Quarteroni, A., Rozza, G. (eds) Modeling, Computations, and Applications. Springer, Berlin (2013)
13. Kaneko, O., Rapisarda, P.: Recursive exact H_∞-identification from impulse-response measurements. Syst. Control. Lett. **49**, 323–334 (2003)
14. Kaneko, O., Rapisarda, P.: On the Takagi interpolation problem. Linear Algebr. Appl. **425**(2–3), 453–470 (2007)
15. Lefteriu, S., Antoulas, A.C.: A new approach to modeling multiport systems from frequency-domain data. IEEE Trans. CAD **29**, 14–27 (2010)
16. Lefteriu, S., Antoulas, A.C.: Model reduction for circuit simulation. In: Benner, P., Hinze, M., ter Mater, E.J.W. (eds) Topics in Model Order Reduction with Applications to Circuit Simulation. Springer, Berlin (2011)
17. Mayo, A.J., Antoulas, A.C.: A behavioural approach to positive real interpolation. J. Math. Comput. Model. Dyn. Syst. **8**, 445455 (2002)
18. Mayo, A.J., Antoulas, A.C.: A framework for the solution of the generalized realization problem. Linear Algebr. Appl. **425**, 634–662 (2007)
19. Polderman, J.W., Willems, J.C.: Introduction to Mathematical System Theory. Springer, New York (1997)
20. Rapisarda, P.:. Linear differential systems. Ph.D. Thesis, University of Groningen (1998)
21. Rapisarda, P., Rao, S.: Realization of lossless systems via constant matrix factorizations. IEEE Trans. Autom. Control. **58**, 2632–2636 (2013)
22. Rapisarda, P., Trentelman, H.L.: Linear Hamiltonian behaviors and bilinear differential forms. SIAM J. Control. Optim. **43**(3), 769–791 (2004)
23. Rapisarda, P., Trentelman, H.L.: Identification and data-driven model reduction of state-space representations of lossless and dissipative systems from noise-free data. Automatica **47**, 1721–1728 (2011)
24. Rapisarda, P., van der Schaft, A.J.: Canonical realizations by factorization of constant matrices. Syst. Control. Lett. **61**(8), 827–833 (2012)
25. Rapisarda, P., van der Schaft, A.J.: Identification and data-driven reduced-order modeling for linear conservative port- and self-adjoint Hamiltonian systems. In Proceeding of the 52nd IEEE CDC, pp. 145–150 (2013)
26. Rapisarda, P., Willems, J.C.: State maps for linear systems. SIAM J. Control. Optim. **35**(3), 1053–1091 (1997)
27. Rapisarda, P., Willems, J.C.: The subspace Nevanlinna interpolation problem and the most powerful unfalsified model. Syst. Control. Lett. **32**, 291–300 (1997)
28. Rapisarda, P., Willems, J.C.: Conserved—and zero-mean quadratic quantities for oscillatory systems. Math. Syst. Signal Control. **17**, 173–200 (2005)
29. van der Schaft, A.J., Rapisarda, P.: State maps from integration by parts. SIAM J. Control. Optim. **49**, 2415–2439 (2011)
30. Willems, J.C.: From time series to linear system—part I. Finite dimensional linear time invariant systems. Automatica **22**, 561–580 (1986)
31. Willems, J.C.: From time series to linear system—part II. Exact modelling. Automatica **22**, 675–694 (1986)
32. Willems, J.C.: Paradigms and puzzles in the theory of dynamical systems. IEEE Trans. Autom. Control. **36**, 259–294 (1991)
33. Willems, J.C., Trentelman, H.L.: On quadratic differential forms. SIAM J. Control. Optim. **36**, 1703–1749 (1998)

Chapter 3
Noninteraction and Triangular Decoupling Using Geometric Control Theory and Transfer Matrices

Jacob van der Woude

Abstract In this chapter we consider linear systems that in addition to a control input and a measurement output also have μ exogenous inputs and μ exogenous outputs. The main topic is then the design of measurement feedback controllers such that in the closed-loop system the transfer matrix between certain specified pairs of exogenous inputs and outputs is zero. Two cases are considered in particular. First, the case that the off-diagonal blocks of the closed-loop system transfer matrix are zero, resulting in a noninteracting behavior. And second, the case that only the blocks above the main diagonal in the closed-loop system transfer matrix are zero, resulting in triangular decoupling. The techniques in this chapter to derive solvability conditions and measurement feedback controllers are based on transfer matrices and the celebrated geometric approach toward system theory. The main results are necessary and sufficient conditions in geometric or transfer matrix terms for the noninteracting control problem and the triangular decoupling problem. Also variations of these problems are treated, like the version with additional stability requirements, or the 'almost' version of the two problems.

3.1 Introduction

In this chapter we present results of joint research with Harry Trentelman that was done in the beginning of his Eindhoven period. It was the time before H_∞- and H_2-control and the behavioral approach toward system theory. More precisely, it was the time that the geometric approach toward system theory still was fully in the spotlight of research in system theory.

J. van der Woude (✉)
Delft Institute of Applied Mathematics, Delft University of Technology,
Mekelweg 4, 2628 CD Delft, The Netherland
e-mail: j.w.vanderwoude@tudelft.nl

M.N. Belur et al. (eds.), *Mathematical Control Theory II*,
Lecture Notes in Control and Information Sciences 462,
DOI 10.1007/978-3-319-21003-2_3

Harry and I started both in 1974 with the study of Mathematics at the Rijksuniversiteit Groningen. About 7 years later, in June 1981, we obtained the MSc degree in Mathematics. Harry was 10 min before me, because his family name comes before mine, alphabetically speaking. Harry still likes to joke that he graduated (a long time) before me.

In our MSc period, around 1979, we were both educated in system theory by Jan Willems, who then was working on invariant and almost invariant subspaces. During the lectures of Jan Willems, we got immersed in the geometric approach toward system theory based on the well-known book by Wonham [12]. The approach is an elegant linear algebra based way of treating fundamental issues in system theory. Although it was not easy, Harry and I got infected by this geometric approach virus. For this reason, we both continued working on invariant and almost invariant subspaces in subsequent PhD projects after our MSc period. Harry directly became PhD student under the supervision of Jan Willems. I first made a detour to industry for about a year before retuning to the university, i.e., to the Eindhoven University of Technology, where I became a PhD student under the supervision of Malo Hautus.

First, we worked separately from each other on challenges in system theory. Unaware of each other, we tried to tackle the same problem on noninteracting control, formulated by Jan Willems in the conference paper [8]. The moment after his PhD that Harry became assistant professor in Eindhoven we learned about each other's research and results on the topic of noninteracting control. Harry had become an expert in all kinds of *exact* and *almost* invariant subspaces, and I had been able to develop some intuition behind the notion of radical, introduced by Wonham.

It was nice that we could complement each other. Together, we could tackle certain aspects of the noninteracting control problem. Jointly, we were definitely more than the two of us alone. The collaboration resulted in a paper for the CDC in Athens and a paper in Linear Algebra and its Applications. Some of our results are also reported in this chapter.

Inspired by our positive results on noninteracting control I continued to study the problem of triangular decoupling and was lucky to find conditions for the solvability of various types of this problem. The results of this research were included in my PhD thesis and some of them are also reported in this chapter.

Working together with Harry in Eindhoven was great. His typical style and humor made it a very nice experience that I never will forget. He also supervised me when I was completing my PhD thesis. It was therefore great that Harry could be co-promoter in my PhD committee, next to the promotors Malo Hautus and Jan Willems.

Harry has a thorough way of working, is very methodically, enthusiastically, with an eye for detail, and has a good sense of humor. I want to thank him for the nice collaboration in Eindhoven, although it lasted for only 2 years or so. I have learned a lot from and by him.

Harry, thank you for the great collaboration during my PhD period in Eindhoven. Perhaps we can do a joint project once more again?

3.2 Setting the Scene

In this chapter we consider the linear system described by

$$\dot{x}(t) = Ax(t) + Bu(t) + \Sigma_{i\in\underline{\mu}} G_i v_i(t), \tag{3.1}$$

$$y(t) = Cx(t), \tag{3.2}$$

$$z_i(t) = H_i x(t), \qquad i \in \underline{\mu}, \tag{3.3}$$

where $\mu \in \mathbb{N}$, $\mu > 1$ and $\underline{\mu} := \{1, 2, \cdots, \mu\}$. In the above, $x(t) \in \mathbb{R}^n$ denotes the state, $u(t) \in \mathbb{R}^m$ the (control) input, $y(t) \in \mathbb{R}^p$ the (measurement) output, and A, B and C are matrices of suitable dimensions. Further, $v_i(t) \in \mathbb{R}^{q_i}$ denotes the ith exogenous input and $z_i(t) \in \mathbb{R}^{r_i}$ the ith exogenous output, where $i \in \underline{\mu}$. The matrices G_i and H_i, for $i \in \underline{\mu}$, are matrices of suitable dimensions.

Throughout this chapter we assume that the system (3.1)–(3.3) is controlled by means of a measurement feedback compensator of the form

$$\dot{w}(t) = Kw(t) + Ly(t), \tag{3.4}$$

$$u(t) = Mx(t) + Ny(t), \tag{3.5}$$

where $w(t) \in \mathbb{R}^k$ denotes the state of the compensator, and K, L, M, and N are matrices of suitable dimensions.

The interconnection of the system (3.1)–(3.3) and compensator (3.4)–(3.5) yields a *closed-loop* system with μ exogenous inputs and μ exogenous outputs. The closed-loop system is described by

$$\dot{x}_e(t) = A_e x_e(t) + \Sigma_{i\in\underline{\mu}} G_{i,e} v_i(t), \tag{3.6}$$

$$z_i(t) = H_{i,e} x_e(t), \qquad i \in \underline{\mu}, \tag{3.7}$$

where

$$x_e(t) = \begin{bmatrix} x(t) \\ w(t) \end{bmatrix}, \quad A_e = \begin{bmatrix} A + BNC & BM \\ LC & K \end{bmatrix}, \tag{3.8}$$

and

$$G_{i,e} = \begin{bmatrix} G_i \\ 0 \end{bmatrix}, \quad H_{i,e} = \begin{bmatrix} H_i & 0 \end{bmatrix}, \quad i \in \underline{\mu}. \tag{3.9}$$

Throughout this chapter we denote the set of rational functions with real coefficients by $\mathbb{R}(s)$. The set of *proper* rational functions with real coefficients and *strictly proper* rational functions with real coefficients is denoted by $\mathbb{R}_0(s)$ and $\mathbb{R}_+(s)$, respectively. We call a vector a ((strictly) proper) rational if all its components are in the set of ((strictly) proper) rational functions, and similarly for matrices.

We assume in this chapter that a stability region $\mathbb{C}_g$ is given a priori. Such a region is a nonempty subset of the complex plane that is 'symmetric' with respect to the real axis, with a nonempty intersection with the real axis when necessary.

We say that a rational function, vector, or matrix is stable if all of its poles are located in $\mathbb{C}_g$. The set of eigenvalues of a real square matrix is denoted by σ. For instance, the eigenvalues of the closed-loop system (3.6)–(3.7) are $\sigma(A_e)$.

Let $T(s)$ denote the transfer matrix of the closed-loop system (3.6)–(3.7). Then $T(s)$ can be partitioned according to the dimensions of the exogenous inputs and outputs as $T(s) = (T_{ij}(s)),\ i, j \in \underline{\mu}$, where $T_{ij}(s) = H_{i,e}(sI - A_e)^{-1}G_{j,e}$ denotes the transfer matrix between the jth exogenous input and the ith exogenous output in the closed-loop system (3.6)–(3.7).

We denote the transfer matrices in the *open loop* system (3.1)–(3.3) by

$$\begin{array}{ll} P(s) = C(sI - A)^{-1}B, & M_j(s) = C(sI - A)^{-1}G_j, \\ L_j(s) = H_i(sI - A)^{-1}B, & K_{ij}(s) = H_i(sI - A)^{-1}G_j, \end{array}$$

where $i, j \in \underline{\mu}$, and the transfer matrix of the compensator (3.4)–(3.5) by

$$F(s) = N + M(sI - K)^{-1}L.$$

An easy calculation shows that in the closed-loop system (3.6)–(3.7)

$$T_{i\,j}(s) = K_{i\,j}(s) + L_i(s)X(s)M_j(s), \quad i, j \in \underline{\mu},$$

where $X(s) = (I - F(s)P(s))^{-1}F(s)$. Note that the inverse in the latter expression exists as a rational matrix, because $I - F(s)P(s)$ is a *bicausal* rational matrix (cf. [2]). A bicausal rational matrix is a proper rational matrix with a proper rational inverse. A proper rational matrix is bicausal if and only if its determinant does not vanish at infinity. It is clear that $X(s)$ is a proper rational matrix and that $F(s) = X(s)(I + P(s)X(s))^{-1}$.

3.3 Problem Formulations and Basic Result

In the spirit of [8] we formulate the following control problems, where we assume that the linear system (3.1)–(3.3) is given together with a stability region $\mathbb{C}_g$. Also, we recall a basic result from [9].

3.3.1 Noninteracting Control

Definition 3.1 The *noninteracting control problem by measurement feedback*, denoted NICPM$_\mu$, consists of finding a measurement feedback compensator (3.4)–(3.5) such that in the closed-loop system (3.6)–(3.7): $T_{i\,j}(s)=0$ for all $i, j \in \underline{\mu}$ with $i \neq j$.

If, in addition to the feature of noninteracting control, we also require internal stabilization, we obtain the following.

Definition 3.2 The *noninteracting control problem by measurement feedback with internal stabilization*, denoted NICPM_μ^s, consists of finding a measurement feedback compensator (3.4)–(3.5) such that in the closed-loop system (3.6)–(3.7): $T_{i\,j}(s) = 0$ for all $i, j \in \underline{\mu}$ with $i \neq j$ and $\sigma(A_e) \subseteq \mathbb{C}_g$.

The almost version, using the H_∞-norm for stable transfer functions, of NICPM_μ reads as follows.

Definition 3.3 The *almost noninteracting control problem by measurement feedback*, denoted ANICPM_μ, consists of finding, for all $\varepsilon > 0$, a measurement feedback compensator (3.4)–(3.5) such that in the closed-loop system (3.6)–(3.7): $||T_{i\,j}(s)||_\infty \leq \varepsilon$ for all $i, j \in \underline{\mu}$ with $i \neq j$.

The following corollary is now easily follows from the previous.

Corollary 3.4

(a) *NICPM_μ is solvable if and only if there exists a proper rational matrix $X(s)$ such that $K_{i\,j}(s) + L_i(s)X(s)M_j(s) = 0$ for all $i, j \in \underline{\mu}$ with $i \neq j$.*
(b) *ANICPM_μ is solvable if and only if for all $\varepsilon > 0$ there exists a proper rational matrix $X(s)$ such that $||K_{i\,j}(s) + L_i(s)X(s)M_j(s)||_\infty \leq \varepsilon$ for all $i, j \in \underline{\mu}$ with $i \neq j$.*

3.3.2 Triangular Decoupling

Definition 3.5 The *triangular decoupling problem by measurement feedback*, denoted TDPM_μ, consists of finding a measurement feedback compensator (3.4)–(3.5) such that in the closed-loop system (3.6)–(3.7): $T_{i\,j}(s) = 0$ for all $i, j \in \underline{\mu}$ with $i < j$.

If we require additional internal stabilization, we obtain the following.

Definition 3.6 The *triangular decoupling problem by measurement feedback with internal stabilization*, denoted TDPM_μ^s, consists of finding a measurement feedback compensator (3.4)–(3.5) such that in the closed-loop system (3.6)–(3.7): $T_{i\,j}(s) = 0$ for all $i, j \in \underline{\mu}$ with $i < j$ and $\sigma(A_e) \subseteq \mathbb{C}_g$.

The almost version of TDPM_μ is given as follows.

Definition 3.7 The *almost triangular decoupling problem by measurement feedback*, denoted ATDPM_μ, consists of finding, for all $\varepsilon > 0$, a measurement feedback compensator (3.4)–(3.5) such that in the closed-loop system (3.6)–(3.7): $||T_{i\,j}(s)||_\infty \leq \varepsilon$ for all $i, j \in \underline{\mu}$ with $i < j$.

Also, the following corollary is immediate.

Corollary 3.8

(a) *$TDPM_\mu$ is solvable if and only if there exists a proper rational matrix $X(s)$ such that $K_{ij}(s) + L_i(s)X(s)M_j(s) = 0$ for all $i, j \in \underline{\mu}$ with $i < j$.*
(b) *$ATDPM_\mu$ is solvable if and only if for all $\varepsilon > 0$ there exists a proper rational matrix $X(s)$ such that $||K_{ij}(s) + L_i(s)X(s)M_j(s)||_\infty \leq \varepsilon$ for all $i, j \in \underline{\mu}$ with $i < j$.*

3.3.3 Basic Result

In order to develop a method by which we can check the solvability of (A)NICPM$_\mu$ and (A)TDPM$_\mu$, we consider the rational matrix equation

$$U(s)X(s) = W(s), \tag{3.10}$$

where $U(s)$ and $W(s)$ are given *strictly proper rational* matrices, $X(s)$ is the unknown *rational* matrix, and where all matrices have suitable dimensions. We say that (3.10) is solvable over $\mathbb{R}(s)$, respectively, over $\mathbb{R}_0(s)$, if there exists a *rational* matrix $X(s)$, respectively, a *proper rational* matrix $X(s)$, such that (3.10) is satisfied.

The next theorem is due to [10] and plays an important role in this chapter.

Theorem 3.9 *For every $\varepsilon > 0$ there exists a proper rational matrix $X_\varepsilon(s)$ such that $||U(s)X_\varepsilon(s) - W(s)||_\infty \leq \varepsilon$ if and only if (3.10) is solvable over $\mathbb{R}(s)$.*

In other words, the solvability of (3.10) over $\mathbb{R}(s)$ is equivalent to the *almost* solvability of (3.10) over $\mathbb{R}_0(s)$.

The proof of the theorem in [10] is quite involved. For an alternative proof, see [13] or [14]. There also is a construction given about how a possibly nonproper rational solution $X(s)$ of (3.10) can be modified into a proper rational matrix $X_\varepsilon(s)$ such that $||U(s)X_\varepsilon(s) - W(s)|| \leq \varepsilon$, where $\varepsilon > 0$ is an a priori given error bound.

Observe that the collection of $\mu^2 - \mu$ equations $K_{ij}(s) + L_i(s)X(s)M_j(s) = 0$ for all $i, j \in \underline{\mu}$ with $i \neq j$, involved in noninteracting control, can be reformulated by Kronecker products into one (large) equation of the form (3.10) (note that in this process the meaning of matrix $X(s)$ is changing). A similar remark holds with respect to triangular decoupling.

Since solvability over $\mathbb{R}(s)$ is equivalent to *almost* solvability over $\mathbb{R}_0(s)$, the following corollary is now obvious.

Corollary 3.10

(a) *$ANICPM_\mu$ is solvable if and only if there exists a rational matrix $X(s)$ such that $K_{ij}(s) + L_i(s)X(s)M_j(s) = 0$ for all $i, j \in \underline{\mu}$ with $i \neq j$.*
(b) *$ATDPM_\mu$ is solvable if and only if there exists a rational matrix $X(s)$ such that $K_{ij}(s) + L_i(s)X(s)M_j(s) = 0$ for all $i, j \in \underline{\mu}$ with $i < j$.*

It is clear that the solvability of ANICPM$_\mu$ is equivalent to the solvability of a certain rational matrix equation of the form (3.10) over $\mathbb{R}(s)$, whereas the solvability of NICPM$_\mu$ is equivalent to the solvability of the same rational matrix equation over $\mathbb{R}_0(s)$. Furthermore, it is clear that any proper rational solution to this equation provides a proper rational $X(s)$ such that $K_{i\,j}(s)+L_i(s)X(s)M_j(s) = 0$ for all $i,\, j \in \underline{\mu}$ with $i \neq j$. The transfer matrix of a compensator that solves NICPM$_\mu$ can then be calculated as $F(s) = X(s)(I+P(s)X(s))^{-1}$. Realizing $F(s)$ as $N+M(sI-K)^{-1}L$ a compensator of type (3.6)–(3.7) is obtained that solves NICPM$_\mu$.

Starting from a rational solution of the equation of type (3.10) mentioned above and a positive ε, a proper rational matrix $X_\varepsilon(s)$ can be computed by the above mentioned construction in [13] or [14], such that $||K_{i\,j}(s)+L_i(s)X(s)M_j(s)||_\infty \leq \varepsilon$ for all $i,\, j \in \underline{\mu}$ with $i \neq j$. Then $F_\varepsilon(s) = X_\varepsilon(s)(I+P(s)X_\varepsilon(s))^{-1}$ will be the transfer matrix of a compensator that achieves almost interaction with accuracy less than $\alpha\varepsilon$, where α is a constant depending on the dimensions of the matrices involved. Realizing $F_\varepsilon(s)$ as $N_\varepsilon + M_\varepsilon(sI - K_\varepsilon)^{-1}L_\varepsilon$ a compensator of type (3.6)–(3.7) is obtained that solves NICPM$_\mu$ approximately.

Similar remarks can be made with respect to TDPM$_\mu$ and ATDPM$_\mu$.

3.4 State Space Approach

The above presented approach is heavily based on finding a common (proper) rational solution to equations of the form $K_{i\,j}(s) + L_i(s)X(s)M_j(s) = 0$ for certain $i,\, j$.

For ANICPM$_\mu$ and ATDPM$_\mu$ it turns out to be possible to express the solvability directly in terms of conditions involving the matrices $K_{i\,j}(s)$, $L_i(s)$, $M_j(s)$, without using Kronecker products! These conditions will be presented later.

For NICPM$_\mu$ and TDPM$_\mu$ the derivation of such 'direct' conditions is much harder and state space methods seem to be more appropriate. To introduce these state space methods, we consider the next linear system

$$\dot{x}(t) = Ax(t) + Bu(t) + Gv(t), \tag{3.11}$$
$$y(t) = Cx(t), \tag{3.12}$$
$$z(t) = Hx(t), \tag{3.13}$$

which can be seen a version of system (3.1)–(3.3) for $\mu = 1$.

Assume that (3.11)–(3.13) is controlled by the feedback compensator (3.4)–(3.5) with $F(s) = N + M(sI - K)^{-1}L$. The closed-loop system is then given by

$$\dot{x}_e(t) = A_e x_e(t) + G_e v(t), \tag{3.14}$$
$$z(t) = H_e x_e(t), \tag{3.15}$$

where A_e is as given in (3.8) and $G_e = \begin{bmatrix} G \\ 0 \end{bmatrix}$, $H_e = \begin{bmatrix} H_i & 0 \end{bmatrix}$. Then, see also before, the transfer matrix between v and z in the closed loop system (3.14)–(3.15) is given by $H_e(sI - A_e)^{-1}G_e$. This transfer matrix is exactly equal to zero if and only if there is a proper rational matrix X such that

$$K(s) + L(s)X(s)M(s) = 0, \text{ where } X(s) = (I - F(s)P(s))^{-1}F(s),$$

with $P(s) = C(sI - A)^{-1}B$ as before, and

$$K(s) = H(sI - A)^{-1}G, \quad L(s) = H(sI - A)^{-1}B, \quad M(s) = C(sI - A)^{-1}G.$$

We then say that the so-called *disturbance decoupling problem by measurement feedback*, abbreviated DDPM, is solved for the system (3.11)–(3.13).

The latter property can also be expressed in state space terms. To derive/present state space conditions for the solvability of DDPM and the control problems in this chapter, we introduce the following concepts, see [1, 4, 5, 9, 11] and [12].

- First, we focus on the system given by $\dot{x} = Ax + Bu$ with initial state $x_0 \in \mathbb{R}^n$.
  - We say that x_0 has a (ξ, ω)-representation if there are rational vectors $\xi(s)$ and $\omega(s)$ of appropriate dimensions such that $x_0 = (sI - A)\xi(s) - B\omega(s)$.
  - We call a (ξ, ω)-representation *regular* when both $\xi(s)$ and $\omega(s)$ are strictly proper rational vectors.
  - We call a (ξ, ω)-representation *stable* if both $\xi(s)$ and $\omega(s)$ are stable rational vectors.
  - We say that a linear subspace $\mathcal{V}$ is a *controlled invariant subspace* if every $x_0 \in \mathcal{V}$ has a regular (ξ, ω)-representation such that $\xi(s) \in \mathcal{V}$.
  - A linear subspace $\mathcal{V}$ is called an *stabilizability subspace* if every $x_0 \in \mathcal{V}$ has a stable regular (ξ, ω)-representation such that $\xi(s) \in \mathcal{V}$.
  - A linear subspace $\mathcal{V}$ is called an *almost controlled invariant subspace* if every $x_0 \in \mathcal{V}$ has a (ξ, ω)-representation such that $\xi(s) \in \mathcal{V}$ (note that the regularity condition is dropped).
- Many properties of the above subspaces are known.
  - For instance, $\mathcal{V}$ is control invariant $\iff A\mathcal{V} \subseteq \mathcal{V} + \text{in}B \iff$ there exists F such that $(A + BF)\mathcal{V} \subseteq \mathcal{V}$.
  - Another useful property is that (A, B) is stabilizable and $\mathcal{V}$ is a stabilizability subspace $\iff$ there exists F such that $(A+BF)\mathcal{V} \subseteq \mathcal{V}$ and $\sigma(A+BF) \subseteq \mathbb{C}_g$.
- It is further well known that the largest of each of the above subspaces contained in $\ker H$ exists and can be computed by means of recursive algorithms only requiring a finite number of iterations.
  - The *largest controlled invariant subspace contained in ker H* is denoted as $\mathcal{V}^*(\ker H)$. The subspace can be described as $\mathcal{V}^*(\ker H) = \{x_0 \in \ker H | x_0$ has a regular (ξ, ω)-representation such that $H\xi(s) = 0\}$.

- The *largest stabilizability subspace contained in ker H* is denoted as $\mathscr{V}_g^*(\ker H)$. The subspace can be described as follows: $\mathscr{V}_g^*(\ker H) = \{x_0 \in \ker H | x_0$ has a stable regular (ξ, ω)-representation such that $H\xi(s) = 0\}$.
- The *largest almost controlled invariant subspace contained in ker H* is denoted as $\mathscr{V}_a^*(\ker H)$. The subspace can be described as follows: $\mathscr{V}_a^*(\ker H) = \{x_0 \in \ker H | x_0$ has a (ξ, ω)-representation such that $H\xi(s) = 0\}$.
- It turns out that the above largest almost controlled invariant subspace does not fit our purposes. The version in which the initial states are not restricted to lie in ker H is more useful in this chapter. This subspace is denoted as $\mathscr{V}_b^*(\ker H)$ and defined as follows: $\mathscr{V}_b^*(\ker H) = \{x_0 \in \mathbb{R}^n | x_0$ has a (ξ, ω)-representation such that $H\xi(s) = 0\}$.

- It is clear that the condition for the celebrated *disturbance decoupling problem* by state feedback, abbreviated DDP, see [12], for the system described by (3.11) and (3.13), given by im $G \subseteq \mathscr{V}^*(\ker H)$ is equivalent to the existence of strictly proper rational matrices $X(s)$ and $U(s)$ such that $I = (sI - A)X(s) - BU(s)$ and $HX(s)G = 0$. This alternative characterization will be useful later on.
- In addition, also the next characterizations are well known, see [1] and [9].
  - im $G \subseteq \mathscr{V}^*(\ker H) \iff$ there exist strictly proper rational matrices $X(s)$ and $U(s)$ such that $(sI - A)X(s) - BU(s) = G$ and $HX(s) = 0 \iff$ there exist a strictly proper rational matrix $V(s)$ such that $H(sI - A)^{-1}BV(s) = H(sI - A)^{-1}G$.
  - im $G \subseteq \mathscr{V}_b^*(\ker H) \iff$ there exist rational matrices $X(s)$ and $U(s)$ such that $(sI - A)X(s) - BU(s) = G$ and $HX(s) = 0 \iff$ there exist a rational matrix $V(s)$ such that $H(sI - A)^{-1}BV(s) = H(sI - A)^{-1}G$.

- Dualizing the above, we can introduce subspaces related to $\dot{x} = Ax$ and $y = Cx$, so relative to the pair (C, A).
  - A subspace S is called *conditioned invariant* (with respect to the pair (C, A)) if $S^\perp$ is controlled invariant with respect to the pair $(A^\top, C^\top)$. It turns out that $\mathscr{S}$ is conditioned invariant $\iff A((\mathscr{S} \cap \ker C) \subseteq \mathscr{S} \iff$ there exists a matrix J such that $(A + JC)\mathscr{S} \subseteq \mathscr{S}$.
  - In the same dual way the *almost conditioned invariant subspace* and the *detectability* subspace can be defined, being the orthogonal complement of the almost controlled invariant subspace and the stabilizability subspace with respect to the pair $(A^\top, C^\top)$.
  - Moreover, given im G there exists a *smallest conditioned invariant subspace that contains im G*. This subspace is denoted by $\mathscr{S}^*(\text{im } G)$ and can be computed by means of recursive algorithms only requiring a finite number of iterations. The same applies to the *smallest almost conditioned invariant subspace containing im G* and the *detectability subspace containing im G*. These subspaces are denoted $\mathscr{S}_b^*(\text{im } G)$ and $\mathscr{S}_g^*(\text{im } G)$, respectively.

In the context of system (3.11)–(3.13) we now have the following results providing the relation between solvability conditions for certain well-known problems in state space terms and in the transfer domain.

Theorem 3.11

(a) *The almost disturbance decoupling problem by measurement feedback, abbreviated ADDPM, see [9]:* $\mathrm{im}\, G \subseteq \mathscr{V}_b^*(\ker H)$, $\mathscr{S}_b^*(\mathrm{im} G) \subseteq \ker H \iff$ *there exists a rational matrix* $X(s)$ *such that* $K(s) + L(s)X(s)M(s) = 0$.

(b) *The disturbance decoupling problem by measurement feedback, abbreviated DDPM, see [5, 11]:* $\mathscr{S}^*(\mathrm{im}\, G) \subseteq \mathscr{V}^*(\ker H) \iff$ *there exists a proper rational matrix* $X(s)$ *such that* $K(s) + L(s)X(s)M(s)=0$.

The following condition in terms of ranks of rational matrices for the solvability of ADDPM will be useful later on.

Corollary 3.12 *There exists a rational matrix* $X(s)$ *such that* $K(s) + L(s)X(s)M(s) = 0 \iff$ *rank* $L(s) =$ *rank* $[L(s), K(s)]$ *and rank* $M(s) =$ *rank* $\begin{bmatrix} M(s) \\ K(s) \end{bmatrix}$.

By Theorem 3.11, the existence of a (proper) rational solution of the equation $K(s) + L(s)X(s)M(s) = 0$ can be expressed in well-known and constructively verifiable state space conditions. It is also possible to include additional stability conditions. This is most easily done in state space terms. Therefore, the following properties, relevant for decoupling and internal stabilization, are presented, see [5] and [13],

Proposition 3.13

(a) *If* $\mathscr{S} \subseteq \mathscr{V}$, $\mathscr{S}$ *is a conditioned invariant subspace and* $\mathscr{V}$ *is a controlled in variant subspace, then there are matrices* F, J *and* N *such that* $(A+BF)\mathscr{V} \subseteq \mathscr{V}$, $(A + JC)\mathscr{S} \subseteq \mathscr{S}$ *and* $(A + BNC)\mathscr{S} \subseteq \mathscr{V}$.

(b) *If* $\mathscr{S} \subseteq \mathscr{V}$, $\mathscr{S}$ *is a detectability subspace,* $\mathscr{V}$ *is a stabilizability subspace,* (A, B) *is stabilizable and* (C, A) *is detectable, then there are matrices* F, J *and* N *such that* $(A + BF)\mathscr{V} \subseteq \mathscr{V}$, $(A + JC)\mathscr{S} \subseteq \mathscr{S}$, $(A + BNC)\mathscr{S} \subseteq \mathscr{V}$, $\sigma(A + BF) \subseteq \mathbb{C}_g$ *and* $\sigma(A + JC) \subseteq \mathbb{C}_g$.

(c) *If* $\mathscr{S}_i \subseteq \mathscr{V}_i$, $\mathscr{S}_{i-1} \subseteq \mathscr{S}_i$, $\mathscr{V}_{i-1} \subseteq \mathscr{V}_i$, $\mathscr{S}_i$ *is a detectability subspace,* $\mathscr{V}_i$ *is a stabilizability subspace, for all* $i \in \underline{\mu}$, *with* $\mathscr{S}_0 = \mathscr{V}_0 = 0$, (A, B) *is stabilizable and* (C, A) *is detectable, then there are matrices* F, J *and* N *such that* $(A + BF)\mathscr{V}_i \subseteq \mathscr{V}_i$, $(A + JC)\mathscr{S}_i \subseteq \mathscr{S}_i$ *and* $(A + BNC)\mathscr{S}_i \subseteq \mathscr{V}_i$, *for all* $i \in \underline{\mu}$, $\sigma(A + BF) \subseteq \mathbb{C}_g$ *and* $\sigma(A + JC) \subseteq \mathbb{C}_g$.

Remark 3.14 The use of Proposition 3.13 lies in the fact that with the matrices F, J and N compensators of the form (3.6)–(3.7) can be derived together with A_e-invariant subspaces in the state space of the closed-loop system such that the noninteracting control or triangular decoupling problems can (be proved to) be solved, see for instance [5] or [11]. Indeed, using the information in part (b), define the matrices

$$K = A + BF + JC - BNC, \quad L = BN - J, \quad M = F - NC$$

and the subspace

$$\mathscr{W} = \left\{ \begin{bmatrix} s \\ 0 \end{bmatrix} + \begin{bmatrix} v \\ v \end{bmatrix} | s \in \mathscr{S}, v \in \mathscr{V} \right\}.$$

Recall that

$$A_e = \begin{bmatrix} A + BNC & BM \\ LC & K \end{bmatrix}$$

and define

$$\mathscr{S} \oplus 0 = \left\{ \begin{bmatrix} s \\ 0 \end{bmatrix} | s \in \mathscr{S} \right\}, \quad \mathscr{V} \oplus \mathbb{R}^n = \left\{ \begin{bmatrix} v \\ w \end{bmatrix} | v \in \mathscr{V}, w \in \mathbb{R}^n \right\}.$$

Then it can be shown easily that

$$\sigma(A_e) \subseteq \mathbb{C}_g, \quad A_e \mathscr{W} \subseteq \mathscr{W}, \quad \mathscr{S} \oplus 0 \subseteq \mathscr{W} \subseteq \mathscr{V} \oplus \mathbb{R}^n.$$

Hence, if im $G \subseteq \mathscr{S} \subseteq \mathscr{V} \subseteq$ ker H, then

$$\sigma(A_e) \subseteq \mathbb{C}_g, \quad A_e \mathscr{W} \subseteq \mathscr{W}, \quad \mathrm{im} G_e \subseteq \mathscr{W} \subseteq \ker H_e,$$

implying that $H_e A_e^k G_e = 0$ for all $k \geq 0$, in turn implying that $T(s) = 0$. Hence, the system is disturbance decoupled by measurement feedback and internally stabilized because $\sigma(A_e)$.

3.5 Noninteracting Control

In this section we will treat the problems of (almost) noninteracting control. It turns out that the version with state feedback can be solved completely. The version with measurement feedback can (so far) only be solved partially from a transfer matrix point of view.

3.5.1 (Almost) Noninteracting Control by State Feedback

In this subsection we shall be dealing with (dynamic) state feedback. This means that in equation (3.2) we assume that $C = I$. Then the interconnection of the feedback compensator (3.4)–(3.5) with the linear system (3.1)–(3.3) results in a closed-loop system described by

$$\dot{x}_e(t) = A'_e x_e(t) + \Sigma_{i \in \underline{\mu}} G_{i,e} v_i(t), \tag{3.16}$$

$$z_i(t) = H_{i,e} x_e(t), \qquad i \in \underline{\mu}, \tag{3.17}$$

where $x_e(t)$, and $G_{i,e}$, $H_{i,e}$, $i \in \underline{\mu}$, are as in (3.8)–(3.9), and

$$A'_e = \begin{bmatrix} A + BN & BM \\ L & K \end{bmatrix}.$$

Let $T'(s)$ denote the transfer matrix of the closed-loop system (3.16)–(3.17) and partition $T'(s) = (T'_{ij}(s))$, $i, j \in \underline{\mu}$, where $T'_{ij}(s) = H_{i,e}(sI - A'_e)^{-1} G_{i,e}$ denotes the transfer matrix between the j-th exogenous input and the i-th exogenous output in the latter closed-loop system.

Denote $P'(s) = (sI - A)^{-1} B$ and $M'_j(s) = (sI - A)^{-1} G_j$, $j \in \underline{\mu}$. In terms of transfer matrices we then have that $T'_{i\,j}(s) = K_{i\,j}(s) + L_i(s) X(s) M'_j(s)$, $i, j \in \underline{\mu}$, where $X(s) = (I - F(s)P'(s))^{-1} F(s)$, and $K_{ij}(s)$, $L_i(s)$ and $F(s)$ are as described in the previous.

We can now formulate the following control problem, where we assume that the system (3.1)–(3.3) with $C = I$ is given.

Definition 3.15 The *noninteracting control problem by state feedback*, denoted NICPS_{μ}, consists of finding a feedback compensator (3.4)–(3.5) such that in the closed-loop system (3.16)–(3.17): $T'_{i\,j}(s) = 0$ for all $i, j \in \underline{\mu}$ with $i \neq j$.

With $\mathbb{C}_g$ as stability region we can formulate the following.

Definition 3.16 The *noninteracting control problem by state feedback with internal stabilization*, denoted NICPS^s_{μ}, consists of finding a feedback compensator (3.4)–(3.5) such that in the closed-loop system (3.16)–(3.17): $T'_{i\,j}(s) = 0$ for all $i, j \in \underline{\mu}$ with $i \neq j$ and $\sigma(A'_e) \subseteq \mathbb{C}_g$.

The almost version reads as follows.

Definition 3.17 The *almost noninteracting control problem by state feedback*, denoted ANICPS_{μ}, consists of finding, for all $\varepsilon > 0$, a feedback compensator (3.4)–(3.5) such that in the closed-loop system (3.16)–(3.17): $||T'_{i\,j}(s)||_{\infty} \leq \varepsilon$ for all $i, j \in \underline{\mu}$ with $i \neq j$.

In order to derive necessary and sufficient conditions in state space terms for the solvability of the control problems defined above, we introduce the following.

Let $\{\mathscr{L}_i | i \in \underline{\mu}\}$ be a family of linear subspaces in $\mathbb{R}^n$ and denote

$$\mathscr{L}_i^{\vee} := \sum_{j \in \underline{\mu}, j \neq i} \mathscr{L}_j, \quad i \in \underline{\mu}.$$

We say that the family is independent if $\mathscr{L}_i \cap \mathscr{L}_i^\vee = 0$ for all $i \in \underline{\mu}$. Define the linear subspace $\mathscr{L}^\vee$ as

$$\mathscr{L}^\vee := \sum_{i\in\underline{\mu}} (\mathscr{L}_i \cap \mathscr{L}_i^\vee).$$

The linear subspace $\mathscr{L}^\vee$ is called the radical of the family $\{\mathscr{L}_i | i \in \underline{\mu}\}$ (see [12]).

It is clear that a family of linear subspaces is independent if and only if its radical is equal to the zero subspace. When the radical is nonzero, it can be shown that the factor spaces $\{(\mathscr{L}_i + \mathscr{L}^\vee)/\mathscr{L}^\vee | i \in \underline{\mu}\}$ are linearly independent. The latter has the following consequence.

Proposition 3.18 *Let $\{\mathscr{L}_i | i \in \underline{\mu}\}$ be a family of linear subspaces in $\mathbb{R}^n$ and let $\{\bar{\mathscr{L}}_i | i \in \underline{\mu}\}$ be a family of linear subspaces such that $\bar{\mathscr{L}}_i \subseteq \mathscr{L}_i$, $\bar{\mathscr{L}}_i \cap \mathscr{L}^\vee = 0$ and $\mathscr{L}_i + \mathscr{L}^\vee = \bar{\mathscr{L}}_i + \mathscr{L}^\vee$ for all $i \in \underline{\mu}$. Then the family $\{\mathscr{L}^\vee, \bar{\mathscr{L}}_1, \bar{\mathscr{L}}_2, \cdots, \bar{\mathscr{L}}_\mu\}$ is independent (in the above sense).*

Proof Straightforward. See, also [7] or [13].

In the following we shall derive necessary and sufficient conditions in state space terms for the solvability of NICPS$_\mu$ and NICPS$^s_\mu$. We denote

$$\mathscr{G}_i := \operatorname{im} G_i, \quad i \in \underline{\mu}, \qquad \mathscr{G}_i^\vee := \sum_{j\in\underline{\mu}, j\neq i} \mathscr{G}_j, \quad i \in \underline{\mu},$$

$$\mathscr{G}^\vee := \sum_{i\in\underline{\mu}} (\mathscr{G}_i \cap \mathscr{G}_i^\vee),$$

$$\mathscr{K}_i := \bigcap_{j\in\underline{\mu}, j\neq i} \ker H_j, \quad i \in \underline{\mu} \qquad \mathscr{K} := \bigcap_{i\in\underline{\mu}} \ker H_i.$$

Let the linear system (3.1)–(3.3) with $C = I$ be given and let $\mathbb{C}_g$ be a given stability region. Then we have the following result.

Theorem 3.19 *NICPS$_\mu$ is solvable if and only if $\mathscr{G}_i \subseteq \mathscr{V}^*(\mathscr{K}_i)$ for all $i \in \underline{\mu}$ and $\mathscr{G}^\vee \subseteq \mathscr{V}^*(\mathscr{K})$.*

Proof For the original proof of the theorem and its extensions, we refer to [7] or [13]. An alternative proof of the necessity of the conditions can be given easily by observing that when NICPS$_\mu$ is solvable there exist A'_e-invariant subspaces $\mathscr{W}_i$ wedged between specific 'input' and 'output' subspaces. The projection of the subspaces $\mathscr{W}_i$ onto the state space of (3.1)–(3.3) are (A, B)-invariant subspaces that contain $\mathscr{G}_i$ and $\mathscr{G}^\vee$, and are contained in $\mathscr{K}_i$ and $\mathscr{K}$, respectively, where $i \in \underline{\mu}$, see [5]. Then taking the largest of these (A, B)-invariant subspaces, the necessity part follows.

The central idea for the sufficiency part of the proof is that because $\mathscr{G}_i \subseteq \mathscr{V}^*(\mathscr{K}_i)$, for all $i \in \underline{\mu}$, there exist strictly proper rational matrices $X_i(s)$ and $U_i(s)$ such that $(sI - A)X_i(s) - BU_i(s) = I$ and $H_j X_i(s) G_i = 0$ for all $i, j \in \underline{\mu}$ with $i \neq j$. Also, because $\mathscr{G}^\vee \subseteq \mathscr{V}^*(\mathscr{K})$, there exist strictly proper rational matrices $X_0(s)$ and

$U_0(s)$ such that $(sI - A)X_0(s) - BU_0(s) = I$ and $H_j X_i(s)\bar{G}_0 = 0$ for all $j \in \underline{\mu}$, where $\bar{G}_0$ is a full column rank matrix such that im $\bar{G}_0 = \mathscr{G}^\vee$. Let further $\bar{\mathscr{G}_i}$ be as indicated in Proposition 3.18, and let $\bar{G}_i$ be a full column rank matrix such that im $\bar{G}_i = \bar{\mathscr{G}_i}$. Then $[\bar{G}_0, \bar{G}_1, \ldots, \bar{G}_\mu]$ is a full column rank matrix. Let $\bar{G}_{\mu+1}$ be such that $[\bar{G}_0, \bar{G}_1, \ldots, \bar{G}_\mu, \bar{G}_{\mu+1}]$ is invertible, and let $X_{\mu+1}(s)$ and $U_{\mu+1}(s)$ be any strictly proper rational matrices such that $(sI - A)X_{\mu+1}(s)\bar{G}_{\mu+1} - BU_{\mu+1}(s)\bar{G}_{\mu+1} = \bar{G}_{\mu+1}$. We can then define $Xs)$ and $U(s)$ through

$$X(s)[\bar{G}_0, \bar{G}_1, \ldots, \bar{G}_{\mu+1}] = [X_0(s)\bar{G}_0, X_1(s)\bar{G}_1, \ldots, X_{\mu+1}(s)\bar{G}_{\mu+1}],$$

$$U(s)[\bar{G}_0, \bar{G}_1, \ldots, \bar{G}_{\mu+1}] = [U_0(s)\bar{G}_0, U_1(s)\bar{G}_1, \ldots, U_{\mu+1}(s)\bar{G}_{\mu+1}].$$

It can be shown that $X(s)$ is invertible and $H_j X(s) G_i = 0$ for all $i, j \in \underline{\mu}$ with $i \neq j$. Further, define $F(s) = U(s)X^{-1}(s)$. It follows easily that $F(s)$ is a proper rational matrix. Realizing $F(s)$ as $N + M(sI - K)^{-1}L$ a compensator of type (3.6)–(3.7) is obtained that yields noninteraction by state feedback, since we assumed that $C = I$, i.e., $y = x$.

In a similar way, the following theorem can be proved.

Theorem 3.20 *NICPS$^s_\mu$ is solvable if and only if $\mathscr{G}_i \subseteq \mathscr{V}_g(\mathscr{K}_i)$ for $i \in \underline{\mu}$, $\mathscr{G}^\vee \subseteq \mathscr{V}_g^*(\mathscr{K})$, and the pair (A, B) is stabilizable.*

We conclude this subsection by presenting necessary and sufficient conditions in state space terms for the solvability of ANICPS$_\mu$, as described previously. In fact, the following theorem can be proved.

Theorem 3.21 *ANICPS$_\mu$ is solvable if and only if $\mathscr{G}_i \subseteq \mathscr{V}_b^*(\mathscr{K}_i)$ for all $i \in \underline{\mu}$ and $\mathscr{G}^\vee \subseteq \mathscr{V}_b^*(\mathscr{K})$.*

The proof of the theorem is completely analogously to the previous theorems by using the following two propositions. In the first proposition, we derive an alternative characterization of the linear subspace $\mathscr{V}_b^*(\ker H)$ (see [6]). See also Theorem 3.9.

Proposition 3.22 $\mathscr{V}_b^*(\ker H) = \{x_0 \in \mathbb{R}^n |$ *for all $\varepsilon > 0$ there exists a regular (ξ, ω)-representation such that $||H\xi(s)||_\infty \leq \varepsilon\}$.*

Proposition 3.23 im $G \subseteq \mathscr{V}_b^*(\ker H)$ *if and only if for all $\varepsilon > 0$ there exist strictly proper rational matrices $X_\varepsilon(s)$ and $U_\varepsilon(s)$ such that $I = (sI - A)X_\varepsilon(s) - BU_\varepsilon(s)$ and $||HX_\varepsilon(s)G||_\infty \leq \varepsilon$.*

3.5.2 Almost Noninteracting Control by Measurement Feedback

In the above the noninteracting control problems by state feedback were considered and nice state space conditions for the solvability of the problems could be derived.

Unfortunately, this is not (yet) the case for the versions of the problem with measurement feedback. The only known result so far for the solvability of ANICPM$_\mu$ is a condition in terms of the transfer matrices of the *open loop* system, i.e., a condition in terms of $K_{i\,j}(s)$, $L_i(s)$ and $M_j(s)$ for the existence of a rational $X(s)$ such that $K_{i\,j}(s) + L_i(s)X(s)M_j(s) = 0$ for all $i, j \in \underline{\mu}$ with $i \neq j$, see also Corollary 3.10.

To present these explicit solvability conditions, we introduce matrices

$$L(s) := \begin{bmatrix} L_1(s) \\ L_2(s) \\ \vdots \\ L_\mu(s) \end{bmatrix}, \quad M(s) := \begin{bmatrix} M_1(s), \; M_2(s), \; \dots, \; M_\mu(s) \end{bmatrix},$$

and for $i \in \underline{\mu}$ we define

$$\check{L}_i(s) := \begin{bmatrix} L_1(s) \\ \vdots \\ L_{i-1}(s) \\ L_{i+1}(s) \\ \vdots \\ L_\mu(s) \end{bmatrix}, \quad \Delta_i(s) := \begin{bmatrix} K_{1i}(s) \\ \vdots \\ K_{i-1\,i}(s) \\ K_{i+1\,i}(s) \\ \vdots \\ K_{\mu\,i}(s) \end{bmatrix},$$

$$\check{M}_i(s) := \begin{bmatrix} M_1(s), \; \dots, \; M_{i-1}(s), \; M_{i+1}(s), \; \dots, \; M_\mu(s), \end{bmatrix},$$

$$\Lambda_i(s) := \begin{bmatrix} K_{i1}(s), \; \dots, \; K_{i\,i-1}(s), \; K_{i\,i+1}(s), \; \dots, \; K_{i\,\mu}(s), \end{bmatrix},$$

$$\Gamma_i(s) := \begin{bmatrix} 0 & \dots & 0 & -K_{1\,i}(s) & 0 & \dots & 0 \\ \vdots & & \vdots & \vdots & \vdots & & \vdots \\ 0 & \dots & 0 & -K_{i-1\,i}(s) & 0 & \dots & 0 \\ K_{i1}(s) & \dots & K_{i\,i-1}(s) & 0 & K_{i\,i+1}(s) & \dots & K_{i\,\mu}(s) \\ 0 & \dots & 0 & -K_{i+1\,i}(s) & 0 & \dots & 0 \\ \vdots & & \vdots & \vdots & \vdots & & \vdots \\ 0 & \dots & 0 & -K_{\mu\,i}(s) & 0 & \dots & 0 \end{bmatrix}.$$

Then the following result can be shown, see [15].

Theorem 3.24 *There is a rational matrix* $X(s)$ *such that* $K_{i\,j}(s) + L_i(s)X(s)M_j(s) = 0$ *for all* $i, j \in \underline{\mu}$ *with* $i \neq j$ *if and only if for all* $i \in \underline{\mu}$

(a) $\operatorname{rank} \check{L}_i(s) = \operatorname{rank}\, [\check{L}_i(s), \Delta_i(s)]$, *(b)* $\operatorname{rank} \check{M}_i(s) = \operatorname{rank} \begin{bmatrix} \check{M}_i(s) \\ \Lambda_i(s) \end{bmatrix}$,

(c) $\operatorname{rank} \begin{bmatrix} 0 & M(s) \\ L(s) & 0 \end{bmatrix} = \operatorname{rank} \begin{bmatrix} 0 & M(s) \\ L(s) & \Gamma_i(s) \end{bmatrix}$

Recall that for $i, j \in \underline{\mu}$

$$K_{ij}(s) = H_i(sI - A)^{-1}G_j, \quad L_i(s) = H_i(sI - A)^{-1}B, \quad M_j(s) = C(sI - A)^{-1}G_j.$$

Hence, for $i \in \underline{\mu}$

$$\begin{aligned} \check{L}_i(s) &= \check{H}_i(sI - A)^{-1}B, \ \Delta_i(s) = \check{H}_i(sI - A)^{-1}G_i, \\ \check{M}_i(s) &= C(sI - A)^{-1}\check{G}_i, \ \Lambda_i(s) = H_i(sI - A)^{-1}\check{G}_i, \end{aligned}$$

where

$$\check{H}_i := \begin{bmatrix} H_1 \\ \vdots \\ H_{i-1} \\ H_{i+1} \\ \vdots \\ H_\mu \end{bmatrix}, \quad \check{G}_i := \begin{bmatrix} G_1, \ldots, G_{i-1}, G_{i+1}, \ldots, G_\mu \end{bmatrix}.$$

Observe that rank $\check{L}_i(s) = \text{rank } [\check{L}_i(s), \Delta_i(s)]$ if and only if there exists a rational matrix $V(s)$ such that $\check{L}_i(s)V(s) = \Delta_i(s)$. Similarly for rank $\check{M}_i(s) = \text{rank} \begin{bmatrix} \check{M}_i(s) \\ \Lambda_i(s) \end{bmatrix}$.

Therefore, the conditions (a) and (b) in Theorem 3.24 can be translated into state space terms as follows:

$$(a) \ \text{im } G_i \subseteq \mathscr{V}_b^*(\ker \check{H}_i) \qquad (b) \ \mathscr{S}_b^*(\text{ im } \check{G}_i) \subseteq \ker H_i,$$

for $i \in \underline{\mu}$

Unfortunately, condition (c) of Theorem 3.24 cannot (so easily) be translated into state space terms. Therefore, conditions for the solvability of ANICPM$_\mu$ that are completely in state space terms are not (yet) known. Hence, for now we have

Theorem 3.25 *ANICPM$_\mu$ is solvable if and only if im $G_i \subseteq \mathscr{V}_b^*(\ker \check{H}_i)$, $\mathscr{S}_b^*(\text{ im } \check{G}_i) \subseteq \ker H_i$, rank $\begin{bmatrix} 0 & M(s) \\ L(s) & 0 \end{bmatrix} = \text{rank} \begin{bmatrix} 0 & M(s) \\ L(s) & \Gamma_i(s) \end{bmatrix}$ for all $i \in \underline{\mu}$.*

3.5.3 Some Extensions

This section is largely based upon [7, 13] and [16], where, in addition to the problems formulated above, also the formulation can be found of the extension of (A)NICPS$_\mu$ toward additional input/output f-stabilization and internal s-stabilization (compare with the formulation of NICPS$_\mu^s$). The latter means that the off-diagonal blocks have to be zero, the diagonal blocks are f-stable (with f for fast), and the underlying system is s-stable (with s for slow). In the formulation of this extension it is assumed that the linear system (3.1)–(3.3) with $C = I$ is given and that $\mathbb{C}_f, \mathbb{C}_s$ with $\mathbb{C}_f \subseteq \mathbb{C}_s$ are given stability regions. In [7] also necessary and sufficient conditions are derived for the solvability of the *almost* version of extended problem. The treatment of this problem gives rise to an analysis that is considerably more involved than its exact counterpart.

3.6 Triangular Decoupling

In this section we present some solvability conditions in the context of (almost) triangular decoupling as introduced before. It turns out that we can tackle the problem directly for measurement feedback (and do not need to focus on state feedback first) and that the conditions can be formulated in state space terms.

Some of the conditions will be obtained immediately, others through a transfer matrix reasoning.

3.6.1 Triangular Decoupling by Measurement Feedback

To present necessary and sufficient conditions in state space terms for solvability of the control problems TDPM$_\mu$ and TDPM$^s_\mu$, let the linear system (3.1)–(3.3) be given and let $\mathbb{C}_g$ be a given stability region. For all $i \in \underline{\mu - 1}$ denote

$$\tilde{\mathscr{G}}_i := \sum_{j=i+1}^{\mu} \text{im } G_j, \qquad \tilde{\mathscr{K}}_i := \bigcap_{j=1}^{i} \ker H_j. \tag{3.18}$$

Then we have the following result.

Theorem 3.26 *TDPM$_\mu$ is solvable if and only if $\mathscr{S}^*(\tilde{\mathscr{G}}_i) \subseteq \mathscr{V}^*(\tilde{\mathscr{K}}_i)$ for all $i \in \underline{\mu - 1}$.*

Theorem 3.27 *TDPM$^s_\mu$ is solvable if and only if $\mathscr{S}^*_g(\tilde{\mathscr{G}}_i) \subseteq \mathscr{V}^*_g(\tilde{\mathscr{K}}_i)$ for all $i \in \underline{\mu - 1}$, the pair (A, B) is stabilizable and the pair (C, A) is detectable.*

Proof The necessity of the subspace inclusions can be given easily by observing that in case TDPM$^s_\mu$ is solvable there exist $\mu - 1$ subspaces $\mathscr{W}_i$ that are A_e-invariant subspaces wedged in between specific 'input' and 'output' subspaces in the state space of the closed-loop system. Following [5], it is clear that for all $i \in \underline{\mu - 1}$

(a) the intersection of $\mathscr{W}_i$ with the state space of the system (3.1)–(3.3) yields a detectability subspace $\mathscr{S}_i$ that contains $\tilde{\mathscr{G}}_i$,
(b) the projection of $\mathscr{W}_i$ onto the state space of the system (3.1)–(3.3) yields a stabilizability subspace $\mathscr{V}_i$ that is contained in $\tilde{\mathscr{K}}_i$,
(c) $\mathscr{S}_i \subseteq \mathscr{V}_i$.

Then taking the smallest of such detectability subspaces and the largest of such stabilizability subspaces, the necessity of the subspace inclusions follows. The necessity of the stabilizability of (A, B) and the detectability of (C, A) is obvious from the fact that $\sigma(A_e) \subseteq \mathbb{C}_g$.

The main idea for the proof of the sufficiency part is the application of Proposition 3.13. Observe that for $\tilde{\mathscr{G}}_{i+1} \subseteq \tilde{\mathscr{G}}_i$ and $\tilde{\mathscr{K}}_{i+1} \subseteq \tilde{\mathscr{K}}_i$, for all $i \in \underline{\mu - 1}$, where we defined $\bar{\mathscr{G}}_\mu = \bar{\mathscr{K}}_\mu = 0$. Hence, it follows that $\mathscr{S}^*(\tilde{\mathscr{G}}_{i+1}) \subseteq \mathscr{S}^*(\tilde{\mathscr{G}}_i)$, $\mathscr{V}^*(\tilde{\mathscr{K}}_{i+1}) \subseteq \mathscr{V}^*(\tilde{\mathscr{K}}_i)$ and $\mathscr{S}^*(\tilde{\mathscr{G}}_i) \subseteq \mathscr{V}^*(\tilde{\mathscr{K}}_i)$, for all $i \in \underline{\mu - 1}$. From Proposition 3.13, it then follows that there exist F, J and N such that $(A+BF)\mathscr{V}^*(\tilde{\mathscr{K}}_i) \subseteq \mathscr{V}^*(\tilde{\mathscr{K}}_i)$, $(A+JC)\mathscr{S}^*(\tilde{\mathscr{G}}_i) \subseteq \mathscr{S}^*(\tilde{\mathscr{G}}_i)$ and $(A + BNC)\mathscr{S}^*(\tilde{\mathscr{G}}_i) \subseteq \mathscr{V}^*(\tilde{\mathscr{K}}_i)$, for all $i \in \underline{\mu - 1}$, $\sigma(A + BF) \subseteq \mathbb{C}_g$ and $\sigma(A + JC) \subseteq \mathbb{C}_g$. Then using a construction as described in Remark 3.14, it follows easily that there exists a measurement feedback compensator of the type (3.6)–(3.7) such that $H_{i,e}A_e^k G_{j,e} = 0$ for all $k \geq 0$ and $1 \leq i < j \leq \mu$. Also it follows that $\sigma(A_e) \subseteq \mathbb{C}_g$.

3.6.2 *Almost Triangular Decoupling by Measurement Feedback*

In this subsection we shall derive necessary and sufficient conditions in state space terms for the solvability of ATDPM$_\mu$. The conditions will not be derived directly using state space methods, but will be a consequence of the result formulated below. Therefore, recall that for $i, j \in \underline{\mu}$

$$K_{ij}(s) = H_i(sI-A)^{-1}G_j, \quad L_i(s) = H_i(sI-A)^{-1}B, \quad M_j(s) = C(sI-A)^{-1}G_j.$$

To present the solvability conditions, we redefine for $i \in \underline{\mu}$

$$\Delta_i(s) := \begin{bmatrix} L_1(s) \\ L_2(s) \\ \vdots \\ L_i(s) \end{bmatrix}, \quad \Lambda_i(s) := \begin{bmatrix} M_{i+1}(s) \; M_{i+2}(s) \; \dots \; M_\mu(s) \end{bmatrix},$$

$$\Gamma_i(s) := \begin{bmatrix} K_{1\,i+1}(s) & K_{1\,i+2}(s) & \dots & K_{1\,\mu}(s) \\ K_{2\,i+1}(s) & K_{2\,i+2}(s) & \dots & K_{1\,\mu}(s) \\ \vdots & \vdots & & \vdots \\ K_{i\ i+1}(s) & K_{ii+2}(s) & \dots & K_{i\mu}(s) \end{bmatrix}.$$

Then the following result can be shown, see [13].

Theorem 3.28 *There is a rational matrix* $X(s)$ *such that* $K_{ij}(s) + L_i(s)X(s)M_j(s) = 0$ *for all* $i, j \in \underline{\mu - 1}$ *with* $i < j$ *if and only if for all* $i \in \underline{\mu - 1}$

$$(a)\ \text{rank}\ \Delta_i(s) = \text{rank}\ [\Delta_i(s), \Gamma_i(s)], \quad (b)\ \text{rank}\ \Lambda_i(s) = \text{rank} \begin{bmatrix} \Lambda_i(s) \\ \Gamma_i(s) \end{bmatrix}.$$

Note that for $i \in \underline{\mu - 1}$

$$\Delta_i(s) = \begin{bmatrix} H_1 \\ H_2 \\ \vdots \\ H_i \end{bmatrix} (sI - A)^{-1}B, \quad \Lambda_i(s) = C(sI - A)^{-1} \begin{bmatrix} G_{i+1} & G_{i+2} & \dots & G_\mu \end{bmatrix},$$

and

$$\Gamma_i(s) = \begin{bmatrix} H_1 \\ H_2 \\ \vdots \\ H_i \end{bmatrix} (sI - A)^{-1} \begin{bmatrix} G_{i+1} & G_{i+2} & \dots & G_\mu \end{bmatrix}.$$

Hence, with the notation introduced in (3.18), it follows that the first rank condition in Theorem 3.28 is equivalent to $\tilde{\mathscr{G}}_i \subseteq \mathscr{V}_b^*(\tilde{\mathscr{K}}_i)$, whereas the second rank condition is equivalent to $\mathscr{S}_b^*(\tilde{\mathscr{G}}_i) \subseteq \tilde{\mathscr{K}}_i$, for all $i \in \underline{\mu - 1}$. Thus, we have proved

Theorem 3.29 *ATDPM*$_\mu$ *is solvable if and only if* $\mathscr{S}_b^*(\tilde{\mathscr{G}}_i) \subseteq \tilde{\mathscr{K}}_i$ *and* $\tilde{\mathscr{G}}_i \subseteq \mathscr{V}_b^*(\tilde{\mathscr{K}}_i)$ *for all* $\in \underline{\mu - 1}$.

3.7 Conclusions

In this chapter we have recalled some results on noninteracting control and triangular decoupling by state and measurement feedback. The results are derived using transfer matrices techniques and the geometric approach toward system theory that was initiated by Wonham in his book [12]. Not all results in this chapter are presented in full generality. More details and results can be found in the list of references, specifically in [7, 13] and [16]. Also in hindsight, some of the results could have been derived easier or in a different way. Further, there are still relevant gaps that can be worked on. For instance, on requirements on the diagonal blocks in both types of problems with respect to rank and/or stability. As the geometric approach has proved to be of limited value for applications, also the H_∞- or H_2-versions of the noninteracting control and triangular decoupling by state and measurement feedback may be worthwhile to look at.

References

1. Hautus, M.L.J.: (A, B)-invariant and stabilizability subspaces, a frequency domain description. Automatica **16**, 703–707 (1980)
2. Hautus, M.L.J., Heymann, M.: Linear feedback - an algebraic approach. SIAM J. Contr. Optimiz. **16**, 83–105 (1979)
3. Morse, A.S., Wonham, W.M.: Status of noninteracting control. IEEE Trans. Autom. Control **AC-16**, 568–581 (1971)
4. Schumacher, J.M.: *(C, A)*-invariant subspaces: Some facts and uses. Wiskundig Seminarium Vrije Universiteit Amsterdam, The Netherlands, Report 110 (1979)
5. Schumacher, J.M.: Compensator synthesis using (C, A, B) pairs. IEEE Trans. Automat. Control **25**, 1133–1138 (1980)
6. Trentelman, H.L.: Almost Invariant Subspaces and High Gain Feedback. CWI tracts 29, Amsterdam (1986)
7. Trentelman, H.L., van der Woude, J.W.: Almost invariance and noninteracting control : a frequency domain analysis. Linear Algebra Appl. **101**, 221–254 (1988)
8. Willems, J.C.: Almost nointeracting control design using dynamic state feedback. In: Proceeding of 4th International Conference Analysis and Optique Systems Versailles. Lecture notes in Control and Information Sciences, vol. 28, pp. 555–561. Springer, Berlin (1980)
9. Willems, J.C.: Almost invariant subspaces: an approach to high feedback design: part I: almost controlled invariant subspaces. IEEE Trans. Automat. Control **AC-26**, 232–252 (1981)
10. Willems, J.C.: Almost invariant subspaces: an approach to high feedback design: part II: almost conditionally invariant subspaces. IEEE Trans. Autom. Control **AC-27**, 1071–1085 (1982)
11. Willems, J.C., Commault, C.: Disturbance decoupling by measurement feedback with stability or pole-placement. SIAM J. Control Optim. **19**, 490–504 (1981)
12. Wonham, W.M.: Linear Multivariable Control: A Geometric Approach, 2-nd edn. Springer, Verlag (1979)
13. van der Woude, J.W.: Feedback decoupling and stabilization for linear systems with multiple exogenous variables. Ph.D. thesis, Eindhoven University of Technology (1987)
14. van der Woude, J.W.: Almost disturbance decoupling by measurement feedback : a frequency domain analysis. IEEE Trans. on Automatic Control **35**, 570–573 (1990)
15. van der Woude, J.W.: On the existence of a common solution X to the matrix equations $A_i X B_j = C_{ij}, (i, j) \in \Gamma$. Linear Algebra Appl. **375**, 135–145 (2003)
16. van der Woude, J.W., Trentelman, H.L.: Non interacting control with internal and input/output stability. In: Proceedings 25th IEEE Conference on Decision and Control, Athens, Greece, pp. 701–702 (1986)

Chapter 4
Simultaneous Stabilization Problem in a Behavioral Framework

Osamu Kaneko

Abstract This article addresses the simultaneous stabilization problem in the behavioral framework. First, simultaneous stabilization problem for two linear systems is addressed. There, a necessary and sufficient condition for two linear systems to be simultaneously stabilizable is a generalization of the result in the standard control theory. In addition, a parameterization of simultaneous stabilizers for a class of a pair of linear systems is also presented. Based on these results, a sufficient condition for three linear systems to be simultaneously stabilizable is provided. Then, a parameterization of simultaneous stabilizer for this case is also presented.

4.1 Introduction

It is a great pleasure to contribute this article to the Festschrift in honor of Harry L. Trentelman on the occasion of his 60th birthday. About 20 years ago, when I was a Ph.D student, I was studying the behavioral system theory. As everyone knows, the behavioral approach was proposed by Jan C. Willems and enables us to view a dynamical system from a more broader perspective than conventional system theory. In 1996, I attended CDC held in Kobe and participated in some organized sessions on behavioral system theory as one of the audience. There Harry gave some talks on quadratic differential forms and control in a behavioral context. I was so impressed that I read and studied papers written by Harry and Jan on behavioral system theory. When I visited Jan in Groningen at the end of the summer of 1999, it was the first time I ever had a talk with Harry. Since then, whenever we met in some international conferences or symposiums, he gave insightful and sharp comments on my results. Luckily, I had a chance to stay in Groningen for about six weeks in the autumn of 2009. At that time, fruitful discussions with Harry were very exciting for me. The next year, he visited Kanzazawa and gave a lecture on rational representations of the

O. Kaneko (✉)
Institute of Science and Engineering, Kanazawa University,
Kakumama-machi, Kanazawa, Ishikawa 920-1192, Japan
e-mail: o-kaneko@ec.t.kanazawa-u.ac.jp

M.N. Belur et al. (eds.), *Mathematical Control Theory II*,
Lecture Notes in Control and Information Sciences 462,
DOI 10.1007/978-3-319-21003-2_4

behaviors, which are very interesting topics (and then he also enjoyed a hot spa). From the above backgrounds, my research centered on behavioral system theory is greatly influenced by Harry. Thus I think that a topic on control in a behavioral context is appropriate for this festschrift.

As one of the issues on behavioral control theory, this article considers simultaneous stabilization problems in a behavioral framework. Simultaneous stabilization is the problem of finding a condition under which there exists a single controller that stabilizes multiple plants. This problem was first proposed and investigated in [22] from the viewpoints of reliable control synthesis. Then there have been many studies in the input/output setting, see [7–9, 16, 17, 21, 23]. In the case of two plants, it was shown that a pair of linear plants is simultaneously stabilizable if and only if there exists a strong stabilizer [26]—the denominator of the stabilizer also has all roots in open left half plane—for the augmented system constructed by using these two systems. In the case of three or more systems, shown in [3–5] by Blondel, it is impossible to find a necessary and sufficient condition that is checkable by a finite number of arithmetic or logical computations. That is, simultaneous stabilization problem for three or more plants is rationally undecidable. Since then, simultaneous stabilization is known as one of open problems in systems and control theory [6, 18]. On the other hand, studies of simultaneous stabilization are expected to provide insightful gradients, which are meaningful for theoretical developments.

J.C. Willems proposed the behavioral approach, which provides a new viewpoint for dynamical system theory [19, 24, 25]. In the behavioral approach, a system is viewed as a set of the trajectories of a system and there is no input/output partition in the variables that interact with the environment while a transfer function as the map from input to output plays a crucial role in the standard system theory. Control in the behavioral approach is regarded as an "interconnection" [15, 20, 25] which corresponds to pick up the desired trajectories by sharing the variables with a controller. Of course, control is not necessarily realized by the feedback architecture. This is also a generalization of the concept of "control" from a broader perspective. Thus, it is expected that the behavioral approach provides new and meaningful insights for an important theoretical issue like the simultaneous stabilization problem.

From these expectations, the author has been studying the simultaneous stabilization problem from the behavioral perspective in [11–14]. This article presents these results with some new remarks and observations. First, we provide an equivalent condition for two linear systems to be simultaneously stabilizable, and then also presents a parameterization of simultaneous stabilizers under the assumption that the interconnection of these two behavior is stable [11, 12]. By using this result, we provide a condition for three linear systems to be simultaneously stabilizable. We show that if one of the behavior stabilizes the other two behaviors, then three behaviors are simultaneously stabilizable [13, 14]. Although this condition corresponds to the result by Blondel in [2], the approach presented here is completely self-contained and is also independent of [2]. Particularly, a parameterization of simultaneous stabilizers in the three systems case is also presented here.

[Notations] Let $\mathbb{R}[\xi]$ denote the set of polynomials with real coefficients and $\mathbb{R}^{\mathrm{p}\times\mathrm{q}}[\xi]$ denote the set of polynomial matrices with real coefficients of size $p\times q$, respectively. Let $\mathbb{R}(\xi)$ denote the set of rational functions with real coefficients and $\mathbb{R}^{\mathrm{p}\times\mathrm{q}}(\xi)$ denote the set of rational matrices with real coefficients of size $p \times q$, respectively. For a nonsingular polynomial matrix $R \in \mathbb{R}^{\mathrm{q}\times\mathrm{q}}[\xi]$, all of the roots of $\det(R)$ are located in the open left half plane, R is said to be Hurwitz. Let $\mathbb{R}_{\mathrm{H}}[\xi]$ and $\mathbb{R}_{\mathrm{H}}^{\mathrm{q}\times\mathrm{q}}[\xi]$ denote the set of Hurwitz polynomials and the set of Hurwitz polynomial matrices of size $q \times q$, respectively.

4.2 Linear Time-Invariant Behaviors

We give brief reviews of behavioral system theory for linear time-invariant systems based on the references [19, 24, 25]. In the behavioral framework, a dynamical system is characterized as the set of the trajectories, i.e., the behavior. Let $\mathcal{P}$ denote the behavior of a system and q denote the number of the variables which interact with its environment. If a system is linear and time-invariant, the behavior $\mathcal{P}$ is representable by $R_N\frac{d^N w}{dt^N}+\cdots+R_1\frac{dw}{dt}+R_0 w=0$, where $R_i \in \mathbb{R}^{\bullet\times\mathrm{q}}$, $i=0,\cdots,N$. This is called a *kernel representation* of $\mathcal{P}$ and the variable w is called a manifest variable. A kernel representation is written as $R(\frac{d}{dt})w = 0$ by using a polynomial matrix $R := R_0 + R_1\xi + \cdots + R_N\xi^N \in \mathbb{R}^{\bullet\times\mathrm{q}}[\xi]$. It should be noted that there is no input/output partition in w. There are many kernel representations for $\mathcal{P}$. Particularly, we call a kernel representation $R(\frac{d}{dt})w = 0$ *minimal* if R has normal full row rank. In the following, the minimal rank of polynomial matrices inducing kernel representations is denoted by p.

$\mathcal{P}$ is said to be *controllable* if for all $w_1, w_2 \in \mathcal{P}$ there exist $w \in \mathcal{P}$ and $T_1, T_2 (\in \mathbb{R})$ such that $w(t) = w_1(t)$ for $t \le T_1$ and $w(t) = w_2(t)$ for $t > T_2$. In the case of linear time invariant behavior, $\mathcal{P}$ is controllable if and only if a minimal kernel representation is induced by a polynomial matrix $R(\xi)$ with the property that $R(\lambda)$ is full row rank for all complex number λ. The controllability of $\mathcal{P}$ is also equivalent to saying that $\mathcal{P}$ is described by $w = M_L\frac{d^L\ell}{dt^L}+\cdots+M_1\frac{d\ell}{dt}+M_0\ell$, where $M_i \in \mathbb{R}^{\mathrm{q}\times\bullet}$, $i=0,\cdots,L$. This is called an *image representation* of $\mathcal{P}$ and ℓ is called a latent variable. Similar to kernel representations, we use the notation $w = M(\frac{d}{dt})\ell$ by using a polynomial matrix $M := M_0+M_1\xi+\cdots+M_L\xi^L \in \mathbb{R}^{\mathrm{q}\times\bullet}[\xi]$. Moreover, there are many image representations for $\mathcal{P}$. Particularly, ℓ is said to be *observable* from w if $w = 0$ implies $\ell = 0$. A latent variable ℓ in $w = M(\frac{d}{dt})\ell$ is observable from w if and only if $M(\lambda)$ is full column rank for all complex number λ.

Note that $RM = 0$ and that there exists a polynomial matrix $Q \in \mathbb{R}^{(\mathrm{q}-\mathrm{p})\times\mathrm{q}}[\xi]$ such that $\left(R^{\mathrm{T}}\ Q^{\mathrm{T}}\right)^{\mathrm{T}}$ is unimodular and $QM = I$. Similarly, there exists a polynomial matrix $N \in \mathbb{R}^{\mathrm{q}\times\mathrm{p}}[\xi]$ such that $(N\ M)$ is unimodular and $RN = I$. Thus,

$$\begin{pmatrix} R \\ Q \end{pmatrix} \begin{pmatrix} N & M \end{pmatrix} = I \tag{4.1}$$

holds, which is referred as a doubly coprime factorization [10]. Throughout this article, we address controllable behaviors and their observable image representations.

$\mathcal{P}$ is said to be stable if $w \in \mathcal{P}$ implies $w(t) \to 0$ as $t \to \infty$. A behavior $\mathcal{P}$ is stable if and only if a minimal kernel representation of $\mathcal{P}$ is induced by a Hurwitz polynomial matrix $R \in \mathbb{R}_{\mathrm{H}}^{\mathtt{q}\times\mathtt{q}}[\xi]$. Control in the behavioral framework can be formalized as "interconnection" [25]. This corresponds to pickup the common trajectories between the behavior of a plant $\mathcal{P}$ and a controller $\mathcal{C}$ by sharing their external variables. The behavior after the interconnection can be restricted as $\mathcal{P} \cap \mathcal{C}$. There exist two kinds of interconnections, one is full interconnection [25] where two systems share all of their variables. The other is partial interconnection [1] where some of their variables are shared by each system. This article focuses on full interconnection.

In order to stabilize the plant $\mathcal{P}$, a controller $\mathcal{C}$, which is described by a kernel representation $Cw = 0$ with $C \in \mathbb{R}^{(\mathtt{q}-\mathtt{p})\times\mathtt{q}}[\xi]$, must be designed so as to satisfy that the behavior $\mathcal{P} \cap \mathcal{C}$ is described by

$$\begin{pmatrix} R\left(\frac{d}{dt}\right) \\ C\left(\frac{d}{dt}\right) \end{pmatrix} w = 0 \tag{4.2}$$

is stable, or equivalently, $\begin{pmatrix} R^{\mathrm{T}} & C^{\mathrm{T}} \end{pmatrix}^{\mathrm{T}}$ must be an element of $\mathbb{R}_{\mathrm{H}}^{\mathtt{q}\times\mathtt{q}}[\xi]$. It was shown in [15] by Kuijper that all of the stabilizing controllers for $\mathcal{P}$ can be induced by polynomial matrices

$$C := \begin{pmatrix} F & B \end{pmatrix} \begin{pmatrix} R \\ Q \end{pmatrix}, \tag{4.3}$$

for arbitrary $B \in \mathbb{R}_{\mathrm{H}}^{(\mathtt{q}-\mathtt{p})\times(\mathtt{q}-\mathtt{p})}[\xi]$ and arbitrary $F \in \mathbb{R}^{(\mathtt{q}-\mathtt{p})\times\mathtt{p}}[\xi]$. This is the behavioral version of the parameterization of all of the stabilizing controllers, which is the extension of the conventional result in [27]. Related to (4.3), it was also shown in [15] that $C \in \mathbb{R}^{(\mathtt{q}-\mathtt{p})\times\mathtt{q}}[\xi]$ induces a stabilizing controller for $\mathcal{P}$ if and only if CM is a Hurwitz polynomial matrix for any observable image representation $w = M\ell$.

4.3 Simultaneous Stabilization Problem for Two Linear Systems in a Behavioral Framework

4.3.1 Problem Formulation

Now we consider the simultaneous stabilization problem for the case of two systems. We are given two linear time-invariant controllable behaviors $\mathcal{P}_1$ and $\mathcal{P}_2$. We assume that the output cardinalities of them are the same, i.e., the ranks of their minimal kernel representations are the same. Let $R_i \in \mathbb{R}^{(\mathtt{q}-\mathtt{p})\times\mathtt{q}}[\xi]$ induce a kernel representation of $\mathcal{P}_i$ for $i = 1, 2$. Then, the problem we consider here is to find a condition under which there exists a single controller $\mathcal{C}$ such that $\mathcal{P}_1 \cap \mathcal{C}$ and $\mathcal{P}_2 \cap \mathcal{C}$ are stable. In terms of polynomial matrices, this problem can be equivalently formalized as finding a condition under which there exists a polynomial matrix $C \in \mathbb{R}^{(\mathtt{q}-\mathtt{p})\times\mathtt{q}}[\xi]$ such that $\begin{pmatrix} R_i(\xi)^{\mathrm{T}} & C(\xi)^{\mathrm{T}} \end{pmatrix}^{\mathrm{T}}$ is an element of $\mathbb{R}_{\mathrm{H}}^{\mathtt{q}\times\mathtt{q}}[\xi]$ for $i = 1, 2$.

In the following, let $M_i \in \mathbb{R}^{(\mathtt{q}-\mathtt{p})\times(\mathtt{q}-\mathtt{p})}[\xi]$ denote a polynomial matrix inducing an observable image representation of $\mathcal{P}_i$ for each $i = 1, 2$. Similarly, let $Q_i \in \mathbb{R}^{(\mathtt{q}-\mathtt{p})\times(\mathtt{q}-\mathtt{p})}[\xi]$ and $N_i \in \mathbb{R}^{(\mathtt{q}-\mathtt{p})\times(\mathtt{q}-\mathtt{p})}[\xi]$ denote polynomial matrices which satisfy the doubly coprime factorization (4.1).

4.3.2 A Necessary and Sufficient Condition for Two Linear Systems to be Simultaneous Stabilizable

In the standard control theory, it is well known that the simultaneous stabilizability of two linear systems is equal to the strong stabilizability [27] of the augmented system which consists of the given two systems [23]. Here, we show the corresponding result in the behavioral framework.

For $\mathcal{P}_1$ and $\mathcal{P}_2$ with involved polynomial matrices, we introduce the augmented behavior $\mathcal{P}_{12}$ described by a kernel representation

$$R_2(\tfrac{d}{dt})\begin{pmatrix} N_1(\tfrac{d}{dt}) & M_1(\tfrac{d}{dt}) \end{pmatrix} w = 0. \tag{4.4}$$

It is easy to see that an observable image representation of $\mathcal{P}_{12}$ is described by

$$w = \begin{pmatrix} R_1(\tfrac{d}{dt}) \\ Q_1(\tfrac{d}{dt}) \end{pmatrix} M_2\left(\frac{d}{dt}\right)\ell. \tag{4.5}$$

Then, we can obtain the following theorem [11].

Theorem 4.1 *Two behaviors $\mathcal{P}_1$ and $\mathcal{P}_2$ are simultaneously stabilizable if and only if there exist $F_{12} \in \mathbb{R}^{(\mathtt{q}-\mathtt{p})\times\mathtt{q}}[\xi]$ and $H_{12} \in \mathbb{R}_{\mathrm{H}}^{(\mathtt{q}-\mathtt{p})\times(\mathtt{q}-\mathtt{p})}[\xi]$ such that*

$$C_{\mathrm{a}12} := \begin{pmatrix} F_{12} & H_{12} \end{pmatrix} \in \mathbb{R}^{(\mathtt{q}-\mathtt{p})\times\mathtt{q}}[\xi] \tag{4.6}$$

induces a stabilizer for $\mathcal{P}_{12}$. □

Here, we give a brief review of the proof of this theorem. We define

$$C_{12} := C_{\mathrm{a}12}\begin{pmatrix} R_1 \\ Q_1 \end{pmatrix} = \begin{pmatrix} F_{12} & H_{12} \end{pmatrix}\begin{pmatrix} R_1 \\ Q_1 \end{pmatrix}. \tag{4.7}$$

It follows from the parameterization (4.3) by Kuijper [15] and $H_{12} \in \mathbb{R}_{\mathrm{H}}^{(\mathtt{q}-\mathtt{p})\times(\mathtt{q}-\mathtt{p})}[\xi]$ that C_{12} described by (4.7) induces a stabilizing controller for $\mathcal{P}_1$. In addition, by using Kuijper's result on the relationship between an observable image representation and a controller as reviewed in the previous section, we see that $C_{12}M_2$ is an element of $\mathbb{R}_{\mathrm{H}}^{(q-p)\times(q-p)}[\xi]$. Thus, $\mathcal{P}_1$ and $\mathcal{P}_2$ are simultaneously stabilizable. Conversely, we assume that $\mathcal{P}_1$ and $\mathcal{P}_2$ are simultaneously stabilizable. In other words, a stabilizer for $\mathcal{P}_1$ induced by

$$\begin{pmatrix} F_{12} & H_{12} \end{pmatrix} \begin{pmatrix} R_1 \\ Q_1 \end{pmatrix} \tag{4.8}$$

for $H_{12} \in \mathbb{R}_{\mathrm{H}}^{(q-p)\times(q-p)}[\xi]$ and $F_{12} \in \mathbb{R}_{\mathrm{H}}^{(q-p)\times p}[\xi]$ also stabilizes $\mathcal{P}_2$. This implies that

$$\underbrace{\begin{pmatrix} F_{12} & H_{12} \end{pmatrix}}_{C_{\mathrm{a}12}} \begin{pmatrix} R_1 \\ Q_1 \end{pmatrix} M_2 =: B_{12} \tag{4.9}$$

is also Hurwitz. This is equivalent to saying that the above $C_{\mathrm{a}12}$ with Hurwitz H_{12} is also a stabilizer for $\mathcal{P}_{12}$. For more detailed proof, see [11].

Here, we have two points to be mentioned. One is on the symmetric structure on simultaneous stabilizers. The above theorem can be equivalently applied to another augmented behavior obtained by the exchange of $\mathcal{P}_1$ and $\mathcal{P}_2$. We denote it by $\mathcal{P}_{21}$. This is described by a kernel representation

$$R_1(\tfrac{d}{dt})\begin{pmatrix} N_2(\tfrac{d}{dt}) & M_2(\tfrac{d}{dt}) \end{pmatrix} w = 0 \tag{4.10}$$

or an observable image representation described by

$$w = \begin{pmatrix} R_2(\tfrac{d}{dt}) \\ Q_2(\tfrac{d}{dt}) \end{pmatrix} M_1\left(\frac{d}{dt}\right)\ell. \tag{4.11}$$

It is easy to see that $\mathcal{P}_1$ and $\mathcal{P}_2$ are simultaneously stabilizable if and only if there exist $F_{21} \in \mathbb{R}^{(\mathrm{q}-\mathrm{p})\times\mathrm{p}}[\xi]$ and $H_{21} \in \mathbb{R}_{\mathrm{H}}^{(\mathrm{q}-\mathrm{p})\times(\mathrm{q}-\mathrm{p})}[\xi]$ such that $\mathcal{P}_{21}$ is stabilized by $C_{\mathrm{a}21}$ induced by

$$C_{\mathrm{a}21} := \begin{pmatrix} F_{21} & H_{21} \end{pmatrix} \in \mathbb{R}^{(\mathrm{q}-\mathrm{p})\times\mathrm{q}}[\xi]. \tag{4.12}$$

The proof of this statement is similar to the above theorem. Thus, we see that

$$\underbrace{\begin{pmatrix} F_{21} & H_{21} \end{pmatrix}}_{C_{\mathrm{a}21}} \begin{pmatrix} R_2 \\ Q_2 \end{pmatrix} M_1 =: B_{21} \tag{4.13}$$

is also Hurwitz. We also define

$$A_{21} := \begin{pmatrix} F_{21} & H_{21} \end{pmatrix} \begin{pmatrix} R_2 \\ Q_2 \end{pmatrix} N_1. \tag{4.14}$$

At this point, we see that $\begin{pmatrix} A_{21} & B_{21} \end{pmatrix}$ also induces a strong stabilizer for $\mathcal{P}_{12}$. Similarly, we also see that $\begin{pmatrix} A_{12} & B_{12} \end{pmatrix}$ also induces a strong stabilizer for $\mathcal{P}_{21}$ where A_{12} is defined by the exchange of $\mathcal{P}_1$ and $\mathcal{P}_2$ in (4.14). If we use the terminologies on the transfer functions in the conventional control theory, the interpretation of this observation is that the pole of the denominator of a strong stabilizer for $\mathcal{P}_{12}$ can be the pole of the closed loop between $\mathcal{P}_{21}$ and C_{a21} and vice versa. Since a strong stabilizer for $\mathcal{P}_{12}$ ($\mathcal{P}_{21}$) is proper while that for $\mathcal{P}_{21}$ ($\mathcal{P}_{12}$, respectively) is nonproper, such an observation on the symmetric structure cannot be obtained in the transfer function setting.

The other point to be mentioned is on the case of multiple plants. Although we treat the case of two linear systems, the above equivalent condition can be quickly extended to the case of multiple linear systems. For $\mathcal{P}_i$ $(i = 1, 2 \cdots, n)$, we introduce the augmented behavior $\mathcal{P}_{1i}$ which is described by a kernel representation induced by

$$R_1 \begin{pmatrix} N_i & M_i \end{pmatrix} \quad \text{for } i = 2, 3, \cdots, n. \tag{4.15}$$

Then, multiple systems $\mathcal{P}_i$ $(i = 1, 2 \cdots, n)$ are simultaneous stabilizable if and only if there exist $F_{1i} \in \mathbb{R}^{(\mathtt{q}-\mathtt{p})\times \mathtt{q}}[\xi]$ and $H_{1i} \in \mathbb{R}_{\mathrm{H}}^{(\mathtt{q}-\mathtt{p})\times(\mathtt{q}-\mathtt{p})}[\xi]$ such that

$$C_{a1i} := \begin{pmatrix} F_{1i} & H_{1i} \end{pmatrix} \in \mathbb{R}^{(\mathtt{q}-\mathtt{p})\times \mathtt{q}}[\xi] \tag{4.16}$$

induces a stabilizer for $\mathcal{P}_{1i}$ for $i = 2, 3, \cdots, n$. This is also a generalization of the result in [22] for the transfer function setting.

4.3.3 The Behavioral Version of the Strong Stabilizability

In the input/output setting, (4.6) can be also described by the transfer function $H_{12}^{-1} F_{12}$. From this, we can regard that the H_{12} corresponds to the "denominator" in the case where $\mathcal{C}_{a12}$ is described by a standard transfer function, so (4.6) corresponds to "strong stabilizer" for $\mathcal{P}_{12}$ in the behavioral setting. It is well known that the strong stabilizability for single-input and single-output systems is characterized as the parity interlacing property (p.i.p) condition on unstable real zeros with ∞ and unstable poles on the real axis [26]. In the behavioral setting, the well-known p.i.p condition can be slightly moderated.

Consider the strong stabilization of single-input and single-output systems in the behavioral framework. Let $\begin{pmatrix} d & n \end{pmatrix} \in \mathbb{R}^{1\times 2}[\xi]$ be a kernel representation which can be also described by the transfer function $-\frac{n}{d}$. The strong stabilizability of this system is equivalent to saying that there exists a $\begin{pmatrix} c_1 & c_2 \end{pmatrix} \in \mathbb{R}^{1\times 2}[\xi]$ with $c_2 \in \mathbb{R}_{\mathrm{H}}[\xi]$ such that $h := c_1 n + c_2 d$ is also Hurwitz. This condition can be also rewritten as that there exist $c_2, h \in \mathbb{R}_{\mathrm{H}}[\xi]$ and $c_1 \in \mathbb{R}[\xi]$ such that

$$d + \frac{c_1}{c_2} n = \frac{h}{c_2} \tag{4.17}$$

holds. Note that $\frac{c_1}{c_2}$ and $\frac{h}{c_2}$ are not restricted to be proper rational functions in the behavioral setting while the obtained result in [23, 27] for the standard control theory requires that these rational functions is to be proper. In addition, we have no restriction on the properness of $\frac{n}{d}$. From these discussions, we can obtain a moderated p.i.p condition as follows.

Theorem 4.2 *Let $d \in \mathbb{R}[\xi]$ and $n \in \mathbb{R}[\xi]$ be coprime polynomials. Let σ_i $(i = 1, 2, \cdots, m)$ denote nonnegative real roots of the numerator n such that $0 \leq \sigma_1 \leq \sigma_2 \leq \cdots \leq \sigma_m < \infty$. Then the following two statement are equivalent.*

1. *There exists $r \in \mathbb{R}(\xi)$ such that the denominator and numerator of $d + rn \in \mathbb{R}(\xi)$ are stable and the denominator of r is stable.*
2. *For every interval (σ_i, σ_{i+1}) $i = 1, 2, \cdots, m-1$, there exists even numbers of roots of d with multiplicity.*

The difference between the above theorem and the well-known p.i.p condition [23, 27] is only the point that the above theorem requires the intervals $0 \leq \cdots \leq \sigma_m < \infty$ while the condition in [23, 27] requires the $0 \leq \cdots \leq \sigma_m \leq \infty$. That is, the above p.i.p condition is moderated in the sense that the interval on the infinite zeros of $\frac{n}{d}$ is eliminated. The proof is straightforward from [23, 27].

We return to the simultaneous stabilization problem by illustrating an example. Consider the following two systems $\mathcal{P}_1$ and $\mathcal{P}_2$ described by

$$R_1 = \begin{pmatrix} -(\xi - 1)(\xi + 1) & \xi - 2 \end{pmatrix} \in \mathbb{R}^{1\times 2}[\xi]$$
$$R_2 = \begin{pmatrix} -(\xi - 2)(\xi + 1) & \xi - 1 \end{pmatrix} \in \mathbb{R}^{1\times 2}[\xi],$$

respectively. For $\mathcal{P}_1$, we compute $N_1 = \begin{pmatrix} -\frac{1}{3} & \frac{1}{3}(\xi + 2) \end{pmatrix}^{\mathrm{T}}$. By using N_1, M_1 and R_2, a kernel representation of the augmented behavior $\mathcal{P}_{12}$ can be induced by

$$R_{12} := \begin{pmatrix} -\frac{2}{3}\xi & (\xi + 1)(2\xi - 3) \end{pmatrix} \in \mathbb{R}^{1\times 2}[\xi].$$

It is easy to see that $R_{12} =: \begin{pmatrix} d_{12} & n_{12} \end{pmatrix}$ satisfies the moderated p.i.p condition (while it does not satisfy the p.i.p condition in the standard setting [23, 27]). In fact, one of the strong stabilizers for $\mathcal{P}_{12}$ can be obtained as a kernel representation induced by

$$C_{a12} := \begin{pmatrix} -\frac{1}{3}\xi & -\xi - \frac{5}{2} \end{pmatrix}.$$

Using this, one of the simultaneous stabilizers of $\mathcal{P}_1$ and $\mathcal{P}_2$ can be obtained as a kernel representation induced by

$$C_{12} := \begin{pmatrix} 2 + \frac{2}{3}\xi & -\frac{3}{2} \end{pmatrix}. \tag{4.18}$$

In fact, it is easy to check that both $\begin{pmatrix} R_1^{\mathrm{T}} & C_{12}^{\mathrm{T}} \end{pmatrix}^{\mathrm{T}}$ and $\begin{pmatrix} R_2^{\mathrm{T}} & C_{12}^{\mathrm{T}} \end{pmatrix}^{\mathrm{T}}$ are Hurwitz. Moreover, (4.18) can be also described by the transfer function as $-\frac{4}{3} + \frac{4}{9}\xi$ which is nonproper. On the other hand, it follows from the standard p.i.p condition in [23, 27] that these two systems cannot be simultaneously stabilizable. Thus, these two systems can be simultaneously stabilizable by a nonproper controller while they cannot be simultaneously stabilizable by any proper controller.

4.3.4 Parameterization of Simultaneous Stabilizers for a Class of Pairs of Linear Systems

We provide a parameterization of simultaneous stabilizers for two linear systems in the behavioral framework. Such a parameterization has not been provided in the standard control theory.

Here, we suppose that $q = 2$ and $p = 1$. In addition, as a crucial assumption, we assume that $\mathcal{P}_1 \cap \mathcal{P}_2$ is stable, which is equivalent to saying that one is a stabilizer of the other. From the viewpoints of polynomials, this assumption is also equivalent to saying that $R_1 M_2$ and $R_2 M_1$ are elements of $\mathbb{R}_H[\xi]$.

From the observation in the previous subsection, $\mathcal{P}_1$ and $\mathcal{P}_2$ are simultaneously stabilizable if and only if there exist Hurwitz H_{12} and H_{21} such that

$$\begin{pmatrix} F_{12} & H_{12} \end{pmatrix} \begin{pmatrix} R_1 \\ Q_1 \end{pmatrix} M_2 = H_{21}. \tag{4.19}$$

Thus, we obtain the following equivalent condition for two linear systems to be simultaneously stabilizable with respect to the solvability of a polynomial equation as follows.

Theorem 4.3 *Let $\mathcal{P}_1$ and $\mathcal{P}_2$ denote linear time invariant controllable behaviors. Then, $\mathcal{P}_1$ and $\mathcal{P}_2$ are simultaneously stabilizable if and only if the following polynomial equation*

$$H_{12} Q_1 M_2 - H_{21} = -F_{12} R_1 M_2 \tag{4.20}$$

is solvable with respect to $H_{12}, H_{21} \in \mathbb{R}_H[\xi]$, and $F_{12} \in \mathbb{R}[\xi]$.

In (4.19), we focus on the part which was already introduced in (4.8). Since H_{12} is Hurwitz, it follows from the parameterization of all stabilizing controllers in (4.3) that (4.8) induces a kernel representation of the stabilizer for $\mathcal{P}_1$ as stated in the previous subsection. In addition, (4.20) implies that (4.8) also induces a kernel representation of the stabilizer for $\mathcal{P}_2$. Namely, if F_{12} and H_{12} are solutions of (4.19), then (4.8) with these solutions induces a kernel representation of the simultaneous stabilizer for $\mathcal{P}_1$ and $\mathcal{P}_2$. Thus, we obtain the following theorem on the parameterization of simultaneous stabilizers for two linear systems.

Theorem 4.4 *Assume that (4.20) is solvable with respect to $H_{12}, H_{21} \in \mathbb{R}_H[\xi]$, and $F_{12} \in \mathbb{R}[\xi]$. Then (4.8) with the solutions H_{12} and F_{12} of (4.20) induces a kernel representation of a simultaneous stabilizer of $\mathcal{P}_1$ and $\mathcal{P}_2$.*

Next, we consider the condition under which (4.20) is solvable. From an assumption that $\mathcal{P}_1 \cap \mathcal{P}_2$ is stable, we define $R_1 M_2 = R_2 M_1 =: B_{12} \in \mathbb{R}_{\mathrm{H}}[\xi]$. Then, (4.20) can be described as

$$H_{12} Q_1 M_2 - H_{21} = -F_{12} B_{12}. \tag{4.21}$$

Note that H_{12} and H_{21} should be Hurwitz polynomials. If they includes B_{12} as a common factor, (4.21) can be rewritten as

$$B_{12} H_{12}' Q_1 M_2 - B_{12} H_{21}' = -F_{12} B_{12} \tag{4.22}$$

where $H_{12}', H_{21}' \in \mathbb{R}_{\mathrm{H}}[\xi]$. The above (4.22) is always solvable with respect to Hurwitz polynomials H_{12}' and H_{21}', and a polynomial F_{12}, because,

$$F_{12} = H_{12}' Q_1 M_2 - H_{21}' \tag{4.23}$$

is a solution of (4.22) for arbitrary Hurwitz polynomials H_{12}' and H_{21}'. Hence, under the assumption that $\mathcal{P}_1 \cap \mathcal{P}_2$ is stable, Theorem 4.4 can be modified as follows:

Theorem 4.5 *Assume that $\mathcal{P}_1 \cap \mathcal{P}_2$ is stable and define a Hurwitz polynomial $B_{12} := R_1 M_2 = R_2 M_1$. Then, a kernel representation of a simultaneous stabilizer for $\mathcal{P}_1$ and $\mathcal{P}_2$ is induced by*

$$C_{12} := \begin{pmatrix} H_{12}' Q_1 M_2 - H_{21}' & B_{12} H_{12}' \end{pmatrix} \begin{pmatrix} R_1 \\ Q_1 \end{pmatrix} \tag{4.24}$$

for arbitrary H_{12}' and $H_{21}' \in \mathbb{R}_{\mathrm{H}}[\xi]$.

The above theorem provides a parameterization of simultaneous stabilizers under the assumption that $\mathcal{P}_1 \cap \mathcal{P}_2$ is stable.

4.4 Simultaneous Stabilization of Three Linear Systems in a Behavioral Framework

4.4.1 A Sufficient Condition for Three Linear Systems to Be Simultaneously Stabilizable

Let $\mathcal{P}_1$, $\mathcal{P}_2$, and $\mathcal{P}_3$ denote the behaviors of three linear systems. We also assume that $\mathcal{P}_1 \cap \mathcal{P}_2$ and $\mathcal{P}_1 \cap \mathcal{P}_3$ are stable. In other words, one of three behaviors stabilizes the other two behaviors. Similar to $B_{12} = R_1 M_2 = R_2 M_1$ for $\mathcal{P}_1$ and $\mathcal{P}_2$, we also define a Hurwitz polynomial

$$B_{13} := R_1 M_3 = R_3 M_1. \tag{4.25}$$

Applying Theorem 4.5 to $\mathcal{P}_1$ and $\mathcal{P}_3$ yields that a kernel representation of a simultaneous stabilizer for $\mathcal{P}_1$ and $\mathcal{P}_3$ is induced by

$$C_{13} := \begin{pmatrix} H'_{13}Q_1M_3 - H'_{31} & B_{13}H'_{13} \end{pmatrix} \begin{pmatrix} R_1 \\ Q_1 \end{pmatrix} \tag{4.26}$$

for arbitrary Hurwitz polynomials H'_{13} and H'_{31}. Compare the simultaneous stabilizer for $\mathcal{P}_1$ and $\mathcal{P}_2$ described by (4.24) with the simultaneous stabilizer for $\mathcal{P}_1$ and $\mathcal{P}_3$ described by (4.26). We see that if the following two identical equations:

$$H'_{12}Q_1M_2 - H'_{21} = H'_{13}Q_1M_3 - H'_{31} \tag{4.27}$$

$$B_{12}H'_{12} = B_{13}H'_{13} \tag{4.28}$$

hold, a simultaneous stabilizers for $\mathcal{P}_1$ and $\mathcal{P}_2$ is equal to that for $\mathcal{P}_1$ and $\mathcal{P}_3$. That is, (4.24) or (4.26) with the identical Eqs. (4.27) and (4.28) induces a simultaneous stabilizer for $\mathcal{P}_1$, $\mathcal{P}_2$, and $\mathcal{P}_3$. Thus, we focus on the problem of whether (4.27) and (4.28) are solvable with respect to Hurwitz polynomials H'_{12}, H'_{21}, H'_{13}, and H'_{31}.

First, we focus on the solvability of (4.28). For this, since we just have to guarantee that H'_{12} and H'_{13} are Hurwitz, we select them as

$$H'_{12} = B_{13}H \tag{4.29}$$

$$H'_{13} = B_{12}H \tag{4.30}$$

where $H \in \mathbb{R}_H[\xi]$ can be arbitrary given. Next, we focus on the solvability of (4.27). Since we have already determined H'_{12} and H'_{13} so as to satisfy Eqs. (4.29) and (4.30), (4.27) is written as

$$HB_{13}Q_1M_2 - H'_{21} = HB_{12}Q_1M_3 - H'_{31} \tag{4.31}$$

or equivalently

$$HB_{13}Q_1M_2 - HB_{12}Q_1M_3 = H'_{21} - H'_{31}. \tag{4.32}$$

For simplicity, we restrict $H = 1$ in the following. The left-hand side of (4.32) has already been determined. This implies that the problem is to show whether the polynomial $B_{13}Q_1M_2 - B_{12}Q_1M_3$ can be described by the subtraction of Hurwitz matrices. In the case where $B_{13}Q_1M_2 - B_{12}Q_1M_3$ is Hurwitz, for instance, we can choose

$$H_{21} := \alpha B_{13}Q_1M_2 - B_{12}Q_1M_3$$

$$H_{31} := (\alpha - 1)B_{13}Q_1M_2 - B_{12}Q_1M_3$$

for an arbitrary $\alpha \in \mathbb{R}$. In general, $B_{13}Q_1M_2 - B_{12}Q_1M_3$ is not necessarily Hurwitz. In such a case, this is described as the product of the Hurwitz polynomial and the anti-Hurwitz one. The Hurwitz part can be included as the factor of both H'_{21} and H'_{31}. Thus, the problem is to check whether the anti-Hurwitz part can be decomposed as the subtraction of Hurwitz polynomials. As for this problem, we can obtain the following lemma.

Lemma 4.6 *Let $a \in \mathbb{R}[\xi]$ be an arbitrary anti-Hurwitz polynomial. Then there exist Hurwitz polynomials b_1 and $b_2 \in \mathbb{R}_{\mathrm{H}}[\xi]$ such that*

$$a = b_1 - b_2. \tag{4.33}$$

The proof will be contained in the future publications by the author. Based on Lemma 4.6, it is possible to guarantee that there exist Hurwitz H'_{21} and H'_{31} such that

$$B_{13}Q_1M_2 - B_{12}Q_1M_3 = H'_{21} - H'_{31}. \tag{4.34}$$

Therefore, if $\mathcal{P}_1 \cap \mathcal{P}_2$ and $\mathcal{P}_1 \cap \mathcal{P}_3$ are stable, then $\mathcal{P}_1$, $\mathcal{P}_2$ and $\mathcal{P}_3$ are simultaneous stabilizable. Moreover, together with Theorem 4.5, we can obtain a parameterization of simultaneous stabilizers for three linear systems under this assumption.

Theorem 4.7 *Assume that $\mathcal{P}_1 \cap \mathcal{P}_2$ and $\mathcal{P}_1 \cap \mathcal{P}_3$ are stable. Define Hurwitz polynomial $B_{12} := R_1M_2 = R_2M_1$ and $B_{13} := R_1M_3 = R_3M_1$. Then, a kernel representation of a simultaneous stabilizer for $\mathcal{P}_1$, $\mathcal{P}_2$ and $\mathcal{P}_3$ is induced by*

$$\begin{pmatrix} HB_{13}Q_1M_2 - H'_{21} & B_{12}B_{13} \end{pmatrix} \begin{pmatrix} R_1 \\ Q_1 \end{pmatrix} \tag{4.35}$$

or

$$\begin{pmatrix} HB_{12}Q_1M_3 - H'_{31} & B_{13}B_{12} \end{pmatrix} \begin{pmatrix} R_1 \\ Q_1 \end{pmatrix} \tag{4.36}$$

where H'_{21} and H'_{31} are Hurwitz polynomials obtained by the decomposition as

$$HB_{13}Q_1M_2 - HB_{12}Q_1M_3 = H'_{21} - H'_{31} \tag{4.37}$$

and H is arbitrary Hurwitz polynomial.

The same condition holds if the role of $\mathcal{P}_1$ and $\mathcal{P}_2$ (or $\mathcal{P}_3$) are replaced. Consequently, we can also obtain the following theorem.

Theorem 4.8 *Let $\mathcal{P}_1$, $\mathcal{P}_2$, $\mathcal{P}_3$ denote the behaviors of three linear systems. If one of them stabilizes the other two behaviors then these three behaviors are simultaneously stabilizable.*

In [2], the similar theorem has already been obtained in the transfer function setting. The approach addressed here is completely self-contained and is derived independently from [2]. Particularly, the approach in the behavioral setting also gives a parameterization of a simultaneous stabilizer, which are different point from the result in [2].

4.4.2 Example

We give a simple example [13] in order to illustrate how the result in this article works in the simultaneous stabilization. Let $\mathcal{P}_1$, $\mathcal{P}_2$, and $\mathcal{P}_3$ be described as kernel representations induced by

$$R_1 = \begin{bmatrix} -2 & \xi - 1 \end{bmatrix} \in \mathbb{R}^{1\times 2}[\xi], \tag{4.38}$$

$$R_2 = \begin{bmatrix} \xi & 2\xi + 1 \end{bmatrix} \in \mathbb{R}^{1\times 2}[\xi], \tag{4.39}$$

$$R_3 = \begin{bmatrix} \xi - 3 & 3\xi - 1 \end{bmatrix} \in \mathbb{R}^{1\times 2}[\xi], \tag{4.40}$$

respectively. A matrix Q_1 such that $[R_1^{\mathrm{T}}\ Q_1^{\mathrm{T}}]^{\mathrm{T}}$ is unimodular can be easily computed as

$$Q_1 = \begin{bmatrix} 0 & \frac{1}{2} \end{bmatrix} \in \mathbb{R}^{1\times 2}[\xi]. \tag{4.41}$$

First, we check whether $\mathcal{P}_1 \cap \mathcal{P}_2$ and $\mathcal{P}_1 \cap \mathcal{P}_3$ are stable. Actually, from simple computations, we see that

$$B_{12} := R_1 M_2 = -\xi^2 - 2\xi - 2 \tag{4.42}$$

$$B_{13} := R_1 M_3 = -\xi^2 - 2\xi - 1. \tag{4.43}$$

Thus, the sufficient condition for a triple of the behaviors to be simultaneously stabilizable is satisfied. We compute

$$B'_{13} Q_1 M_2 - B'_{12} Q_1 M_3 = -\xi^2 - 4\xi - 3. \tag{4.44}$$

One of the candidates of the decomposition described by (4.44) is

$$\begin{aligned} H'_{21} &= 2(-\xi^2 - 4\xi - 3) \\ H'_{31} &= -H'_{21}. \end{aligned}$$

Then, we obtain

$$B'_{13}Q_1M_2 - H'_{21} = B'_{12}Q_1M_3 - H'_{31} = \frac{1}{2}\xi^3 - \xi^2 - \frac{15}{2}\xi - 6.$$

Hence, the behavior of a simultaneous stabilizer, say $\mathcal{C}$, is induced by

$$\begin{aligned} C &:= \begin{pmatrix} B'_{13}Q_1M_2 - H'_{21} & B'_{12}B'_{13} \end{pmatrix} \begin{pmatrix} R_1 \\ Q_1 \end{pmatrix} \\ &= \begin{pmatrix} -\xi^3 + 2\xi + 15\xi + 12 & -4\xi^3 - 11\xi - 2\xi + 5 \end{pmatrix} \in \mathbb{R}^{1\times 2}[\xi]. \end{aligned}$$

Indeed, we can validate that C induces simultaneous stabilizers for $\mathcal{P}_1$, $\mathcal{P}_2$, and $\mathcal{P}_3$. As for $\mathcal{P}_1 \cap \mathcal{C}$, we see that $CM_1 = B'_{12}B'_{13}$, which is stable. The roots of polynomials of CM_2 and CM_3 are $\{-1, -1, -1, -3\}$ and $\{-1, -1, -2, -3\}$ respectively, which implies that they are also Hurwitz. Thus, we see that C described by (4.37) induces a simultaneous stabilizers for $\mathcal{P}_1$, $\mathcal{P}_2$, and $\mathcal{P}_3$.

4.5 Conclusions and Future Works

In this article, we have addressed the simultaneous stabilization problem in the behavioral framework. We have explained a necessary and sufficient condition for a pair of linear systems to be simultaneously stabilizable. We have also discussed the relationship between the strong stabilizability in the behavioral setting and that in the standard setting. We have also provided a parameterization of simultaneous stabilizer under the assumption that the interconnection of the two systems are stable. Then we have also considered the case of three systems. For this problem, we have provided a sufficient condition for three systems to be simultaneously stable. This condition is that one of the three systems stabilizes the other two systems. Although this condition coincides with the results in [2] by Blondel, the derivation is completely self-contained and our result yields a representation of simultaneous stabilizers. Finally, we have also provided a parameterization of simultaneous stabilizers under the assumption that one of the three behaviors stabilizes the other two behaviors.

There are many issues to be looked into in the future. In this article, we suppose that simultaneous stabilization can be done by regular interconnection. On the other hand, there exist many cases where a kernel representation of $\mathcal{P}_i$ is not minimal, that is, the output cardinality is not the same. In such cases, a necessary and/or sufficient condition should be provided. To address this problem, we believe that the results by Harry and Praagman in [20] could be very useful.

In addition, we have explained necessary and sufficient conditions for a class of pairs/triples of linear systems. These results should be extended to larger classes of pairs/triples. Finally, as shown in [5, 6], the simultaneous stabilization problem for a triple of linear systems is undecidable. Then, the problem on what kind of triples is decidable might be interesting from the theoretical points of view.

References

1. Belur, N.M., Trentelman, H.L.: Stabilization, pole placement and regular implementability. IEEE Trans. Autom. Control **AC-47**, 735–744 (2002)
2. Blondel, V.D., Campion, G., Gevers, M.: A sufficient condition for simultaneous stabilization. IEEE Trans. Autom. Control **AC-38**, 1264–1266 (1993)
3. Blondel, V.D.: Simultaneous stabilization of linear systems. Springer, Berlin (1994)
4. Blondel, V.D., Gevers, M., Mortini, R., Rupp, R.: Simultaneous stabilization of three or more systems: Conditions on the real axis does not suffice. SIAM J. Control Optim. **32**, 572–590 (1994)
5. Blondel, V.D., Gevers, M.: Simultaneous stabilization of three linear systems is rationally undecidable. Math. Control Signals Syst. **6**, 135–145 (1994)
6. Blondel, V.D.: Simultaneous stabilization of linear systems and interpolation with rational function. In: Blondel, V.D., Sontag, E., Vidyasagar, M., Willems, J.C. (eds.) Open Problems in Mathematical Systems and Control Theory, pp. 53–60. Springer Verlag, London (1999)
7. Ghosh, B.: An approach to simultaneous system design: part 1. SIAM J. Control Optim. **24**, 480–496 (1986)
8. Ghosh, B.: Transcendental and interpolation methods in simultaneous stabilization and simultaneous partial pole placement problems. SIAM J. Control Optim. **24**, 1091–1109 (1986)
9. Ghosh, B.: An approach to simultaneous system design: part 2. SIAM J. Control Optim. **26**, 919–963 (1988)
10. Kailath, T.: Linear Systems. Prentice-Hall, Englewood Cliffs, NJ (1980)
11. Kaneko, O., Mori, K., Yoshida, K., Fujii, T.: A parameterization of simultaneous stabilizers for a pair of linear systems in a behavioral framework. In: Proceedings of the 43rd IEEE Conference on Decision and Control, pp. 329–334 (2004)
12. Kaneko, O., Fujii, T.: A sufficient condition for a triple of linear systems to be simultaneously stabilizable within a behavioral framework. In: Proceedings of the 44th IEEE Conference on Decision and Control, pp. 149–154 (2005)
13. Kaneko, O.: The behavioral approach to simultaneous stabilization. In: Proceedings of the 19th International Symposium on Mathematical Theory of Network and Systems, pp. 697–701 (2010)
14. Kaneko, O.: Cyclic condition of simultaneous stabilizability for three linear systems in a behavioral framework. In: Proceedings of IFAC Joint conference: 5th Symposium on System Structure and Control, 11th Workshop on Time-Delay Systems, and 6th Workshop on Fractional Differentiation and Its Applications, pp. 90–95 (2013)
15. Kuijper, M.: Why do stabilizing controllers stabilize? Automatica **33**, 621–625 (1995)
16. Maeda, H., Vidyasagar, M.: Some results on simultaneous stabilization. Syst. Control Lett. **5**, 205–208 (1984)
17. Obinata, G., Moore, J.B.: Characterization of controllers in simultaneous stabilization. Syst. Control Lett. **10**, 333–340 (1989)
18. Patel, V.V.: Solution to the "Champagne problem" on the simultaneous stabilization of three plants. Syst. Control Lett. **37**, 173–175 (1999)
19. Polderman, J.W., Willems, J.C.: Introduction to Mathematical Systems Theory -A Behavioral Approach. Springer, Berlin (1997)
20. Praagman, C., Trentelman, H.L., Yoe, R.Z.: On the parameterization of all regularly implementing and stabilizing controllers. SIAM Journal on Control and Optimization **45**, 2035–2053 (2007)
21. Saeks, R., Murray, J.: Fractional representation, algebraic geometry and the simultaneous stabilization. IEEE Trans. Autom. Control **AC-27**, 895–903 (1982)
22. Vidyasagar, M., Viswanadham, N.: Algebraic design techniques for reliable stabilization. IEEE. Trans. Autom. Control **AC-27**, 1085–1095 (1982)
23. Vidyasagar, M.: Control System Synthesis, A Factorization Approach. MIT Press, Cambridge MA (1985)

24. Willems, J.C.: Paradigms and puzzles in the theory of dynamical systems. IEEE Trans. Autom. Control **AC-36**, 259–294 (1991)
25. Willems, J.C.: On interconnections, control, and feedback. IEEE Trans. Autom. Control **AC-42**, 326–339 (1997)
26. Youla, D.C., Bongiorno Jr, J.J., Lu, C.N.: Single-loop feedback-stabilization of linear multivariable dynamical plants. Automatica **10**, 159–173 (1974)
27. Youla, D.C., Jabr, H.A., Bongiorno Jr, J.J.: Modern Wiener-Hopf design of optimal controllers, part 2: the multivariable case. IEEE Trans. Autom. Control **AC–21**, 319–338 (1976)

Chapter 5
New Properties of ARE Solutions for Strictly Dissipative and Lossless Systems

Chayan Bhawal, Sandeep Kumar, Debasattam Pal and Madhu N. Belur

Abstract Algebraic Riccati Equation (ARE) solutions play an important role in many optimal/suboptimal control problems. However, a key assumption in formulation and solution of the ARE is a certain 'regularity condition' on the feedthrough term D of the system. For example, formulation of the ARE requires nonsingularity of $D + D^T$ in positive real dissipative systems and, in the case of bounded real dissipative systems, one requires nonsingularity of $I - D^T D$. Note that for lossless systems $D + D^T = 0$, while for all-pass systems $I - D^T D = 0$; this rules out the formulation of the ARE. Noting that the ARE solutions are also extremal "storage functions" for dissipative systems, one can speak of storage function for the lossless case too. This contributed chapter formulates new properties of the ARE solution; we then show that this property is satisfied by the storage function for the lossless case too. The formulation of this property is via the set of trajectories of minimal dissipation. We show that the states in a first-order representation of this set satisfy *static* relations that are closely linked to ARE solutions; this property holds for the lossless case too. Using this property, we propose an algorithm to compute the storage function for the lossless case.

With best wishes to Harry L. Trentelman on the occasion of his 60th birthday.
The last author adds: "to my Ph.D. supervisor, who always has been very friendly, and imparted a sense of discipline and rigour in teaching and research during the course of my Ph.D. Through your actions, I learnt the meaning of 'zero tolerance' to careless and hasty work. Thanks very much for all the skills I imbibed from you: both technical and non-technical."

C. Bhawal · S. Kumar · D. Pal · M.N. Belur (✉)
Department of Electrical Engineering, Indian Institute of Technology Bombay, Mumbai, India
belur@iitb.ac.in

C. Bhawal
chayanbhawal@ee.iitb.ac.in

S. Kumar
sandeepkumar.iitb@gmail.com

D. Pal
debasattam@ee.iitb.ac.in

M.N. Belur et al. (eds.), *Mathematical Control Theory II*,
Lecture Notes in Control and Information Sciences 462,
DOI 10.1007/978-3-319-21003-2_5

5.1 Introduction

The algebraic Riccati equation (ARE) has found widespread application in many optimal and suboptimal control/estimation problems. For example, Kalman filter, LQ control, H_∞ and H_2 control; see [1, 11], for example. Since its introduction in control theory by Kalman, many conceptual and numerical methods to solve ARE have been developed [3, 11] for instance. In the context of dissipative systems, the ARE solutions are extremal storage functions of the system. More about the link between storage functions, dissipative systems and solvability of AREs can be found in [16, 18]. However, for a special class of dissipative systems, namely, conservative systems, the ARE does not exist. This happens due to the formulation of the ARE depending on a suitable regularity condition on the feedthrough term D of any input-state-output representation of a system. The precise form of the regularity condition depends on the supply rate function, with respect to which dissipativity holds. For example, in case of the "positive real supply rate," $u^T y$, where u is the input and y is the output of the system, existence of the corresponding ARE requires nonsingularity of $D + D^T$. Similarly, for the "bounded real supply rate," $u^T u - y^T y$, nonsingularity of $I - D^T D$ is required for existence of the corresponding ARE. Contrary to this regularity condition, systems that are conservative with respect to the positive real supply rate and the bounded real supply rate have $D + D^T = 0$ and $I - D^T D = 0$, respectively.[1] Hence, for such systems the regularity conditions are violated, and consequently, the corresponding ARE does not exist. In this chapter, we formulate new properties of the ARE solution in terms of the set of trajectories of "minimal dissipation" as formulated recently in [17]: for reasons we will elaborate later, we will call this set "a Hamiltonian system." We show that the ARE solution is closely linked to the static relations that hold between the states in a first-order representation of this set. We then show that this property is satisfied for the storage function for the conservative case too, though the ARE does not exist in this case. We use this result to develop an algorithm to compute the unique storage function for the conservative systems case.

We now elaborate further on the key property that the ARE solution satisfies: which we extend to the lossless case. The property is based on an observation concerning the relation between ARE solutions and Hamiltonian systems. It is well known that when the feedthrough term satisfies the regularity conditions, that is, when the ARE exists, the solutions to the ARE can be found using suitable invariant subspaces of a corresponding Hamiltonian matrix. Note that, in the singular cases (lossless/all-pass), the Hamiltonian matrix does not exist. Consequently, this method involving the invariant subspace fails to work for the singular cases. However, this same method, when viewed from a different perspective opens up a new way of computing the

[1] Lossless systems, with u input and y output, are conservative with respect to the "positive real supply rate" $u^T y$ and have $D + D^T = 0$. Similarly, all-pass systems are conservative with respect to the "bounded real supply rate" $u^T u - y^T y$. For all-pass systems $I - D^T D = 0$. Hence, all arguments about ARE solutions and storage functions made for lossless systems are applicable to all-pass systems as well.

ARE solutions, which extends naturally to the singular case, too. This new point of view stems from the fact that the first-order system defined by the Hamiltonian matrix associated to an ARE is nothing but a state representation of a system comprised of the "trajectories of minimal dissipation." Consequently, choosing an invariant subspace im $\begin{bmatrix} I \\ K \end{bmatrix}$ of the Hamiltonian matrix to get K as a solution to the ARE, can be viewed as obtaining a subsystem of the Hamiltonian system by *restricting* the trajectories to satisfy an extra set of equations as $z = Kx$, where x, z are state variables of the original system and its 'dual', respectively. The crucial fact about this new view-point is that, although, the Hamiltonian *matrix* and the ARE do not exist in the singular case, the Hamiltonian *system*, comprising of the trajectories of minimal system does exist. We show in this chapter that, in such cases too, the strategy of putting static relation $z = Kx$ leads to a storage function $x^T Kx$ to the original system.

The notation used in the chapter is standard. The set $\mathbb{R}$ and $\mathbb{C}$ denote the fields of real and complex numbers, respectively. The set $\mathbb{R}[\xi]$ denotes the ring of polynomials in ξ with real coefficients. The set $\mathbb{R}^{\mathtt{w}\times\mathtt{p}}[\xi]$ denotes all $\mathtt{w} \times \mathtt{p}$ matrices with entries from $\mathbb{R}[\xi]$. We use $\bullet$ when a dimension need not be specified: for example, $\mathbb{R}^{\mathtt{w}\times\bullet}$ denotes the set of real constant matrices having $\mathtt{w}$ rows. $\mathbb{R}[\zeta, \eta]$ denotes the set of real polynomials in two indeterminates: ζ and η. The set of $\mathtt{w} \times \mathtt{w}$ matrices with entries in $\mathbb{R}[\zeta, \eta]$ is denoted by $\mathbb{R}^{\mathtt{w}\times\mathtt{w}}[\zeta, \eta]$. The space $\mathfrak{C}^\infty(\mathbb{R}, \mathbb{R}^{\mathtt{w}})$ stands for the space of all infinitely often differentiable functions from $\mathbb{R}$ to $\mathbb{R}^{\mathtt{w}}$, and $\mathfrak{D}(\mathbb{R}, \mathbb{R}^{\mathtt{w}})$ stands for the subspace of all compactly supported trajectories in $\mathfrak{C}^\infty(\mathbb{R}, \mathbb{R}^{\mathtt{w}})$.

This chapter is structured as follows: Sect. 5.2 summarizes preliminaries required in the chapter. New properties of ARE solutions are presented in Sect. 5.3. In Sect. 5.4, we formulate and prove new results that help computation of storage function K for conservative behaviors based on the notion of "trajectories of minimal dissipation". Section 5.5 uses the main result in Sect. 5.4 and proposes a numerical algorithm to compute storage function of conservative systems. Section 5.6 contains numerical examples to illustrate the main results. Some concluding remark is presented in Sect. 5.7.

5.2 Preliminaries

In this section, we give a brief introduction to various concepts that are required to formulate and solve the problem addressed in the chapter.

5.2.1 Behavior

We start with some essential preliminaries of the behavioral approach: a detailed exposition can be found in [12].

Definition 5.1 A linear differential behavior $\mathfrak{B}$ is defined as the subspace of infinitely often differentiable functions $\mathfrak{C}^{\infty}(\mathbb{R}, \mathbb{R}^{\mathtt{w}})$ consisting of all solutions to a set of linear ordinary differential equations with constant coefficients, i.e., for $R(\xi) \in \mathbb{R}^{\bullet \times \mathtt{w}}[\xi]$

$$\mathfrak{B} := \left\{ w \in \mathfrak{C}^{\infty}(\mathbb{R}, \mathbb{R}^{\mathtt{w}}) \mid R\left(\frac{d}{dt}\right) w = 0 \right\}. \tag{5.1}$$

The variable w in Eq. (5.1) is called the *manifest variable* of the behavior $\mathfrak{B}$. We denote linear differential behaviors with $\mathtt{w}$ number of manifest variables as $\mathfrak{L}^{\mathtt{w}}$. Equation (5.1) is what we call a *kernel representation* of the behavior $\mathfrak{B} \in \mathfrak{L}^{\mathtt{w}}$ and we sometimes also write $\mathfrak{B} = \ker R(\frac{d}{dt})$. We assume the polynomial matrix $R(\xi)$ has full row rank without loss of generality (see [12, Chap. 6]). This assumption guarantees existence of a nonsingular block $P(\xi)$ (after a permutation of columns, if necessary, with a corresponding permutation of the components of w) such that $R(\xi) = \begin{bmatrix} P(\xi) & Q(\xi) \end{bmatrix}$. Conforming to this partition of $R(\xi)$, partition w into $w = \left[\begin{smallmatrix} y \\ u \end{smallmatrix}\right]$, where it has been shown that u, y are the input and output of the behavior $\mathfrak{B}$ respectively: note that this partition is not unique. Such a partition is called an *input-output* partition of the behavior. The input-output partition is called *proper* if $P^{-1}Q$ is a matrix of *proper* rational functions. Although there are a number of ways in which the manifest variables can be partitioned as input and output, the number of components of the input depends only on $\mathfrak{B}$: we denote this number as $\mathtt{m}(\mathfrak{B})$, and call it the *input cardinality* of the behavior. The number of components in the output is called the *output cardinality* represented as $\mathtt{p}(\mathfrak{B})$. It is well known that $\mathtt{p}(\mathfrak{B}) = \operatorname{rank} R(\xi)$ and $\mathtt{m}(\mathfrak{B}) = \mathtt{w} - \mathtt{p}(\mathfrak{B})$.

In the behavioral approach, a system is nothing but its behavior: we use the terms behavior/system interchangeably. There are various ways of representing a behavior depending on how the system is modeled: a useful one is the *latent variable representation*: for $R(\xi) \in \mathbb{R}^{\bullet \times \mathtt{w}}$ and $M(\xi) \in \mathbb{R}^{\bullet \times \mathtt{m}}[\xi]$,

$$\mathfrak{B} := \left\{ w \in \mathfrak{C}^{\infty}(\mathbb{R}, \mathbb{R}^{\mathtt{w}}) \mid \text{there exists } \ell \in \mathfrak{C}^{\infty}(\mathbb{R}, \mathbb{R}^{\mathtt{m}}) \text{ such that } R\left(\frac{d}{dt}\right) w = M\left(\frac{d}{dt}\right) \ell \right\}.$$

Here ℓ is called a latent variable.

Controllability is another important concept required for this chapter.

Definition 5.2 A behavior $\mathfrak{B}$ is said to be *controllable* if for every pair of trajectories $w_1, w_2 \in \mathfrak{B}$ there exists $w_3 \in \mathfrak{B}$ and $\tau > 0$ such that

$$w_3(t) = \begin{cases} w_1(t) & \text{for } t \leqslant 0, \\ w_2(t) & \text{for } t \geqslant \tau. \end{cases}$$

We represent the set of all controllable behaviors with $\mathtt{w}$ variables as $\mathfrak{L}^{\mathtt{w}}_{\text{cont}}$. The familiar PBH rank test for controllability has been generalized: a behavior $\mathfrak{B}$ with minimal kernel representation $\mathfrak{B} = \ker R(\frac{d}{dt})$ is controllable if and only if $R(\lambda)$ has constant rank for all $\lambda \in \mathbb{C}$. One of the ways by which a behavior $\mathfrak{B}$ can be represented if (and only if) $\mathfrak{B}$ is controllable is the *image representation*: for $M(\xi) \in \mathbb{R}^{\mathtt{w} \times \mathtt{m}}[\xi]$

$$\mathfrak{B} := \left\{ w \in \mathfrak{C}^{\infty}(\mathbb{R}, \mathbb{R}^{\mathrm{w}}) \mid \text{there exists } \ell \in \mathfrak{C}^{\infty}(\mathbb{R}, \mathbb{R}^{\mathrm{m}}) \text{ such that } w = M\left(\frac{d}{dt}\right)\ell \right\}. \tag{5.2}$$

If $M(\xi)$ is such that $M(\lambda)$ has full column rank for all $\lambda \in \mathbb{C}$, then the image representation is said to be an *observable* image representation: this can be assumed without loss of generality (see [12, Sect. 6.6]).

5.2.2 Quadratic Differential Forms and Dissipativity

This section contains a brief review of Quadratic Differential Forms (QDFs): more on QDFs can be found in [18]. We often encounter quadratic expressions of derivatives of the manifest and/or latent variables of the behavior $\mathfrak{B}$. Two-variable polynomial matrices can be associated with such quadratic forms. Consider a two-variable polynomial matrix $\phi(\zeta, \eta) := \sum_{j,k} \phi_{jk} \zeta^j \eta^k \in \mathbb{R}^{\mathrm{w} \times \mathrm{w}}[\zeta, \eta]$ where $\phi_{jk} \in \mathbb{R}^{\mathrm{w} \times \mathrm{w}}$. The QDF Q_{ϕ} induced by $\phi(\zeta, \eta)$ is a map $Q_{\phi} : \mathfrak{C}^{\infty}(\mathbb{R}, \mathbb{R}^{\mathrm{w}}) \to \mathfrak{C}^{\infty}(\mathbb{R}, \mathbb{R})$ defined as

$$Q_{\phi}(w) := \sum_{j,k} \left(\frac{d^j w}{dt^j}\right)^T \phi_{jk} \left(\frac{d^k w}{dt^k}\right).$$

Of course, when $\Sigma \in \mathbb{R}^{\mathrm{w} \times \mathrm{w}}$, then $Q_{\Sigma}(w) = w^T \Sigma w$. Using the definition of QDFs, we next define a dissipative system.

Definition 5.3 Consider $\Sigma = \Sigma^T \in \mathbb{R}^{\mathrm{w} \times \mathrm{w}}$ and controllable $\mathfrak{B} \in \mathfrak{L}^{\mathrm{w}}_{\mathrm{cont}}$. The system $\mathfrak{B}$ is said to be Σ-dissipative if

$$\int_{\mathbb{R}} Q_{\Sigma}(w)\, dt \geqslant 0 \quad \text{for every} \quad w \in \mathfrak{B} \cap \mathfrak{D}. \tag{5.3}$$

The function $Q_{\Sigma}(w)$ is also called the supply rate: it is the rate of supply of energy to the system. For simplicity, we also call Σ the supply rate. Equation (5.3) formalizes the notion that dissipative systems are such that the net energy exchange is always an *absorption* when the trajectories considered are those that start-from-rest and end-at-rest, i.e. compactly supported. The link with existence of a storage function is well known for the controllable system case: a controllable behavior $\mathfrak{B} \in \mathfrak{L}^{\mathrm{w}}_{\mathrm{cont}}$ is dissipative with respect to Σ if and only if there exists a quadratic differential form $Q_{\psi}(w)$ such that

$$\frac{d}{dt} Q_{\psi}(w) \leqslant Q_{\Sigma}(w) \text{ for all } w \in \mathfrak{B}.$$

The QDF Q_{ψ} is called a storage function for $\mathfrak{B}$ with respect to the supply rate Σ.

The notion of a storage function captures the intuition that the rate of increase of stored energy in a dissipative system is at most the power supplied. In this chapter, we shall be dealing with supply rates Q_Σ induced by real symmetric constant nonsingular matrices Σ only. We need a count of the number of positive eigenvalues (with multiplicities) of the symmetric matrix Σ: call this number the *positive signature* of the matrix Σ and denote it by $\sigma_+(\Sigma)$.

For a Σ-dissipative system, $\mathtt{m}(\mathfrak{B})$, the input cardinality of the behavior, cannot exceed the positive signature $\sigma_+(\Sigma)$ of the supply rate Σ i.e. $\mathtt{m}(\mathfrak{B}) \leqslant \sigma_+(\Sigma)$ (details in [18, Remark 5.11] and [19]). For this chapter, we restrict ourselves to the so-called *maximum input cardinality condition*, i.e.

$$\mathtt{m}(\mathfrak{B}) = \sigma_+(\Sigma). \tag{5.4}$$

Given $\Sigma \in \mathbb{R}^{\mathtt{w}\times\mathtt{w}}$ and a system described by the observable image representation $w = M(\frac{d}{dt})\ell$, the QDF $Q_\Sigma(w)$ can also be expressed as $Q_\Phi(\ell)$ in the latent variables induced by $\Phi(\zeta,\eta) \in \mathbb{R}^{\mathtt{m}\times\mathtt{m}}[\zeta,\eta]$ is given by

$$\Phi(\zeta,\eta) := M(\zeta)^T \Sigma M(\eta).$$

Conservative systems are a special class of dissipative systems and this work focusses on the conservative systems' case: this is when the algebraic Riccati equation does *not* exist.

Definition 5.4 Consider a symmetric and nonsingular matrix $\Sigma \in \mathbb{R}^{\mathtt{w}\times\mathtt{w}}$ and a behavior $\mathfrak{B} \in \mathfrak{L}^{\mathtt{w}}_{\text{cont}}$. The system $\mathfrak{B}$ is called Σ-conservative if

$$\int_{\mathbb{R}} Q_\Sigma(w)dt = 0 \text{ for all } w \in \mathfrak{B} \cap \mathfrak{D}.$$

In order to simplify the exposition in this chapter, we shall be using the positive real supply rate $2u^T y$ i.e.

$$Q_\Sigma = \begin{bmatrix} u \\ y \end{bmatrix}^T \Sigma \begin{bmatrix} u \\ y \end{bmatrix} \text{induced by } \Sigma = \begin{bmatrix} 0 & I \\ I & 0 \end{bmatrix} \tag{5.5}$$

where u and y are the input and output of the system respectively. Systems conservative with respect to the positive real supply rate are known in the literature as lossless systems: see Footnote 1. We will be dealing with only lossless systems in this chapter. However, the results in the chapter can be extended to system conservative with respect to other supply rates also.

5.2.3 State Representation and Trimness

A *state variable representation* of a behavior $\mathfrak{B}$ is a latent variable representation where the latent variable x satisfies the *axiom of state:* whenever $(w_1, x_1),\ (w_2, x_2) \in$

$\mathfrak{B}_{\text{full}}$ and $x_1(0) = x_2(0)$, the *concatenation*[2] $(w_1, x_1) \wedge_0 (w_2, x_2)$ at $t = 0$ also satisfies the equations of $\mathfrak{B}_{\text{full}}$ in a weak/distributional sense. For such a system, we have a first-order description, called the state-space description:

$$E\frac{dx}{dt} + Fx + Gw = 0 \text{ where } E, F, G \text{ are constant real matrices.} \tag{5.6}$$

A state-space description is said to be *minimal* if the number of components in the state x is the minimum amongst all possible state representations. The number of states corresponding to a minimal state representation of $\mathfrak{B}$ is called the *McMillan degree* of the behavior $\mathfrak{B}$. When the state x is not minimal (but is observable from the system variable w), it is known that one or more components in x satisfy a static relation and the states are said to be *nontrim*. A more formal definition of *state trim* is presented next.

Definition 5.5 The state x in Eq. (5.6) is said to be trim if for every $a \in \mathbb{R}^{\mathtt{n}}$ there exist a $w \in \mathfrak{B}$ such that $x(0) = a$ and (w, x) satisfies Eq. (5.6).

The algorithm proposed in this chapter is based on this notion of state trimness. The static relation between the state x of the given lossless system and the "dual state" z of the adjoint system are used to find the unique storage function for the lossless case: see Theorem 5.13 below.

5.2.4 Minimal Polynomial Basis

This section contains a review of the notion of a minimal polynomial basis.

The *degree* of a polynomial vector $r(s) \in \mathbb{R}^{\mathtt{n}}[s]$ is the maximum degree among the $\mathtt{n}$ components of the vector. The degree of the zero polynomial and the zero vector in $\mathbb{R}^{\mathtt{n}}[s]$ is defined as $-\infty$.

For $R(s) \in \mathbb{R}^{\mathtt{n}\times\mathtt{m}}[s]$, the set of all polynomial vectors $v(s) \in \mathbb{R}^{\mathtt{m}}[s]$ that satisfy $R(s)v(s) = 0$ forms a vector space over the field of scalar rational functions. It is known from the literature that such a vector space admits a polynomial basis called the right nullspace basis of the polynomial matrix $R(s)$: see [8, Sect. 6.5.4]. There is a special nullspace basis called the *minimal polynomial basis* of the polynomial matrix $R(s)$ which is of importance to us in this chapter. Consider the polynomial matrix $R(s) \in \mathbb{R}^{\mathtt{n}\times\mathtt{m}}[s]$ of rank $\mathtt{n}$. Let the set $\{p_1(s), p_2(s), \ldots, p_{\mathtt{m}-\mathtt{n}}(s)\}$ be a nullspace basis of $R(s)$ ordered by their degrees $d_1 \leqslant d_2 \leqslant \cdots \leqslant d_{\mathtt{m}-\mathtt{n}}$. The set $\{p_1(s), p_2(s), \ldots, p_{\mathtt{m}-\mathtt{n}}(s)\}$ is said to be a *minimal polynomial basis* of $R(s)$ if every other nullspace basis $\{q_1(s), q_2(s), \ldots, q_{\mathtt{m}-\mathtt{n}}(s)\}$ with degree $c_1 \leqslant c_2 \leqslant$

---

[2]For trajectories (w_1, x_1) and (w_2, x_2), their *concatenation at* t_0, denoted by $(w_1, x_1)\wedge_{t_0}(w_2, x_2)$, is defined as

$$(w_1, x_1) \wedge_{t_0} (w_2, x_2)(t) := \begin{cases} (w_1, x_1)(t) & \text{for } t < t_0 \\ (w_2, x_2)(t) & \text{for } t \geqslant t_0. \end{cases}$$

.

$\cdots \leqslant c_{\mathtt{m}-\mathtt{n}}$ is such that $d_i \leqslant c_i$, for$i = 1, 2, \ldots \mathtt{m} - \mathtt{n}$. The degrees of the vectors of minimal polynomial basis of $R(s)$ are called the *(Forney invariant) minimal indices* or *Kronecker indices* (more details in [8, Sect. 6.5.4]).

5.3 The Algebraic Riccati Equation (ARE) and Hamiltonian Systems

With a proper input-output partition (u, y), a controllable dissipative behavior $\mathfrak{B}$ admits the following minimal i/s/o representation.

$$\dot{x} = Ax + Bu, \qquad y = Cx + Du, \qquad A \in \mathbb{R}^{\mathtt{n}\times\mathtt{n}}, \quad B,\ C^T \in \mathbb{R}^{\mathtt{n}\times\mathtt{p}} \text{ and } D \in \mathbb{R}^{\mathtt{p}\times\mathtt{p}} \tag{5.7}$$

with (C, A) observable. We assume here that the number of input $\mathtt{m}(\mathfrak{B})$ = number of output $\mathtt{p}(\mathfrak{B})$: this assumption is in view of the maximum input cardinality condition and $\Sigma = \begin{bmatrix} 0 & I \\ I & 0 \end{bmatrix}$. The storage functions of a dissipative behavior are closely related to the algebraic Riccati inequality (ARI) and the Hamiltonian matrix. One of the results relating LMI, controllable behavior and storage function is the *Kalman–Yakubovich–Popov (KYP)* lemma: details in [6, Sect. 5.6]. For easy reference, we present the *KYP* lemma, in a behavioral context, as a proposition next.

Proposition 5.6 *A behavior $\mathfrak{B} \in \mathfrak{L}^{\mathtt{w}}_{\text{cont}}$, with a controllable and observable minimal i/s/o representation as in Eq.* (5.7)*, is Σ-dissipative if and only if there exists a solution $K = K^T \in \mathbb{R}^{\mathtt{n}\times\mathtt{n}}$ to the LMI*

$$\begin{bmatrix} A^TK + KA & KB - C^T \\ B^TK - C & -(D + D^T) \end{bmatrix} \leqslant 0. \tag{5.8}$$

For systems with $D + D^T > 0$, the Schur complement with respect to $D + D^T$ in LMI (5.8) results in the algebraic Riccati inequality

$$A^TK + KA + (KB - C^T)(D + D^T)^{-1}(B^TK - C) \leqslant 0. \tag{5.9}$$

The corresponding equation to the inequality (5.9) is called the algebraic Riccati equation (ARE). Symmetric solutions to the ARE have a one-to-one correspondence to $\mathtt{n}$-dimensional invariant subspaces of the matrix below (details in [10, Theorem 3.1.1]).

$$\mathscr{H} = \begin{bmatrix} A - B(D + D^T)^{-1}C & B(D + D^T)^{-1}B^T \\ -C^T(D + D^T)^{-1}C & -A^T + C^T(D + D^T)^{-1}B^T \end{bmatrix} \tag{5.10}$$

The matrix $\mathscr{H}$ is known as the Hamiltonian matrix. Every $\mathtt{n}$-dimensional $\mathscr{H}$ invariant subspace spanned by columns of $\begin{bmatrix} I \\ K \end{bmatrix}$ corresponding to a suitably chosen set of eigenvalues of $\mathscr{H}$, provides a solution K to the ARE.

The detailed procedure to find the solution to the ARE from $\mathtt{n}$-dimensional eigenspaces of the Hamiltonian matrix can be found in [4]. We provide a brief review of the procedure next. In the lines of [10] and [13, Definition 5.1.1], we define a Lambda set (Λ) to define the partition of eigenvalues of the Hamiltonian matrix $\mathscr{H}$. $\bar{\Lambda}$ denotes the set of complex conjugates of the elements in Λ.

Definition 5.7 Consider an even polynomial $p(\xi) \in \mathbb{R}[\xi]$ with no roots on the imaginary axis. A set of complex numbers $\Lambda \subset$ roots (p) is called a Lambda set of the roots of p if the following conditions are satisfied:

1. $\Lambda = \bar{\Lambda}$
2. $\Lambda \cap (-\Lambda) = \emptyset$
3. $\Lambda \cup (-\Lambda) =$ roots of $p(\xi)$ (counted with multiplicity)

Condition 1 in Definition 5.7 implies that the Lambda set should contain conjugate pairs of complex roots of $p(\xi)$. By condition 2, polynomial $p(\xi)$ should not have any roots on the imaginary axis.

In this chapter, we use the word Lambda set with respect to the eigenvalues of a matrix to mean the Lambda set corresponding to the roots of the characteristic polynomial of the matrix. Constructing Lambda set from the set of eigenvalues of $\mathscr{H}$ (spec($\mathscr{H}$)), we find the solutions to the ARE. This is a well-known result in the literature [10] and we present it as a proposition here.

Proposition 5.8 *Consider a minimal i/s/o system given by Eq.* (5.7) *and the algebraic Riccati equation* $A^T K + KA + (KB - C^T)(D + D^T)^{-1}(B^T K - C) = 0$. *The corresponding Hamiltonian matrix* $\mathscr{H} \in \mathbb{R}^{2\mathtt{n}\times 2\mathtt{n}}$ *is given by Eq.* (5.10). *Assume that the Hamiltonian matrix* $\mathscr{H}$ *has no eigenvalues on the imaginary axis and define* Λ *to be a Lambda set of spec* ($\mathscr{H}$). *Let the* $\mathtt{n}$*-dimensional* $\mathscr{H}$*-invariant subspace corresponding to the Lambda set* Λ *be*

$$S_\Lambda := im \begin{bmatrix} X_1 \\ X_2 \end{bmatrix}, \text{ where } X_1, X_2 \in \mathbb{R}^{\mathtt{n}\times\mathtt{n}}$$

Then, X_1 *is invertible and* $K := X_2 X_1^{-1}$ *is a real symmetric solution to the ARE.*

The solutions to the ARE are storage functions $x^T Kx$ of the behavior $\mathfrak{B}$ with x the state in i/s/o representation (Eq. (5.7)).

In order to describe the algorithm and the main results of the chapter, we need the definition of the orthogonal complement of a behavior $\mathfrak{B}$.

Definition 5.9 Consider a controllable behavior $\mathfrak{B} \in \mathfrak{L}^{\mathtt{w}}_{\text{cont}}$ and a symmetric $\Sigma \in \mathbb{R}^{\mathtt{w}\times\mathtt{w}}$. The Σ-orthogonal complement behavior $\mathfrak{B}^{\perp_\Sigma}$ of $\mathfrak{B}$ is defined as

$$\mathfrak{B}^{\perp_\Sigma} := \left\{ v \in \mathfrak{C}^\infty(\mathbb{R}, \mathbb{R}^{\mathtt{w}}) \mid \int_{-\infty}^{\infty} v^T \Sigma w \, dt = 0 \text{ for all } w \in \mathfrak{B} \cap \mathfrak{D} \right\}.$$

It is well known that an i/s/o representation of $\mathfrak{B}$ (with $w = (u, y) \in \mathfrak{B}$) gives one for $\mathfrak{B}^{\perp_\Sigma}$: see [18, Sect. 10]. If $\dot{x} = Ax + Bu$, $y = Cx + Du$ is a minimal i/s/o representation of $\mathfrak{B}$, then (with respect to the positive real supply rate), a minimal i/s/o representation $\mathfrak{B}^{\perp_\Sigma}$ (with $v \in \mathfrak{B}^{\perp_\Sigma}$, $v = (u, y)$) is

$$\dot{z} = -A^T z + C^T u \quad \text{and} \quad y = B^T z - D^T u. \tag{5.11}$$

For a given behavior $\mathfrak{B} \in \mathfrak{L}^{\mathtt{w}}_{\text{cont}}$ and supply rate Σ, we call $\mathfrak{B} \cap \mathfrak{B}^{\perp_\Sigma}$ a *Hamiltonian system* and denote it by $\mathfrak{B}_{\text{Ham}}$: see Remark 5.10 below for a brief background. It has been shown in [17] that these trajectories are *trajectories of minimal dissipation* for the given supply rate. The first-order representation for this set has a good structure: this has been used in [15] for example. Define $E := \begin{bmatrix} I_n & 0 & 0 \\ 0 & I_n & 0 \\ 0 & 0 & 0 \end{bmatrix}$ and $H := \begin{bmatrix} A & 0 & B \\ 0 & -A^T & C^T \\ C & -B^T & D + D^T \end{bmatrix}$. A (possibly nonminimal) first-order representation of $\mathfrak{B}_{\text{Ham}}$ is given by

$$\left(\frac{d}{dt} E - H \right) \begin{bmatrix} x \\ z \\ y \end{bmatrix} = 0. \tag{5.12}$$

Define $R(\xi) := (\xi E - H)$; we call $R(\xi)$ a 'Hamiltonian pencil'.

Remark 5.10 In classical optimal control theory, given a quadratic cost functional, the system of trajectories satisfying the corresponding Euler–Lagrange (EL) equation can be considered a Hamiltonian system. Further, the trajectories are called stationary with respect to this cost: see [14, Sect. 4] for example. The EL equation with respect to the integral of QDF Q_Σ turns out to be $\partial\Phi(\frac{d}{dt})\ell := M(-\frac{d}{dt})^T \Sigma M(\frac{d}{dt})\ell = 0$ with $\ell \in \mathfrak{C}^\infty(\mathbb{R}, \mathbb{R}^{\mathtt{m}})$. For $\ell^\star$ satisfying this system of equations, $w^\star := M(\frac{d}{dt})\ell^\star$ turns out to be stationary with respect to $w^T \Sigma w$: see [14, Proposition 4.1]. For a behavior $\mathfrak{B} \in \mathfrak{L}^{\mathtt{w}}_{\text{cont}}$ and its orthogonal complement $\mathfrak{B}^{\perp_\Sigma}$, it is shown in [7, Theorem 3.3] that $\mathfrak{B} \cap \mathfrak{B}^{\perp_\Sigma} = M(\frac{d}{dt})\ker\partial\Phi(\frac{d}{dt})$; with this background, we call $\mathfrak{B} \cap \mathfrak{B}^{\perp_\Sigma}$ a Hamiltonian behavior and the matrix pencil $R(\xi)$ related to the first-order representation of $\mathfrak{B}_{\text{Ham}}$, a *Hamiltonian pencil.*

Corresponding to a Λ-set of the eigenvalues of $\mathscr{H}$, we associate a behavior $(\mathfrak{B}_{\text{Ham}})_\Lambda \in \mathfrak{L}^{\mathtt{w}}$ such that $(\mathfrak{B}_{\text{Ham}})_\Lambda$ contains (possibly polynomial times) exponential trajectories with the time-exponent λ_i an element in Λ. Further $(\mathfrak{B}_{\text{Ham}})_\Lambda$ is a sub-behavior of $\mathfrak{B}_{\text{Ham}}$, i.e., all the trajectories in $(\mathfrak{B}_{\text{Ham}})_\Lambda$ are trajectories in $\mathfrak{B}_{\text{Ham}}$. This notion has been used elsewhere too. For example, for Λ-set corresponding to the $\mathtt{n}$ eigenvalues of $\mathscr{H}$ in $\mathbb{C}^+$, the corresponding $(\mathfrak{B}_{\text{Ham}})_\Lambda = (\mathfrak{B}_{\text{Ham}})_{antistab}$ as defined in [17, Theorem 3.6]. The same notion has also been used in [15, Sect. 3]. We present a theorem next which shows the relations between Hamiltonian systems and storage functions of a behavior. Some of the equivalences are known. This theorem is the one we extend to the lossless case in Theorem 5.13 below.

Theorem 5.11 *Consider a controllable, strictly dissipative behavior* $\mathfrak{B} \in \mathfrak{L}^{\mathtt{w}}_{\text{cont}}$ *with minimal state-space representation as in Eq.* (5.7) *and McMillan degree* $\mathtt{n}$. *The corresponding Hamiltonian behavior* $\mathfrak{B}_{Ham} = \ker R(\frac{d}{dt})$ *where* $R(\xi) := \xi E - H \in \mathbb{R}^{(2\mathtt{n}+\mathtt{p})\times(2\mathtt{n}+\mathtt{p})}$ *is the Hamiltonian pencil defined in Eq.* (5.12). *Suppose* $K \in \mathbb{R}^{\mathtt{n}\times\mathtt{n}}$ *is a solution to the ARE corresponding to the behavior* $\mathfrak{B}$. *Then, the following statements hold.*

1. *The Hamiltonian behavior* $\mathfrak{B}_{Ham}$ *is autonomous, i.e.* $\det R(\xi) \neq 0$.
   In fact $\deg \det R(\xi) = 2\mathtt{n}$.
2. $\frac{d}{dt} x^T K x = 2u^T y \quad$ *for all* $\begin{bmatrix} u \\ y \end{bmatrix} \in (\mathfrak{B}_{Ham})_\Lambda$.
3. *rank* $\begin{bmatrix} & R(\xi) & \\ -K & I & 0 \end{bmatrix} =$ *rank* $R(\xi) = 2\mathtt{n} + \mathtt{p}$.
4. *rank* $\begin{bmatrix} & R(\lambda) & \\ -K & I & 0 \end{bmatrix} =$ *rank* $R(\lambda) < 2\mathtt{n} + \mathtt{p}$ *for each* $\lambda \in \Lambda(\text{roots} \det R(\xi))$.

Proof Statement **1** is trivial and so we do not dwell on it further: see [18, Sect. 4]. The polynomial matrix $R(\xi)$ is full row rank and hence **3** is true. Statement **2** has been proved in [18, Theorem 4.8]. Hence, we proceed to prove Statement 4.

4: In order to prove 4 of Theorem 5.11, we first prove that

$$\ker \begin{bmatrix} & R(\lambda) & \\ -K & I & 0 \end{bmatrix} = \ker R(\lambda) \text{ for any } \lambda \in \Lambda\,(\text{roots} \det R(\xi)) = \Lambda(\text{spec}(\mathscr{H})).$$

Of course $\ker \begin{bmatrix} & R(\lambda) & \\ -K & I & 0 \end{bmatrix} \subseteq \ker R(\lambda)$ holds and the reverse inclusion requires to be proved.

Conversely, let $v \in \ker R(\lambda)$. Hence v is an eigenvector[3] of $R(\xi)$ corresponding to eigenvalue λ. By Proposition 5.8 we have $\begin{bmatrix} I \\ K \\ 0 \end{bmatrix}$ spans the eigenspace of $R(\xi)$. Hence $v \in \text{span}\begin{bmatrix} I \\ K \\ 0 \end{bmatrix}$. It is obvious that $\begin{bmatrix} -K \\ I \\ 0 \end{bmatrix}$ is orthogonal to $\begin{bmatrix} I \\ K \\ 0 \end{bmatrix}$. Hence $[\,-K \; I \; 0\,]\, v = 0$. Thus we conclude that $\ker R(\lambda) \subseteq \ker \begin{bmatrix} & R(\lambda) & \\ -K & I & 0 \end{bmatrix}$, and this proves *equality* of the kernels. This proves that the ranks are equal. Hence 4 follows. This completes the proof of Theorem 5.11. □

[3] As in [5], for a square and nonsingular polynomial matrix $R(s)$, we call the values of $\lambda \in \mathbb{C}$ at which rank of $R(\lambda)$ drops the *eigenvalues* of the polynomial matrix $R(s)$ and we call the vectors in the nullspace of $R(\lambda)$ the *eigenvectors* of $R(s)$ corresponding to λ.

5.4 Storage Functions for Lossless Systems

Due to the condition $D + D^T = 0$ for lossless systems, Proposition 5.8 cannot be used to find storage functions of lossless systems. However, for lossless systems, the LMI (5.8) still exists with equality and solution to this LME can be interpreted as storage functions even in the absence of the ARI and Hamiltonian matrix. The LME is equivalent to solving the following matrix equations.

$$A^T K + KA = 0 \quad \text{and} \quad B^T K - C = 0 \tag{5.13}$$

For a lossless behavior $\mathfrak{B}$, the first-order representation of the Hamiltonian system $\mathfrak{B}_{\text{Ham}}$ is

$$\begin{bmatrix} \xi I_n - A & 0 & -B \\ 0 & \xi I_n + A^T & -C^T \\ -C & B^T & 0 \end{bmatrix} \begin{bmatrix} x \\ z \\ y \end{bmatrix} = 0. \tag{5.14}$$

Our main result (Theorem 5.13) below uses the nontrimness aspect in the states above. A special case of [2, Lemma 11] relates to trimness: we state this as a proposition below for easy reference.

Proposition 5.12 *Consider a Σ-dissipative behavior $\mathfrak{B} \in \mathfrak{L}^{\mathtt{w}}_{\text{cont}}$ and its orthogonal complement behavior $\mathfrak{B}^{\perp_\Sigma}$ with supply rate induced by the nonsingular matrix Σ of Eq.* (5.5) *(i.e. the positive real supply rate). Assume the behavior satisfies the maximum input cardinality (Eq.* (5.4)*). Then the following are equivalent.*

1. *$\mathfrak{B}$ is lossless.*
2. $\mathfrak{B} = \mathfrak{B} \cap \mathfrak{B}^{\perp_\Sigma} = \mathfrak{B}^{\perp_\Sigma}$

Since the McMillan degree of $\mathfrak{B}$ is $\mathtt{n}$, from Proposition 5.12, we infer that McMillan degree of the Hamiltonian behavior $\mathfrak{B}_{\text{Ham}}$ is also $\mathtt{n}$. However, the Hamiltonian behavior in Eq. (5.14) has $2\mathtt{n}$ states and hence $\mathfrak{B} \cap \mathfrak{B}^{\perp_\Sigma} = \mathfrak{B}_{\text{Ham}}$ is not state trim, i.e., there is a static relationship between state x and the dual state z. The next theorem helps extract the static relations of the first-order representation (5.14) of behavior $\mathfrak{B}_{\text{Ham}}$ and in the process yields the unique storage function for the lossless behavior $\mathfrak{B}$.

Theorem 5.13 *Consider a controllable, lossless behavior $\mathfrak{B} \in \mathfrak{L}^{\mathtt{w}}_{\text{cont}}$ with minimal state-space representation as in Eq.* (5.7)*. The McMillan degree of $\mathfrak{B}$ is $\mathtt{n}$. The corresponding Hamiltonian behavior $\mathfrak{B}_{Ham} = \ker R(\frac{d}{dt})$ where $R(\xi) := \xi E - H$ is the Hamiltonian pencil described in Eq.* (5.12) *with $D + D^T = 0$. Then the following statements hold.*

1. *The Hamiltonian behavior $\mathfrak{B}_{Ham}$ is not autonomous, i.e. det $R(\xi) = 0$.*
2. *There exists a unique symmetric matrix $K \in \mathbb{R}^{\mathtt{n}\times\mathtt{n}}$ that satisfies*

$$\frac{d}{dt} x^T K x = 2u^T y \quad \textit{for all} \quad \begin{bmatrix} u \\ y \end{bmatrix} \in \mathfrak{B}_{Ham} = \mathfrak{B}. \tag{5.15}$$

3. *There exists a unique symmetric matrix* $K \in \mathbb{R}^{n \times n}$ *that satisfies*

$$rank \begin{bmatrix} & R(\xi) & \\ -K & I & 0 \end{bmatrix} = rank\, R(\xi). \tag{5.16}$$

Proof Statement 1 is well known and details on it can be found in [7, 14] for example. Statement 2 shows the existence of a storage function and this has been dealt with in [18, Remark 5.9]. Hence we prove 3 next.

3: We prove Eq. (5.16) of Theorem 5.13 here.
Using 2 of Theorem 5.13, we have

$$\frac{d}{dt} x^T K x = 2u^T y \quad \text{i.e.} \quad \dot{x}^T K x + x^T K \dot{x} = 2u^T y$$

Using system Eq. (5.7) of behavior $\mathfrak{B}$, we have

$$\begin{bmatrix} x \\ u \end{bmatrix}^T \begin{bmatrix} A^T K + KA & KB - C^T \\ B^T K - C & 0 \end{bmatrix} \begin{bmatrix} x \\ u \end{bmatrix} = 0 \text{ for each} (x, u) \text{ satisfying system equations.}$$

Since (A, B) is controllable and u is input to the system, there is a system trajectory (x, u) that passes through each (x_0, u_0) for $x_0 \in \mathbb{R}^n$ and $u_0 \in \mathbb{R}^m$. Hence

$$\begin{bmatrix} A^T K + KA & KB - C^T \\ B^T K - C & 0 \end{bmatrix} = 0$$

Therefore, we infer that

$$A^T K + KA = 0 \quad \text{and} \quad B^T K - C = 0 \tag{5.17}$$

It is known from [18, Sect. 10] that

$$\frac{d}{dt} x^T z = 2u^T y = \frac{d}{dt} x^T K x \text{ which evaluates to } \dot{x}^T z + x^T \dot{z} - \dot{x}^T K x - x^T K \dot{x} = 0.$$

Using the Eqs. (5.7) and (5.11), we have

$$\begin{aligned} (Ax + Bu)^T z + x^T(-A^T z + C^T u) - (Ax + Bu)^T K x - x^T K(Ax + Bu) = 0 \\ \text{i.e.} \quad u^T B^T z + x^T C^T u - x^T (A^T K + KA)x - u^T B^T K x - u^T B^T K x = 0 \end{aligned}$$

Using Eq. (5.17), we have

$$2u^T B^T (z - Kx) = 0 \tag{5.18}$$

Equation (5.18) is satisfied for all system trajectories and at every time instant. This proves that $B^T(z - Kx) = 0$. We crucially use (A, B) controllability and (C, A) observability, together with Eq. (5.17) to conclude that $z - Kx = 0$ is satisfied over *all* system trajectories. Thus we proved that adding the laws $\begin{bmatrix} -K & I & 0 \end{bmatrix} \begin{bmatrix} x \\ z \\ y \end{bmatrix}$ to the system equations $R(\frac{d}{dt}) \begin{bmatrix} x \\ z \\ y \end{bmatrix}$ imposes *no further restriction* on $\mathfrak{B}$. This proves that $\text{rank} \begin{bmatrix} R(\xi) \\ -K \quad I \quad 0 \end{bmatrix} = \text{rank}\, R(\xi)$, and thus completes the proof of Theorem 5.13.

□

The next corollary states that Conditions 2 and 3 of Theorem 5.13 are equivalent. This equivalence condition is used to develop an algorithm to compute the storage function of a lossless behavior $\mathfrak{B}$.

Corollary 5.14 *Consider a controllable, lossless behavior $\mathfrak{B} \in \mathfrak{L}^{\mathtt{w}}_{\text{cont}}$ with minimal state-space representation as in Eq.* (5.7). *Let the McMillan degree of $\mathfrak{B}$ be* $\mathtt{n}$. *Consider the corresponding Hamiltonian behavior $\mathfrak{B}_{Ham} = \ker R(\frac{d}{dt})$ where $R(\xi) := \xi E - H$ is the Hamiltonian pencil described in Eq.* (5.12) *with* $D + D^T = 0$. *Then a necessary and sufficient condition for $K = K^T \in \mathbb{R}^{\mathtt{n}\times\mathtt{n}}$ to be a storage function for $\mathfrak{B}$ is*

$$\text{rank} \begin{bmatrix} R(\xi) \\ -K \quad I \quad 0 \end{bmatrix} = \text{rank}\, R(\xi). \tag{5.19}$$

Proof (Necessity) This follows from Statements 2 and 3 of Theorem 5.13.
(Sufficiency) We assume a symmetric matrix $K \in \mathbb{R}^{\mathtt{n}\times\mathtt{n}}$ satisfies Eq. (5.19) and show that K satisfies Eq. (5.15) i.e. K induces the storage function for $\mathfrak{B}$. Using Eq. (5.19), behavior $\mathfrak{B}_{\text{Ham}}$ has trajectories that satisfy $z = Kx$. By definition of "dual states," the relation between "states" and its "dual states" is

$$\frac{d}{dt} x^T z = 2u^T y \quad \text{i.e.} \quad \frac{d}{dt} x^T K x = 2u^T y.$$

Hence K satisfies Eq. (5.15) if and only if K satisfies Eq. (5.19). This completes the proof of Corollary 5.14.

□

Using Corollary 5.14, we conclude that $[-K \quad I \quad 0]$ is in the row span of the polynomial matrix $R(\xi)$. The corollary guarantees that $K \in \mathbb{R}^{\mathtt{n}\times\mathtt{n}}$ serves as the unique storage function of the lossless behavior $\mathfrak{B}$. In the next section, we present an algorithm to find the unique storage function of the lossless behavior $\mathfrak{B}$ using the fact that $[-K \quad I \quad 0]$ is in the row span of $R(\xi)$.

5.5 Lossless System's Storage Function: Algorithmic Aspects

Algorithm 5.5.1 is based on extraction of static relations in first order representation of the Hamiltonian behavior $\mathfrak{B}_{\mathrm{Ham}}$ described in Sect. 5.4. The Hamiltonian pencil $R(\xi)$ is an input to the algorithm and a unique symmetric matrix K that induces storage function of the lossless behavior is the output.

Algorithm 5.5.1 Static relations extraction-based algorithm.

Input: $R(\xi) := \xi E - H \in \mathbb{R}[\xi]^{(2\mathrm{n}+\mathrm{p})\times(2\mathrm{n}+\mathrm{p})}$, a rank 2n polynomial matrix.
Output: $K \in \mathbb{R}^{\mathrm{n}\times\mathrm{n}}$ with $x^T K x$ the storage function.

1: Calculate a minimal polynomial nullspace basis of $R(\xi)$.
2: *Result*: A full column rank polynomial matrix $M(\xi) \in \mathbb{R}[\xi]^{(2\mathrm{n}+\mathrm{p})\times\mathrm{p}}$.
3: Partition $M(\xi)$ as $\begin{bmatrix} M_1(\xi) \\ M_2(s) \end{bmatrix}$ where $M_1(\xi) \in \mathbb{R}[\xi]^{2\mathrm{n}\times\mathrm{p}}$.
4: Calculate a minimal polynomial nullspace basis of $M_1(\xi)^T$.
5: *Result*: A full column rank polynomial matrix $N(\xi) \in \mathbb{R}[\xi]^{2\mathrm{n}\times(2\mathrm{n}-\mathrm{p})}$.
6: Partition $N(\xi) = \begin{bmatrix} N_{11} & N_{12}(\xi) \\ N_{21} & N_{22}(\xi) \end{bmatrix}$ with $N_{11},\ N_{21} \in \mathbb{R}^{\mathrm{n}\times\mathrm{n}}$. (See Theorem 5.15 below)
7: The storage function $x^T K x$ induced by the symmetric matrix K is given by

$$K := -N_{11}N_{21}^{-1} \in \mathbb{R}^{\mathrm{n}\times\mathrm{n}}$$

Using the partition of the various matrices in the Algorithm 5.5.1, we state the following result about the unique storage function for a lossless behavior.

Theorem 5.15 *Consider* $R(\xi) := \xi E - H \in \mathbb{R}[\xi]^{(2\mathrm{n}+\mathrm{p})\times(2\mathrm{n}+\mathrm{p})}$ *as defined in Eq.* (5.12) *constructed for the lossless behavior* $\mathfrak{B} \in \mathfrak{L}^{2\mathrm{p}}_{\mathrm{cont}}$. *Let* $M(\xi) \in \mathbb{R}[\xi]^{(2\mathrm{n}+\mathrm{p})\times\mathrm{p}}$ *be any minimal polynomial nullspace basis (MPB) for* $R(\xi)$. *Partition* $M = \begin{bmatrix} M_1(\xi) \\ M_2(\xi) \end{bmatrix}$ *with* $M_1 \in \mathbb{R}[\xi]^{2\mathrm{n}\times\mathrm{p}}$. *Let* $N(\xi)$ *be any MPB for* $M_1(\xi)^T$. *Then, the following statements are true.*

1. *The first n (Forney invariant) minimal indices of* $N(\xi)$ *are* 0, *i.e. first n columns of* $N(\xi)$ *are constant vectors.*
2. *Partition N into* $\begin{bmatrix} N_1 & N_2(\xi) \end{bmatrix}$ *with* $N_1 \in \mathbb{R}^{2\mathrm{n}\times\mathrm{n}}$ *and further partition* $N_1 = \begin{bmatrix} N_{11} \\ N_{21} \end{bmatrix}$ *with* $N_{21} \in \mathbb{R}^{\mathrm{n}\times\mathrm{n}}$. *Then,*
   a. N_{21} *is invertible.*
   b. $K := -N_{11}N_{21}^{-1}$ *is symmetric.*
   c. $x^T K x$ *is the unique storage function for* $\mathfrak{B}$, *i.e.* $\frac{d}{dt}x^T K x = 2u^T y$ *for all system trajectories.*

Proof **1**: Using Statement 1 of Theorem 5.13, we have det $R(\xi) = 0$. Hence there exists a nullspace $M(\xi)$ of $R(\xi)$. Since rank $R(\xi) = 2\mathtt{n}$ where $\mathtt{n}$ is the McMillan degree of the behavior $\mathfrak{B}$ and $R(\xi) \in \mathbb{R}^{(2\mathtt{n}+\mathtt{p})\times(2\mathtt{n}+\mathtt{p})}[\xi]$, we have that the minimal polynomial basis $M(\xi) \in \mathbb{R}^{(2\mathtt{n}+\mathtt{p})\times\mathtt{p}}[\xi]$.
Using Corollary 5.14, we have $\begin{bmatrix}-K & I & 0\end{bmatrix}$ is in the row span of $R(\xi)$. Therefore,

$$\begin{bmatrix}-K & I & 0\end{bmatrix} M(\xi) = 0 \qquad \text{i.e.} \qquad \begin{bmatrix}-K & I & 0\end{bmatrix}\begin{bmatrix}M_1(\xi)\\ M_2(\xi)\end{bmatrix} = 0, \text{ where } M_1 \in \mathbb{R}[\xi]^{2\mathtt{n}\times\mathtt{p}}$$

This implies that

$$\begin{bmatrix}-K & I\end{bmatrix}\begin{bmatrix}M_1(\xi)\end{bmatrix} = 0 \qquad \text{i.e.} \qquad M_1(\xi)^T\begin{bmatrix}-K\\ I\end{bmatrix} = 0$$

The nullspace of $M_1(\xi)^T$ must have $\mathtt{n}$ constant polynomial vectors. Hence the first $\mathtt{n}$ (Forney invariant) minimal indices are 0. This proves 1 of Theorem 5.15.
2: Here we prove 2 of Theorem 5.15.
The minimal nullspace basis of $M_1(\xi)^T$ is the columns of $N(\xi) \in \mathbb{R}[\xi]^{2\mathtt{n}\times(2\mathtt{n}-\mathtt{p})}$. Partition N into $\begin{bmatrix}N_1 & N_2(\xi)\end{bmatrix}$ with $N_1 \in \mathbb{R}^{2\mathtt{n}\times\mathtt{n}}$ and further partition $N_1 = \begin{bmatrix}N_{11}\\ N_{21}\end{bmatrix}$ with $N_{21} \in \mathbb{R}^{\mathtt{n}\times\mathtt{n}}$. Further

$$\text{span}\begin{bmatrix}N_{11}\\ N_{21}\end{bmatrix} = \text{span}\begin{bmatrix}-K\\ I\end{bmatrix}.$$

This proves that N_{21} is invertible and therefore $K = -N_{11}N_{21}^{-1}$. The entire proof is based on Theorem 5.13 and Corollary 5.14, hence the symmetric matrix K found by Algorithm 5.5.1 induces storage function of the lossless behavior $\mathfrak{B}$ i.e. $\frac{d}{dt}x^TKx = 2u^Ty$ for all system trajectories. Hence 2 of Theorem 5.15 follows. This completes the proof of Theorem 5.15. □

Algorithm 5.5.1 is based on computation of nullspace basis of polynomial matrices. Computation of nullspace basis of a polynomial matrix can be done by block Toeplitz matrix algorithm: more details can be found in [9, 20].

5.6 Examples

In this section, we consider two examples: one in which we have strict dissipativity and another in which we have losslessness. We use Algorithm 5.5.1 for calculating K for the lossless case.

Example 5.16 In this example, we illustrate the conditions in Theorem 5.11. Consider a strictly dissipative behavior $\mathfrak{B}$ with transfer function $G(s) = \frac{s+2}{s+1}$. A minimal i/s/o representation of the system is $\dot{x} = -x + u$ and $y = x + u$. The Hamiltonian pencil for the behavior $\mathfrak{B}$ as obtained from Eq. (5.12) is

$$R(\xi) = \begin{bmatrix} \xi + 1 & 0 & -1 \\ 0 & \xi - 1 & -1 \\ -1 & 1 & -2 \end{bmatrix}$$

Hence $\det R(\xi) = 4 - 2\xi^2 \neq 0$, $\deg \det R(\xi) = 2$ and $R(\xi) \in \mathbb{R}^{3\times 3}[\xi]$ i.e. Hamiltonian system is autonomous. The roots of $\det R(\xi) = \{-\sqrt{2}, \sqrt{2}\}$. Following Definition 5.7, two Lambda sets can be formed $\Lambda_1 = \{-\sqrt{2}\}$ and $\Lambda_2 = \{\sqrt{2}\}$. For Λ_1, the storage function $K_{\Lambda_1} = 0.171$. Notice that

$$\text{rank} \begin{bmatrix} -\sqrt{2}+1 & 0 & -1 \\ 0 & -\sqrt{2}-1 & -1 \\ -1 & 1 & -2 \end{bmatrix} = 2 \text{ and rank} \begin{bmatrix} -\sqrt{2}+1 & 0 & -1 \\ 0 & -\sqrt{2}-1 & -1 \\ -1 & 1 & -2 \\ -0.171 & 1 & 0 \end{bmatrix} = 2.$$

It can be verified that the storage function for Lambda set Λ_2 is $K_{\Lambda_2} = 5.828$ and it also satisfies the conditions in Theorem 5.11. Consider any other arbitrary value of K which is not a solution to the ARE corresponding to the behavior $\mathfrak{B}$. Say $K = 1$ then

$$\text{rank} \begin{bmatrix} -\sqrt{2}+1 & 0 & -1 \\ 0 & -\sqrt{2}-1 & -1 \\ -1 & 1 & -2 \\ -1 & 1 & 0 \end{bmatrix} = 3.$$

Hence for any other arbitrary value of K, rank $\begin{bmatrix} & R(\xi) & \\ -K & I & 0 \end{bmatrix} \neq \text{rank } R(\xi)$.

Next we consider transfer function of a *lossless* behavior $\mathfrak{B}$ that brings out the use of Theorem 5.13. In order to calculate the storage function K we use Algorithm 5.5.1.

Example 5.17 Consider a lossless behavior $\mathfrak{B}$ with transfer function $G(s) = \frac{s}{s^2+1}$. A minimal i/s/o representation of the behavior is

$$\dot{x} = \begin{bmatrix} 0 & 1 \\ -1 & 0 \end{bmatrix} x + \begin{bmatrix} 0 \\ 1 \end{bmatrix} u \quad \text{and} \quad y = \begin{bmatrix} 0 & 1 \end{bmatrix} x + 0\,u$$

The Hamiltonian pencil for the behavior $\mathfrak{B}$ as obtained from Eq. (5.12) is

$$R(\xi) = \begin{bmatrix} \xi & -1 & 0 & 0 & 0 \\ 1 & \xi & 0 & 0 & -1 \\ 0 & 0 & \xi & -1 & 0 \\ 0 & 0 & 1 & \xi & -1 \\ 0 & -1 & 0 & 1 & 0 \end{bmatrix} \quad \text{and one can check that } \det R(\xi) = 0.$$

Thus the behavior $\mathfrak{B}_{\text{Ham}}$ is not autonomous. We next calculate the storage function using Algorithm 5.5.1.

1. A minimal polynomial nullspace basis (MPB) of $R(\xi)$ is $M(\xi) = \begin{bmatrix} 1 \\ \xi \\ 1 \\ \xi \\ 1+\xi^2 \end{bmatrix}$.
2. Partitioning $M(\xi)$ by Step 3 of Algorithm 5.5.1, we have: $M_1(\xi) = \begin{bmatrix} 1 \\ \xi \\ 1 \\ \xi \end{bmatrix}$.
3. MPB of $M_1(\xi)^T$ is $N(\xi) = \begin{bmatrix} -4 & -\sqrt{2} & -3\xi \\ \sqrt{2} & -4 & 3 \\ 4 & \sqrt{2} & -3\xi \\ -\sqrt{2} & 4 & 3 \end{bmatrix}$.
4. Using Step 6 of the same algorithm, we partition $N(\xi)$. Hence $N_{11} = \begin{bmatrix} -4 & -\sqrt{2} \\ \sqrt{2} & -4 \end{bmatrix}$ and $N_{21} = \begin{bmatrix} 4 & \sqrt{2} \\ -\sqrt{2} & 4 \end{bmatrix}$.
5. Therefore, the matrix $K = -N_{11}N_{21}^{-1} = \begin{bmatrix} 1 & 0 \\ 0 & 1 \end{bmatrix}$ induces the storage function of the lossless behavior $\mathfrak{B}$.

It can be verified that rank $\begin{bmatrix} & R(\xi) & \\ -K & I & 0 \end{bmatrix}$ = rank $R(\xi) = 4$.

With any arbitrary $K = \begin{bmatrix} 2 & 0 \\ 0 & 1 \end{bmatrix}$ (say), we will have rank $\begin{bmatrix} & R(\xi) & \\ -K & I & 0 \end{bmatrix} = 5$. Hence for arbitrary K, rank $\begin{bmatrix} & R(\xi) & \\ -K & I & 0 \end{bmatrix} \neq$ rank $R(\xi)$.

5.7 Concluding Remarks

This chapter dealt with the formulation of new properties of the ARE solution for the case when the equation exists: namely, when regularity conditions on the feedthrough term are satisfied. These results were extended to the case when the ARE does not exist: for example, the lossless case. For this case, the "ARE" solution is the storage function, which is unique for the lossless case. We formulated an algorithm that computes this storage function. The algorithm was developed exploiting the fact that the states in the Hamiltonian system (corresponding to a conservative behavior) are not trim. Static relations of the form $z = Kx$ helped to extract this nontrimness and hence led to a storage function $x^T Kx$ to the original system.

Acknowledgments This work was supported in part by SERB-DST, IRCC (IIT Bombay) and BRNS, India.

References

1. Antoulas, A.C.: Approximation of Large-scale Dynamical Systems. Advances in Design and Control. SIAM, Philadelphia (2005)
2. Belur, M.N., Pillai, H.K., Trentelman, H.L.: Synthesis of dissipative systems: a linear algebraic approach. Linear Algebr. Appl. **425**(2–3), 739–756 (2007)
3. Bini, D.A., Iannazzo, B., Meini, B.: Numerical Solution of Algebraic Riccati Equations. SIAM, Philadelphia (2012)
4. Bittanti, S., Laub, A.J., Willems, J.C. (eds.): The Riccati Equation. Springer, New York (1991)
5. Gohberg, I., Lancaster, P., Rodman, L.: Matrix Polynomials. Academic Press, New York (2009)
6. Haddad, W.M., Chellaboina, V.: Nonlinear Dynamical Systems and Control: A Lyapunov-Based Approach. Princeton University Press, Princeton (2008)
7. Jugade, S.C., Pal, D., Kalaimani, R.K., Belur, M.N.: Stationary trajectories, singular Hamiltonian systems and ill-posed interconnection. In: European Control Conference (ECC), Zurich, Switzerland, pp. 1740–1745 (2013)
8. Kailath, T.: Linear Systems. Prentice-Hall, Englewood Cliffs (1980)
9. Khare, S.R., Pillai, H.K., Belur, M.N.: An algorithm to compute a minimal nullspace basis of a polynomial matrix. In: Proceedings of the Mathematical Theory of Networks and Systems (MTNS), Budapest, Hungary, vol. 5, no. 9, pp. 219–224 (2010)
10. Kučera, V.: Algebraic Riccati equation: Hermitian and definite solutions. In: Bittanti, S., Laub, A.J., Willems, J.C. (eds.) The Riccati Equation, pp. 53–88. Springer, Berlin (1991)
11. Lancaster, P., Rodman, L.: Algebraic Riccati Equations. Oxford University Press, Oxford (1995)
12. Polderman, J.W., Willems, J.C.: Introduction to Mathematical Systems Theory: A Behavioral Approach. Springer, Heidelberg (1998)
13. Rapisarda, P.: Linear differential systems. Ph.D. thesis, University of Groningen, The Netherlands (1998)
14. Rapisarda, P., Trentelman, H.L.: Linear Hamiltonian behaviors and bilinear differential forms. SIAM J. Control Optim. **43**(3), 769–791 (2004)
15. Sorensen, D.C.: Passivity preserving model reduction via interpolation of spectral zeros. Syst. Control Lett. **54**(4), 347–360 (2005)
16. Trentelman, H.L., Willems, J.C.: The dissipation inequality and the algebraic Riccati equation. In: Bittanti, S., Laub, A.J., Willems, J.C. (eds.) The Riccati Equation, pp. 197–242. Springer, Berlin (1991)
17. Trentelman, H.L., Minh, H.B., Rapisarda, P.: Dissipativity preserving model reduction by retention of trajectories of minimal dissipation. Math. Control Signal. Syst. **21**(3), 171–201 (2009)
18. Willems, J.C., Trentelman, H.L.: On quadratic differential forms. SIAM J. Control Optim. **36**(5), 1703–1749 (1998)
19. Willems, J.C., Trentelman, H.L.: Synthesis of dissipative systems using quadratic differential forms: Parts I and II. IEEE Trans. Autom. Control **47**(1), 53–69 and 70–86 (2002)
20. Anaya, J.C.Z., Henrion, D.: An improved Toeplitz algorithm for polynomial matrix null-space computation. Appl. Math. Comput. **207**(1), 256–272 (2009)

Chapter 6
Stochastic Almost Output Synchronization for Time-Varying Networks of Nonidentical and Non-introspective Agents Under External Stochastic Disturbances and Disturbances with Known Frequencies

Meirong Zhang, Anton A. Stoorvogel and Ali Saberi

Abstract We consider stochastic almost output synchronization for time-varying directed networks of nonidentical and non-introspective (i.e., agents have no access to their own states or outputs) agents under external stochastic disturbances. The network experiences switches at unknown moments in time without chattering. A purely decentralized (i.e., the additional communication channel among agents is dispensed) time-invariant protocol based on a low- and high-gain method is designed for each agent to achieve stochastic almost output synchronization, while reducing the impact of stochastic disturbances. Moreover, we extend the problem to the case where stochastic disturbances can have nonzero mean or other disturbances are present with known frequencies. Another purely decentralized protocol is designed to completely reject the impact of disturbances with known frequencies on the synchronization error.

6.1 Introduction

Almost disturbance decoupling has a long history. It was the main topic of the Ph.D. thesis of Harry Trentelman. Anton Stoorvogel was, as a Ph.D. student of Harry, also looking at almost disturbance decoupling in connection to H_2 and H_∞ control. Ali

M. Zhang · A. Saberi
School of Electrical Engineering and Computer Science,
Washington State University, Pullman, WA 99164-2752, USA
e-mail: meirong.zhang@email.wsu.edu

A. Saberi
e-mail: saberi@eecs.wsu.edu

A.A. Stoorvogel (✉)
Department of Electrical Engineering, Mathematics and Computer Science,
University of Twente, P.O. Box 217, Enschede, The Netherlands
e-mail: A.A.Stoorvogel@utwente.nl

M.N. Belur et al. (eds.), *Mathematical Control Theory II*,
Lecture Notes in Control and Information Sciences 462,
DOI 10.1007/978-3-319-21003-2_6

Saberi was in this period working on the same class of problems. This paper looks at a version of almost disturbance decoupling in the context of multiagent systems.

In the last decade, the topic of synchronization in a multiagent system has received considerable attention. Its potential applications can be seen in cooperative control on autonomous vehicles, distributed sensor network, swarming and flocking, and others. The objective of synchronization is to secure an asymptotic agreement on a common state or output trajectory through decentralized control protocols (see [1, 12, 18, 28]). Research has mainly focused on the state synchronization based on full-state/partial-state coupling in a homogeneous network (i.e., agents have identical dynamics), where the agent dynamics progress from single- and double-integrator dynamics to more general dynamics (e.g., [7, 14, 15, 21, 24–26, 29]). The counterpart of state synchronization is output synchronization, which is mostly done on heterogeneous networks (i.e., agents are nonidentical). When the agents have access to part of their own states it is frequently referred to as introspective and, otherwise, non-introspective. Quite a few of the recent works on output synchronization have assumed agents are introspective (e.g., [3, 6, 27, 30]) while few have considered non-introspective agents. For non-introspective agents, the paper [5] addressed the output synchronization for heterogeneous networks.

In [7] for homogeneous networks a controller structure was introduced which included not only sharing the relative outputs over the network but also sharing the relative states of the protocol over the network. This was also used in our earlier work such as [5, 16, 17]. This type of additional communication is not always natural. Some papers such as [21] (homogeneous network) and [6] (heterogeneous network but introspective) already avoided this additional communication of controller states.

Almost synchronization is a notion that was brought up by Peymani and his coworkers in [17] (introspective) and [16] (homogeneous, non-introspective), where it deals with agents that are affected by external disturbances. The goal of their work is to reduce the impact of disturbances on the synchronization error to an arbitrary degree of accuracy (expressed in the $\mathscr{H}_\infty$ norm). But they assume availability of an additional communication channel to exchange information about internal controller or observer states between neighboring agents. The earlier work on almost synchronization for introspective, heterogeneous networks was extended in [31] to design a dynamic protocol to avoid exchange of controller states.

The majority of the works assumes that the topology associated with the network is fixed. Extensions to time-varying topologies are done in the framework of switching topologies. Synchronization with time-varying topologies is studied utilizing concepts of dwell time and average dwell time (e.g., [11, 22, 23]). It is assumed that time-varying topologies switch among a finite set of topologies. In [32], switching laws are designed to achieve synchronization.

This paper also aims to solve the almost regulated output synchronization problem for heterogeneous networks of non-introspective agents under switching graphs. However, instead of deterministic disturbances with finite power, we consider stochastic disturbances with bounded variance. We name this problem as stochastic almost regulated output synchronization. Moreover, we extend this problem by allowing nonzero mean stochastic disturbances and other disturbances with known frequencies.

6.1.1 Notations and Definitions

Given a matrix A, A' denotes its conjugate transpose and $\|A\|$ is the induced 2-norm. For square matrices, $\lambda_i(A)$ denotes its i'th eigenvalue, and it is said to be Hurwitz stable if all eigenvalues are in the open left half complex plane. We denote by $\mathrm{blkdiag}\{A_i\}$, a block diagonal matrix with $A_1, \ldots, A_N$ as the diagonal elements, and by $\mathrm{col}\{x_i\}$ or $[x_1; \ldots; x_N]$, a column vector with $x_1, \ldots, x_N$ stacked together, where the range of index i can be identified from the context. $A \otimes B$ depicts the Kronecker product between A and B. I_n denotes the n-dimensional identity matrix and 0_n denotes the $n \times n$ zero matrix; sometimes we drop the subscript if the dimension is clear from the context. Finally, the $\mathscr{H}_\infty$ norm of a transfer function T is indicated by $\|T\|_\infty$.

A *weighted directed graph* $\mathscr{G}$ is defined by a triple $(\mathscr{V}, \mathscr{E}, \mathscr{A})$ where $\mathscr{V} = \{1, \ldots, N\}$ is a node set, $\mathscr{E}$ is a set of pairs of nodes indicating connections among nodes, and $\mathscr{A} = [a_{ij}] \in \mathbb{R}^{N\times N}$ is the weighting matrix, and $a_{ij} > 0$ iff $(i, j) \in \mathscr{E}$. Each pair in $\mathscr{E}$ is called an *edge*. A *path* from node i_1 to i_k is a sequence of nodes $\{i_1, \ldots, i_k\}$ such that $(i_j, i_{j+1}) \in \mathscr{E}$ for $j = 1, \ldots, k-1$. A *directed tree* with *root* r is a subset of nodes of the graph $\mathscr{G}$ such that a path exists between r and every other node in this subset. A *directed spanning tree* is a directed tree containing all the nodes of the graph. For a weighted graph $\mathscr{G}$, a matrix $L = [\ell_{ij}]$ with

$$\ell_{ij} = \begin{cases} \sum_{k=1}^{N} a_{ik}, & i = j, \\ -a_{ij}, & i \neq j, \end{cases}$$

is called the *Laplacian matrix* associated with the graph $\mathscr{G}$. Since our graph $\mathscr{G}$ has nonnegative weights, we know that L has all its eigenvalues in the closed right half plane and at least one eigenvalue at zero associated with right eigenvector $\mathbf{1}$.

Definition 6.1 Let $\mathscr{L}_N \subset \mathbb{R}^{N\times N}$ be the family of all possible Laplacian matrices associated to a graph with N nodes. We denote by $\mathscr{G}_L$ the graph associated with a Laplacian matrix $L \in \mathscr{L}_N$. A time-varying graph $\mathscr{G}(t)$ with N nodes is such that

$$\mathscr{G}(t) = \mathscr{G}_{\sigma(t)}$$

where $\sigma : \mathbb{R} \to \mathscr{L}_N$ is a piecewise constant, right-continuous function with minimal dwell time τ (see [8]), i.e., $\sigma(t)$ remains fixed for $t \in [t_k, t_{k+1})$, $k \in \mathbb{Z}$ and switches at $t = t_k, k = 1, 2, \ldots$ where $t_{k+1} - t_k \geq \tau$ for $k = 0, 1, \ldots$. For ease of presentation we assume $t_0 = 0$.

Definition 6.2 A matrix pair (A, C) is said to contain the matrix pair (S, R) if there exists a matrix Π such that $\Pi S = A\Pi$ and $C\Pi = R$.

Remark 6.3 Definition 6.2 implies that for any initial condition $\omega(0)$ of the system

$$\dot{\omega} = S\omega, \qquad y_r = R\omega,$$

there exists an initial condition $x(0)$ of the system

$$\dot{x} = Ax, \qquad y = Cx,$$

such that $y(t) = y_r(t)$, for all $t \geq 0$ (see [10]).

6.2 Stochastic Disturbances

In this section, we consider the problem of almost output synchronization for time-varying networks (i.e., multiagent systems) with nonidentical and non-introspective agents under stochastic disturbances. The time-varying network is constrained in the sense that we exclude chattering by imposing an, arbitrary small, minimal dwell time. Our agents need not be the same and are non-introspective (i.e., they have no access to any of their own states). We will achieve stochastic almost output synchronization in such a way that outputs of agents are asymptotically regulated to a reference trajectory generated by an autonomous system.

6.2.1 Multiagent System Description

Suppose a multiagent system/network consists of N nonidentical, non-introspective agents $\tilde{\Sigma}_i$ with $i \in \{1, \ldots, N\}$ described by the stochastic differential equation:

$$\tilde{\Sigma}_i : \begin{cases} \mathrm{d}\tilde{x}_i = \tilde{A}_i\tilde{x}_i\mathrm{d}t + \tilde{B}_i\tilde{u}_i\mathrm{d}t + \tilde{G}_i\mathrm{d}w_i, & \tilde{x}_i(0) = \tilde{x}_{i0}, \\ y_i = \tilde{C}_i\tilde{x}_i, \end{cases} \tag{6.1}$$

where $\tilde{x}_i \in \mathbb{R}^{\tilde{n}_i}$, $\tilde{u}_i \in \mathbb{R}^{m_i}$, and $y_i \in \mathbb{R}^p$ are the state, input, and output of agent i, and $w = \mathrm{col}\{w_i\}$ is a Wiener process (a Brownian motion) with mean 0 and rate Q_0, that is, $\mathrm{Cov}[w(t)] = tQ_0$ and the initial condition $\tilde{x}_{i0}$ of (6.1) is a Gaussian random vector which is independent of w_i. Here Q_0 is block diagonal such that w_i and w_j are independent for any $i \neq j$. Its solution $\tilde{x}_i$ is rigorously defined through Wiener integrals and is a Gauss–Markov process. See, for instance, [13]. We denote by $\tilde{\rho}_i$ the maximal order of the infinite zeros of (6.1) with input $\tilde{u}_i$ and output y_i.

Remark 6.4 If we have an agent described by:

$$\breve{\Sigma}_i : \begin{cases} \dot{\breve{x}}_i = \breve{A}_i \breve{x}_i + \breve{B}_i \tilde{u}_i + \breve{G}_i \breve{w}_i, \\ y_i = \breve{C}_i \breve{x}_i, \end{cases} \tag{6.2}$$

with $\breve{w}_i$ stochastic colored noise, then we assume that $\breve{w}_i$ can be (approximately) modeled by a linear model:

$$\breve{\Sigma}_{wi} : \begin{cases} \mathrm{d}\breve{p}_i = \breve{A}_{wi} \breve{p}_i \mathrm{d}t + \breve{G}_{wi} \mathrm{d}w_i, \\ \breve{w}_i = \breve{C}_{wi} \breve{p}_i. \end{cases} \tag{6.3}$$

Combining (6.2) and (6.3) we get a model of the form (6.1).

The time-varying network provides each agent with a linear combination of its own output relative to those of other neighboring agents, that is, agent $i \in \mathscr{V}$, has access to the quantity

$$\zeta_i(t) = \sum_{j=1}^{N} a_{ij}(t)(y_i(t) - y_j(t)), \tag{6.4}$$

which is equivalent to

$$\zeta_i(t) = \sum_{j=1}^{N} \ell_{ij}(t) y_j(t). \tag{6.5}$$

We make the following assumption on the agent dynamics.

Assumption 6.5 For each agent $i \in \mathscr{V}$, we have:

- $(\tilde{A}_i, \tilde{B}_i, \tilde{C}_i)$ is right-invertible and minimum-phase;
- $(\tilde{A}_i, \tilde{B}_i)$ is stabilizable, and $(\tilde{A}_i, \tilde{C}_i)$ is detectable;

Remark 6.6 Right invertibility of a triple $(\tilde{A}_i, \tilde{B}_i, \tilde{C}_i)$ means that, given a reference output $y_r(t)$, there exist an initial condition $\tilde{x}_i(0)$ and an input $\tilde{u}_i(t)$ such that $y_i(t) = y_r(t)$, for all $t \geq 0$.

6.2.2 Problem Formulation

As described at the beginning of this section, the outputs of agents will be asymptotically regulated to a given reference trajectory in the presence of external stochastic disturbances. The reference trajectory is generated by an autonomous system

$$\Sigma_0 : \begin{cases} \dot{x}_r = S_r x_r, \qquad x_r(0) = x_{r0}, \\ y_r = R_r x_r, \end{cases} \tag{6.6}$$

where $x_r \in \mathbb{R}^{n_r}$, $y_r \in \mathbb{R}^p$. Moreover, we assume that (S_r, R_r) is observable and all eigenvalues of S_r are in the closed right half complex plane.

Define $e_i := y_i - y_r$ as the regulated output synchronization error for agent $i \in \mathscr{V}$ and $\mathbf{e} = \text{col}\{e_i\}$ for the complete network. In order to achieve our goal, it is clear that a nonempty subset π of agents must have knowledge of their output relative to the reference trajectory y_r generated by the reference system. Specially, each agent has access to the quantity

$$\psi_i = \iota_i(y_i - y_r), \qquad \iota_i = \begin{cases} 1, & i \in \pi, \\ 0, & i \notin \pi, \end{cases} \tag{6.7}$$

where π is a subset of $\mathscr{V}$.

Assumption 6.7 Every node of the network graph $\mathscr{G}$ is a member of a directed tree with the root contained in π.

In the following, we will refer to the node set π as *root set* in view of Assumption 6.7 (A special case is when π consists of a single element corresponding to the root of a directed spanning tree of $\mathscr{G}$).

Based on the Laplacian matrix $L(t)$ of our time-varying network graph $\mathscr{G}(t)$, we define the expanded Laplacian matrix as

$$\bar{L}(t) = L(t) + \text{blkdiag}\{\iota_i\} = [\bar{\ell}_{ij}(t)].$$

Note that $\bar{L}(t)$ is clearly not a Laplacian matrix associated to some graph since it does not have a zero row sum. From [5, Lemma 7], all eigenvalues of $\bar{L}(t)$ are in the open right half complex plane for all $t \in \mathbb{R}$.

It should be noted that, in practice, perfect information of the communication topology is usually not available for controller design and only some rough characterization of the network can be obtained. Next, we will define a set of time-varying graphs based on some rough information of the graph. Before doing so, we first define a set of fixed graphs, based on which the set of time-varying graphs will be defined.

Definition 6.8 For given root set $\pi, \alpha, \beta, \varphi > 0$ and positive integer N, $\mathbb{G}^{\varphi,N}_{\alpha,\beta,\pi}$ is the set of directed graphs $\mathscr{G}$ composed of N agents satisfying the following properties:

- The eigenvalues of the associated expanded Laplacian $\bar{L}$, denoted by $\lambda_1, \ldots, \lambda_N$, satisfy $\text{Re}\{\lambda_i\} > \beta$ and $|\lambda_i| < \alpha$.
- The condition number[1] of the expanded Laplacian $\bar{L}$ is less than φ.

[1] In this context, we mean by condition number the minimum of $\|U\|\|U^{-1}\|$ over all possible matrices U whose columns are the (generalized) eigenvectors of the expanded Laplacian matrix $\bar{L}$.

Remark 6.9 In order to achieve regulated output synchronization for all agents, the first condition is obviously necessary.

Note that for undirected graphs the condition number of the Laplacian matrix is always bounded. Moreover, if we have a *finite* set of possible directed graphs each of which has a spanning tree then there always exists a set of the graph $\mathbb{G}^{\varphi,N}_{\alpha,\beta,\pi}$ for suitable α, β, $\varphi > 0$ and N containing these graphs. The only limitation is that we cannot find **one** protocol for a sequence of graphs converging to a graph without a spanning tree or whose Laplacian matrix either diverges or approaches some ill-conditioned matrix.

Definition 6.10 Given a root set π, α, β, φ, $\tau > 0$ and positive integer N, we define the set of time-varying network graphs $\tilde{\mathbb{G}}^{\varphi,\tau,N}_{\alpha,\beta,\pi}$ as the set of all time-varying graphs $\mathscr{G}$ with minimal dwell time τ for which

$$\mathscr{G}(t) = \mathscr{G}_{\sigma(t)} \in \mathbb{G}^{\varphi,N}_{\alpha,\beta,\pi}$$

for all $t \in \mathbb{R}$.

Remark 6.11 Note that a minimal dwell time is assumed to avoid chattering problems. However, it can be arbitrarily small.

We will define the stochastic almost regulated output synchronization problem under switching graphs as follows.

Problem 6.12 Consider a multiagent system (6.1) and (6.4) under Assumption 6.5, and reference system (6.6) and (6.7) under Assumption 6.7. For any given root set π, α, β, φ, $\tau > 0$ and positive integer N defining a set of time-varying network graphs $\tilde{\mathbb{G}}^{\varphi,\tau,N}_{\alpha,\beta,\pi}$, the *stochastic almost regulated output synchronization* problem is to find, if possible, for any $\gamma > 0$, a linear time-invariant dynamic protocol such that, for any $\mathscr{G} \in \tilde{\mathbb{G}}^{\varphi,\tau,N}_{\alpha,\beta,\pi}$, for all initial conditions of agents and reference system, the stochastic almost regulated output synchronization error satisfies

$$\begin{aligned}
&\lim_{t\to\infty} \mathbb{E}\mathbf{e}(t) = 0, \\
&\limsup_{t\to\infty} \mathrm{Var}[\mathbf{e}(t)] = \limsup_{t\to\infty} \mathbb{E}\mathbf{e}'(t)\mathbf{e}(t) < \gamma \operatorname{trace} Q_0.
\end{aligned} \tag{6.8}$$

Remark 6.13 Clearly, we can also define (6.8) in terms of the RMS, (see, e.g., [2]) as:

$$\limsup_{T\to\infty} \frac{1}{T}\mathbb{E}\int_0^T \mathbf{e}(t)'\mathbf{e}(t)\,\mathrm{d}t < \gamma \operatorname{trace} Q_0.$$

Remark 6.14 Note that because of the time-varying graph the complete system is time variant and hence the variance of the error signal might not converge as time tends to infinity. Hence we use in the above a lim sup instead of a regular limit.

6.2.3 Distributed Protocol Design

The main result in this section is given in the following theorem.

Theorem 6.15 *Consider a multiagent system* (6.1) *and* (6.4), *and reference system* (6.6) *and* (6.7). *Let a root set* π, α, β, φ, $\tau > 0$ *and positive integer* N *be given, and hence a set of time-varying network graphs* $\tilde{\mathbb{G}}_{\alpha,\beta,\pi}^{\varphi,\tau,N}$ *be defined.*

Under Assumptions 6.5 and 6.7, the stochastic almost regulated output synchronization problem is solvable, i.e., for any given $\gamma > 0$, *there exists a family of distributed dynamic protocols, parametrized in terms of low- and high-gain parameters* δ, ε, *of the form*

$$\begin{cases} \dot{\chi}_i = \mathscr{A}_i(\delta, \varepsilon)\chi_i + \mathscr{B}_i(\delta, \varepsilon)\begin{pmatrix}\zeta_i \\ \psi_i\end{pmatrix} \\ \tilde{u}_i = \mathscr{C}_i(\delta, \varepsilon)\chi_i + \mathscr{D}_i(\delta, \varepsilon)\begin{pmatrix}\zeta_i \\ \psi_i\end{pmatrix} \end{cases}, \quad i \in \mathscr{V} \tag{6.9}$$

where $\chi_i \in \mathbb{R}^{q_i}$, *such that for any time-varying graph* $\mathscr{G} \in \tilde{\mathbb{G}}_{\alpha,\beta,\pi}^{\varphi,\tau,N}$, *for all initial conditions of agents, the stochastic almost regulated output synchronization error satisfies* (6.8).

Specifically, there exits a $\delta^* \in (0, 1]$ *such that for each* $\delta \in (0, \delta^*]$, *there exists an* $\varepsilon^* \in (0, 1]$ *such that for any* $\varepsilon \in (0, \varepsilon^*]$, *the protocol* (6.9) *achieves stochastic almost regulated output synchronization.*

Remark 6.16 In the above, we would like to stress that the initial condition of the reference system is deterministic while the initial conditions of the agents are stochastic. Our protocol yields (6.8) independent of the initial condition of the reference system and independent of the stochastic properties for the agents, i.e., we do not need to impose bounds on the second-order moments.

In the next section, we will present a more general problem after which we will present a joint proof of these two cases in Sect. 6.4.

6.3 Stochastic Disturbances and Disturbances with Known Frequencies

In this section, the agent model (6.1) is modified as follows:

$$\tilde{\Sigma}_i : \begin{cases} d\tilde{x}_i = \tilde{A}_i\tilde{x}_i dt + \tilde{B}_i\tilde{u}_i dt + \tilde{G}_i dw_i + \tilde{H}_i^1 d_i dt, \\ y_i = \tilde{C}_i\tilde{x}_i + \tilde{H}_i^2 d_i, \end{cases} \tag{6.10}$$

where $\tilde{x}_i$, $\tilde{u}_i$, y_i, and w_i are the same as those in (6.1), while $d_i \in \mathbb{R}^{m_{d_i}}$ is an external disturbance with known frequencies, which can be generated by the following exosystem:

$$\begin{aligned} \dot{x}_{id} &= S_{id} x_{id}, \quad x_{id}(0) = x_{id0} \\ d_i &= R_{id} x_{id}, \end{aligned} \tag{6.11}$$

where $x_{id} \in \mathbb{R}^{n_{d_i}}$ and the initial condition x_{id0} can be arbitrarily chosen.

In Remark 6.4 we considered colored noise. However, the model we used in that remark to generate colored noise, clearly cannot incorporate bias terms. This is one of the main motivations of the model (6.10) since the above disturbance term d_i can generate bias terms provided S_{id} has zero eigenvalues. However, it clearly can also handle other cases where we have disturbances with known frequency content.

Note that we have two exosystems (6.6) and (6.11) which generate the reference signal y_r and the disturbance d_i, respectively. We can unify the two in one exosystem:

$$\begin{aligned} \dot{x}_{ie} &= S_i x_{ie}, \qquad x_{ie}(0) = x_{ie0} \\ d_i &= R_{ie} x_{ie}, \\ y_r &= R_{re} x_{ie}, \end{aligned} \tag{6.12}$$

where

$$S_i = \begin{pmatrix} S_{id} & 0 \\ 0 & S_r \end{pmatrix}, \quad R_{ie} = \begin{pmatrix} R_{id} & 0 \end{pmatrix}, \quad R_{re} = \begin{pmatrix} 0 & R_r \end{pmatrix}. \tag{6.13}$$

Note that

$$x_{ie0} = \begin{pmatrix} x_{id0} \\ x_{r0} \end{pmatrix}$$

and therefore the second part of the initial condition has to be the same for each agent while the first part might be different for each agent. Note that in case we have no disturbances with known frequencies (as in the previous section) then we can still use the model (6.12) but with

$$S_i = S_r, \qquad R_{ie} = 0, \qquad R_{re} = R_r \tag{6.14}$$

while $x_{ie0} = x_{r0}$.

The time-varying topology $\mathscr{G}(t)$ has exactly the same structure as in Sect. 6.2, and it also belongs to a set of time-varying graph $\tilde{\mathbb{G}}^{\varphi,\tau,N}_{\alpha,\beta,\pi}$ as defined in Definition 6.10. The network defined by the time-varying topology also provides each agent with the measurement $\zeta_i(t)$ given in (6.4).

6.3.1 Distributed Protocol Design

Here is the main result in this section:

Theorem 6.17 *Consider a multiagent system described by* (6.10), (6.4), (6.7), *and reference system* (6.12). *Let a root set* π, α, β, φ, $\tau > 0$ *and positive integer* N *be given, and hence a set of time-varying network graphs* $\tilde{\mathbb{G}}^{\varphi,\tau,N}_{\alpha,\beta,\pi}$ *be defined.*

Under Assumptions 6.5 and 6.7, the stochastic almost regulated output synchronization problem is solvable, i.e., there exists a family of distributed dynamic protocols, parametrized in terms of low- and high-gain parameters δ, ε, *of the form*

$$\begin{cases} \dot{\chi}_i = \mathscr{A}_i(\delta,\varepsilon)\chi_i + \mathscr{B}_i(\delta,\varepsilon)\begin{pmatrix}\zeta_i\\ \psi_i\end{pmatrix} \\ \tilde{u}_i = \mathscr{C}_i(\delta,\varepsilon)\chi_i + \mathscr{D}_i(\delta,\varepsilon)\begin{pmatrix}\zeta_i\\ \psi_i\end{pmatrix} \end{cases}, \quad i \in \mathscr{V} \tag{6.15}$$

where $\chi_i \in \mathbb{R}^{q_i}$, *such that for any time-varying graph* $\mathscr{G} \in \tilde{\mathbb{G}}^{\varphi,\tau,N}_{\alpha,\beta,\pi}$, *for all initial conditions of agents, the stochastic almost regulated output synchronization error satisfies* (6.8).

Specifically, there exits a $\delta^* \in (0, 1]$ *such that for each* $\delta \in (0, \delta^*]$, *there exists an* $\varepsilon^* \in (0, 1]$ *such that for any* $\varepsilon \in (0, \varepsilon^*]$, *the protocol* (6.15) *solves the stochastic almost regulated output synchronization problem.*

The proof will be presented in a constructive way in the following section.

6.4 Proof of Theorems 6.15 and 6.17

Note that Theorem 6.15 is basically a corollary of Theorem 6.17 if we replace (6.13) by (6.14) and still use exosystem (6.12). In this section, we will present a constructive proof in three steps. As noted, we can concentrate on the proof of Theorem 6.17.

In Step 1, precompensators are designed for each agent to make the interconnection of an agent and its precompensator square, uniform rank (i.e., all the infinite zeros are of the same order) and such that it can generate the reference signal for all possible initial condition of the joint exosystem (6.12). In Step 2, a distributed linear dynamic protocol is designed for each interconnection system obtained from Step 1. Finally, in Step 3, we combine the precompensator from Step 1 and the protocol for the interconnection system in Step 2, and get a protocol of the form (6.15) for each agent in the network (6.10) (if disturbances with known frequencies are present) or a protocol of the form (6.9) for each agent in the network (6.1) (if disturbances with known frequencies are not present).

Step 1: In this step, we augment agent (6.10) with a precompensator in such a way that the interconnection is square, minimum-phase uniform rank such that it can generate the reference signal for all possible initial condition of the joint exosystem (6.12).

To be more specific, we need to find precompensators

$$\begin{cases} \dot{z}_i = A_{ip}z_i + B_{ip}u_i, \\ \tilde{u}_i = C_{ip}z_i + D_{ip}u_i, \end{cases} \tag{6.16}$$

for each agent $i = 1, \ldots, N$, such that agent (6.10) plus precompensator (6.16) can be represented as:

$$\begin{cases} \mathrm{d}x_i = A_i x_i \mathrm{d}t + B_i u_i \mathrm{d}t + G_i \mathrm{d}w_i + H_i^1 d_i \mathrm{d}t, \\ y_i = C_i x_i + H_i^2 d_i, \end{cases} \tag{6.17}$$

where $x_i \in \mathbb{R}^{n_i}$, $u_i \in \mathbb{R}^p$, $y_i \in \mathbb{R}^p$ are states, inputs, and outputs of the interconnection of agent (6.10) and precompensator (6.16). Moreover,

- There exists Π_i such that

$$\begin{aligned} A_i\Pi_i + H_i^1 R_{ie} &= \Pi_i S_i \\ C_i\Pi_i + H_i^2 R_{ie} &= R_{re} \end{aligned} \tag{6.18}$$

- (A_i, B_i, C_i) is square and has uniform rank $\rho \geq 1$.

The first condition implies that for any initial condition of (6.12) there exists an initial condition for (6.17) such that for $u_i = 0$, we have that $\mathbb{E}y_i = y_r$. We could, equivalently, impose $w_i = 0$ in which case, we should have $y_i = y_r$. In the special case where we do not have disturbances with known frequencies (Theorem 6.15) we have $R_{ie} = 0$ and $S_i = S_r$. In that case, the first condition reduces to the condition that (C_i, A_i) contains (S_r, R_r).

For our construction of precompensator (6.16), we first note that the following regulator equation

$$\tilde{A}_i\tilde{\Pi}_i + \tilde{B}_i\tilde{\Gamma}_i + \tilde{H}_i^1 R_{ie} = \tilde{\Pi}_i S_i, \qquad \tilde{C}_i\tilde{\Pi}_i + \tilde{H}_i^2 R_{ie} = R_{re}. \tag{6.19}$$

has a unique solution $\tilde{\Pi}_i$ and $\tilde{\Gamma}_i$ since $(\tilde{A}_i, \tilde{B}_i, \tilde{C}_i)$ is right-invertible and minimum-phase while S_i has no stable eigenvalues (see [19]). Let $\Gamma_{oi}, S_{oi})$ be the observable subsystem of $(\tilde{\Gamma}_i, S_i)$. Then we consider the following precompensator:

$$\dot{p}_{i,1} = S_{oi}p_{i,1} + B_{i,1}u_i^1, \qquad \tilde{u}_i = \Gamma_{oi}p_{i,1} + D_{i,1}u_i^1 \tag{6.20}$$

where $B_{i,1}$ and $D_{i,1}$ are chosen according to the technique presented in [9] to guarantee that the interconnection of (6.10) and (6.20) is minimum-phase and right-invertible. It is not hard to verify that the interconnection of (6.10) and (6.20) is a system of the form (6.17) for which there exists Π_i satisfying (6.18). However, we still need to guarantee that this interconnection is square and uniform rank.

Let ρ_i be the maximal order of the infinite zeros for the interconnection of (6.20) and (6.10). For $i = 1, \ldots, N$ and set $\rho = \max\{\rho_i\}$. According to [20, Theorem 1], a precompensator of the form

$$\begin{aligned} \dot{p}_{i_2} &= A_{ip2} p_{i_2} + B_{ip2} u_i^2, \\ u_i^1 &= C_{ip2} p_{i_2} + D_{ip2} u_i^2, \end{aligned} \tag{6.21}$$

exists such that the interconnection of (6.20), (6.21), and (6.10) is square, minimum-phase, and has uniform rank ρ. This interconnection of (6.20) and (6.21) is of the form (6.16) while the interconnection of (6.20), (6.21), and (6.10) is of the form (6.17) which has the required properties.

Without loss of generality, we assume that (A_i, B_i, C_i) is already in the SCB form, i.e., the system has a specific form where $x_i = [x_{ia}; x_{id}]$, with $x_{ia} \in \mathbb{R}^{n_i - p\rho}$ representing the finite zero structure and $x_{id} \in \mathbb{R}^{p\rho}$ the infinite zero structure. We obtain that (6.17) can be written as:

$$\left\{ \begin{aligned} \mathrm{d}x_{ia} &= A_{ia} x_{ia} \mathrm{d}t + L_{iad} y_i \mathrm{d}t + G_{ia} \mathrm{d}w_i + H_{ia}^1 d_i \mathrm{d}t, \\ \mathrm{d}x_{id} &= A_d x_{id} \mathrm{d}t + B_d (u_i + E_{ida} x_{ia} + E_{idd} x_{id}) \mathrm{d}t + G_{id} \mathrm{d}w_i + H_{id}^1 d_i \mathrm{d}t, \\ y_i &= C_d x_{id} + H_i^2 d_i \mathrm{d}t, \end{aligned} \right. \tag{6.22}$$

for $i = 1, \ldots, N$, where A_d, B_d, and C_d have the special form

$$A_d = \begin{pmatrix} 0 & I_{p(\rho-1)} \\ 0 & 0 \end{pmatrix}, \quad B_d = \begin{pmatrix} 0 \\ I_p \end{pmatrix}, \quad C_d = \begin{pmatrix} I_p & 0 \end{pmatrix}. \tag{6.23}$$

Furthermore, the eigenvalues of A_{ia} are the invariant zeros of (A_i, B_i, C_i), which are all in the open left half complex plane.

Step 2: Each agent after applying the precompensator (6.16) is of the form (6.22). For this system, we will design a purely decentralized controller based on a low- and high-gain method. Let $\delta \in (0, 1]$ be the low-gain parameter and $\varepsilon \in (0, 1]$ be the high-gain parameter as in [4]. First, select K such that $A_d - KC_d$ is Hurwitz stable. Next, choose $F_\delta = -B_d' P_d$ where $P_d' = P_d > 0$ is uniquely determined by the following algebraic Riccati equation:

$$P_d A_d + A_d' P_d - \beta P_d B_d B_d' P_d + \delta I = 0, \tag{6.24}$$

where $\beta > 0$ is the lower bound on the real parts of all eigenvalues of expanded Laplacian matrices $\bar{L}(t)$, for all t. Next, define

$$S_\varepsilon = \mathrm{blkdiag}\{I_p, \varepsilon I_p, \ldots, \varepsilon^{\rho-1} I_p\},$$

$$K_\varepsilon = \varepsilon^{-1} S_\varepsilon^{-1} K \quad \text{and} \quad F_{\delta\varepsilon} = \varepsilon^{-\rho} F_\delta S_\varepsilon.$$

Then, we design the dynamic controller for each agent $i \in \mathscr{V}$:

$$\begin{aligned} \dot{\hat{x}}_{id} &= A_d\hat{x}_{id} + K_\varepsilon(\zeta_i + \psi_i - C_d\hat{x}_{id}), \\ u_i &= F_{\delta\varepsilon}\hat{x}_{id}, \end{aligned} \tag{6.25}$$

where ψ_i is defined in (6.7).

The state $\hat{x}_{id}$ is not an estimator for x_{id} but actually an estimator for

$$\sum_{j=1}^{N} \bar{\ell}_{ij}(t)x_{id}(t). \tag{6.26}$$

The following lemma then provides a constructive proof of Theorem 6.17. However, by replacing (6.13) with (6.14) it also provides a constructive proof of Theorem 6.15.

Lemma 6.18 *Consider the agents in SCB format* (6.22) *obtained after applying the precompensators* (6.16). *For any given* $\gamma > 0$, *there exits a* $\delta^* \in (0, 1]$ *such that, for each* $\delta \in (0, \delta^*]$, *there exists an* $\varepsilon^* \in (0, 1]$ *such that for any* $\varepsilon \in (0, \varepsilon^*]$, *the protocol* (6.25) *solves the stochastic almost regulated output synchronization problem for any time-varying graph* $\mathscr{G} \in \tilde{\mathbb{G}}^{\varphi,\tau,N}_{\alpha,\beta,\pi}$, *for all initial conditions.*

Proof Recall that $x_i = [x_{ia}; x_{id}]$ and that (6.17) is a shorthand notation for (6.22). For each $i \in \mathscr{V}$, let $\bar{x}_i = x_i - \Pi_i x_r$, where Π_i is defined by (6.18). Then

$$\begin{aligned} \mathrm{d}\bar{x}_i &= A_i x_i \mathrm{d}t - \Pi_i S_i x_r \mathrm{d}t + B_i u_i \mathrm{d}t + H_i^1 d_i \mathrm{d}t + G_i \mathrm{d}w_i \\ &= A_i \bar{x}_i \mathrm{d}t + B_i u_i \mathrm{d}t + G_i \mathrm{d}w_i \end{aligned}$$

and

$$e_i = y_i - y_r = C_i x_i + H_i^2 R_{ie} - R_{re} x_r = C_i x_i - C_i \Pi_i x_r = C_i \bar{x}_i.$$

Since the dynamics of the $\bar{x}_i$ system with output e_i is governed by the same dynamics as the dynamics of agent i, we can present $\bar{x}_i$ in the same form as (6.22), with $\bar{x}_i = [\bar{x}_{ia}; \bar{x}_{id}]$, where

$$\begin{aligned} \mathrm{d}\bar{x}_{ia} &= A_{ia}\bar{x}_{ia}\mathrm{d}t + L_{iad}e_i\mathrm{d}t + G_{ia}\mathrm{d}w_i, \\ \mathrm{d}\bar{x}_{id} &= A_d\bar{x}_{id}\mathrm{d}t + B_d(u_i + E_{ida}\bar{x}_{ia} + E_{idd}\bar{x}_{id})\mathrm{d}t + G_{id}\mathrm{d}w_i, \\ e_i &= C_d\bar{x}_{id}. \end{aligned}$$

Define $\xi_{ia} = \bar{x}_{ia}$, $\xi_{id} = S_\varepsilon\bar{x}_{id}$, and $\hat{\xi}_{id} = S_\varepsilon\hat{x}_{id}$. Then

$$\begin{aligned} \mathrm{d}\xi_{ia} &= A_{ia}\xi_{ia}\mathrm{d}t + V_{iad}\xi_{id}\mathrm{d}t + G_{ia}\mathrm{d}w_i, \\ \varepsilon\mathrm{d}\xi_{id} &= A_d\xi_{id}\mathrm{d}t + B_d F_\delta\hat{\xi}_{id}\mathrm{d}t + V^\varepsilon_{ida}\xi_{ia}\mathrm{d}t + V^\varepsilon_{idd}\xi_{id}\mathrm{d}t + \varepsilon G^\varepsilon_{id}\mathrm{d}w_i, \\ e_i &= C_d\xi_{id}, \end{aligned}$$

where $V_{iad} = L_{iad}C_d$, $V^\varepsilon_{ida} = \varepsilon^\rho B_d E_{ida}$, $V^\varepsilon_{idd} = \varepsilon^\rho B_d E_{idd} S_\varepsilon^{-1}$ and $G^\varepsilon_{id} = S_\varepsilon G_{id}$.

Then,

$$\zeta_i + \psi_i = \sum_{j=1}^{N} \ell_{ij}(t)(y_j - y_r) + \iota_i(y_i - y_r) = \sum_{j=1}^{N} \bar{\ell}_{ij}(t)e_j.$$

Similarly, the controller (6.25) can be rewritten as

$$\varepsilon \mathrm{d}\hat{\xi}_{id} = A_d\hat{\xi}_{id}\mathrm{d}t + K\sum_{j=1}^{N} \bar{\ell}_{ij}(t)C_d\xi_{jd}\mathrm{d}t - KC_d\hat{\xi}_{id}\mathrm{d}t.$$

Let

$$\xi_a = \mathrm{col}\{\xi_{ia}\},\ \xi_d = \mathrm{col}\{\xi_{id}\},\ \hat{\xi}_d = \mathrm{col}\{\hat{\xi}_{id}\},\ w = \mathrm{col}\{w_i\}.$$

Then we have,

$$\begin{aligned} \mathrm{d}\xi_a &= A_a\xi_a\mathrm{d}t + V_{ad}\xi_d\mathrm{d}t + G_a\mathrm{d}w, \\ \varepsilon\mathrm{d}\xi_d &= (I_N \otimes A_d)\xi_d\mathrm{d}t + (I_N \otimes B_dF_\delta)\hat{\xi}_d\mathrm{d}t + V^\varepsilon_{da}\xi_a\mathrm{d}t + V^\varepsilon_{dd}\xi_d\mathrm{d}t + \varepsilon G^\varepsilon_d\mathrm{d}w, \\ \varepsilon\mathrm{d}\hat{\xi}_d &= (I_N \otimes A_d)\hat{\xi}_d\mathrm{d}t + (\bar{L}(t) \otimes KC_d)\xi_d\mathrm{d}t - (I_N \otimes KC_d)\hat{\xi}_d\mathrm{d}t, \end{aligned}$$

where $A_a = \mathrm{blkdiag}\{A_{ia}\}$, and V_{ad}, V^ε_{da}, V^ε_{dd}, G_a, G^ε_d are similarly defined.

Define $U_t^{-1}\bar{L}(t)U_t = J_t$, where J_t is the Jordan form of $\bar{L}(t)$, and let

$$v_a = \xi_a, \qquad v_d = (J_tU_t^{-1} \otimes I_{p\rho})\xi_d, \qquad \tilde{v}_d = v_d - (U_t^{-1} \otimes I_{p\rho})\hat{\xi}_d.$$

By our assumptions on the graph, we know that J_t and J_t^{-1} are bounded. Moreover, by the assumption on the condition number we can guarantee that U_t and U_t^{-1} are both bounded as well. Note that when a switching of the network graph occurs then v_d and $\tilde{v}_d$ will in most case experience a discontinuity (because of a sudden change in J_t and U_t) while v_a remains continuous. There exists $m_1, m_2 > 0$ such that we will have:

$$\|v_d(t_k^+)\| \leq m_1\|v_d(t_k^-)\|, \qquad \|\tilde{v}_d(t_k^+)\| \leq m_2\|\tilde{v}_d(t_k^-)\|$$

for any switching time t_k because of our bounds on U_t and J_t. Here

$$f(t^+) = \lim_{h\downarrow 0} f(t+h), \quad f(t^-) = \lim_{h\downarrow 0} f(t-h)$$

Between switching, the behavior of v_a, v_d, and $\tilde{v}_d$ is described by the following stochastic differential equations:

$$\begin{aligned} \mathrm{d}v_a &= A_av_a\mathrm{d}t + W_{ad,t}v_d\mathrm{d}t + G_a\mathrm{d}w, \\ \varepsilon\mathrm{d}v_d &= (I_N \otimes A_d)v_d\mathrm{d}t + (J_t \otimes B_dF_\delta)(v_d - \tilde{v}_d)\mathrm{d}t \\ &\quad + W^\varepsilon_{da,t}v_a\mathrm{d}t + W^\varepsilon_{dd,t}v_d\mathrm{d}t + \varepsilon\bar{G}^\varepsilon_{d,t}\mathrm{d}w, \\ \varepsilon\mathrm{d}\tilde{v}_d &= (I_N \otimes (A_d - KC_d))\tilde{v}_d\mathrm{d}t + (J_t \otimes B_dF_\delta)(v_d - \tilde{v}_d)\mathrm{d}t \\ &\quad + W^\varepsilon_{da,t}v_a\mathrm{d}t + W^\varepsilon_{dd,t}v_d\mathrm{d}t + \varepsilon\bar{G}^\varepsilon_{d,t}\mathrm{d}w, \end{aligned} \tag{6.27}$$

where $W_{ad,t} = V_{ad}(U_t J_t^{-1} \otimes I_{p\rho})$, $W^{\varepsilon}_{da,t} = (J_t U_t^{-1} \otimes I_{p\rho}) V^{\varepsilon}_{da}$, $W^{\varepsilon}_{dd,t} = (J_t U_t^{-1} \otimes I_{p\rho}) V^{\varepsilon}_{dd}(U_t J_t^{-1} \otimes I_{p\rho})$, and $\bar{G}^{\varepsilon}_{d,t} = (J_t U_t^{-1} \otimes I_{p\rho}) G^{\varepsilon}_{d}$.

Finally, let $\eta_a = v_a$, and define N_d such that

$$\eta_d := N_d \begin{pmatrix} v_d \\ \tilde{v}_d \end{pmatrix} = \begin{pmatrix} v_{1d} \\ \tilde{v}_{1d} \\ \vdots \\ v_{Nd} \\ \tilde{v}_{Nd} \end{pmatrix} \text{ where } N_d = \begin{pmatrix} e_1 & 0 \\ 0 & e_1 \\ \vdots & \vdots \\ e_N & 0 \\ 0 & e_N \end{pmatrix} \otimes I_{p\rho},$$

where $e_i \in \mathbb{R}^N$ is the i'th standard basis vector whose elements are all zero except for the i'th element which is equal to 1. Again the switching can only cause discontinuities in η_d (and not in η_a). There exists $m_3 > 0$ such that we will have:

$$\|\eta_d(t_k^+)\| \le m_3 \|\eta_d(t_k^-)\|, \tag{6.28}$$

for any switching time t_k. Between switching the stochastic differential equation (6.27) can be rewritten as:

$$\begin{aligned} \mathrm{d}\eta_a &= A_a \eta_a \mathrm{d}t + \tilde{W}_{ad,t} \eta_d \mathrm{d}t + G_a \mathrm{d}w, \\ \varepsilon \mathrm{d}\eta_d &= \tilde{A}_{\delta,t} \eta_d \mathrm{d}t + \tilde{W}^{\varepsilon}_{da,t} \eta_a \mathrm{d}t + \tilde{W}^{\varepsilon}_{dd,t} \eta_d \mathrm{d}t + \varepsilon \tilde{G}^{\varepsilon}_{d,t} \mathrm{d}w, \end{aligned} \tag{6.29}$$

where

$$\tilde{A}_{\delta,t} = I_N \otimes \begin{pmatrix} A_d & 0 \\ 0 & A_d - K C_d \end{pmatrix} + J_t \otimes \begin{pmatrix} B_d F_\delta & -B_d F_\delta \\ B_d F_\delta & -B_d F_\delta \end{pmatrix}, \tag{6.30}$$

and

$$\tilde{W}_{ad,t} = \begin{pmatrix} W_{ad,t} & 0 \end{pmatrix} N_d^{-1}, \qquad \tilde{G}^{\varepsilon}_{d,t} = N_d \begin{pmatrix} \bar{G}^{\varepsilon}_{d,t} \\ \bar{G}^{\varepsilon}_{d,t} \end{pmatrix},$$

$$\tilde{W}^{\varepsilon}_{da,t} = N_d \begin{pmatrix} W^{\varepsilon}_{da,t} \\ W^{\varepsilon}_{da,t} \end{pmatrix}, \qquad \tilde{W}^{\varepsilon}_{dd,t} = N_d \begin{pmatrix} W^{\varepsilon}_{dd,t} & 0 \\ W^{\varepsilon}_{dd,t} & 0 \end{pmatrix} N_d^{-1}.$$

Lemma 6.19 *Consider the matrix $\tilde{A}_{\delta,t}$ defined in (6.30). For any δ small enough the matrix $\tilde{A}_{\delta,t}$ is asymptotically stable for any Jordan matrix J_t whose eigenvalues satisfy $Re\{\lambda_{ti}\} > \beta$ and $|\lambda_{ti}| < \alpha$. Moreover, there exists $P_\delta > 0$ and $\nu > 0$ such that*

$$\tilde{A}'_{\delta,t} P_\delta + P_\delta \tilde{A}_{\delta,t} \le -\nu P_\delta - 4I \tag{6.31}$$

is satisfied for all possible Jordan matrices J_t and such that there exists $P_a > 0$ for which

$$P_a A_a + A'_a P_a = -\nu P_a - I. \tag{6.32}$$

Proof First note that if ν is small enough such that $A_a + \frac{\nu}{2}I$ is asymptotically stable then there exists $P_a > 0$ satisfying (6.32).

For the existence of P_δ and the stability of $\tilde{A}_{\delta,t}$ we rely on techniques developed earlier in [4, 21]. If we define

$$\bar{A}_{\delta,ti} = \begin{pmatrix} A_d + \lambda_{ti} B_d F_\delta & -\lambda_{ti} B_d F_\delta \\ \lambda_{ti} B_d F_\delta & A_d - K C_d - \lambda_{ti} B_d F_\delta \end{pmatrix}$$

and

$$\bar{B} = \begin{pmatrix} B_d F_\delta & -B_d F_\delta \\ B_d F_\delta & -B_d F_\delta \end{pmatrix}$$

then

$$\tilde{A}_{\delta,t} = \begin{pmatrix} \bar{A}_{\delta,t1} & \mu_1 \bar{B} & 0 & \cdots & 0 \\ 0 & \bar{A}_{\delta,t2} & \ddots & \ddots & \vdots \\ \vdots & \ddots & \ddots & \ddots & 0 \\ \vdots & & \ddots & \ddots & \mu_{N-1}\bar{B} \\ 0 & \cdots & \cdots & 0 & \bar{A}_{\delta,tN} \end{pmatrix},$$

where $\lambda_{t1}, \ldots, \lambda_{tN}$ are the eigenvalues of J_t and $\mu_i \in \{0, 1\}$ is determined by the Jordan structure of J_t. Define

$$\bar{P}_\delta = \rho \begin{pmatrix} P_d & 0 \\ 0 & \sqrt{\|P_d\|}P \end{pmatrix},$$

where P_d is the solution of the Riccati equation (6.24) and P is uniquely defined by the Lyapunov equation:

$$P(A_d - KC_d) + (A_d - KC_d)'P = -I.$$

In the above we choose ρ such that $\rho\delta > 1$ and $\rho\sqrt{\|P_d\|} > 2$. As shown in [4] we then have:

$$\bar{A}_{\delta,ti}'\bar{P}_\delta + \bar{P}_\delta\bar{A}_{\delta,ti} \le -\rho \begin{pmatrix} \delta I & 0 \\ 0 & \frac{1}{2}\sqrt{\|P_d\|}I \end{pmatrix} \le -I.$$

Via Schur complement, it is easy to verify that if matrices $A_{11} < -kI$, $A_{22} < 0$ and A_{12} are given then there exists μ sufficiently large such that the matrix

$$\begin{pmatrix} A_{11} & A_{12} \\ A_{12}' & \mu A_{22} \end{pmatrix} < -(k-1)I.$$

Define the matrix:

$$P_\delta = \begin{pmatrix} \alpha_1 \bar{P}_\delta & 0 & \cdots & 0 \\ 0 & \ddots & \ddots & \vdots \\ \vdots & \ddots & \ddots & 0 \\ 0 & \cdots & 0 & \alpha_N \bar{P}_\delta \end{pmatrix}.$$

Then we have that $P_\delta \tilde{A}_{\delta,t} + \tilde{A}_{\delta,t}' P_\delta$ is less than or equal to:

$$\begin{pmatrix} -\alpha_1 I & \alpha_1 \mu_1 \bar{P}_\delta \bar{B} & 0 & \cdots & 0 \\ \alpha_1 \mu_1 \bar{B}' \bar{P}_\delta & \ddots & \ddots & \ddots & \vdots \\ 0 & \ddots & \ddots & \ddots & 0 \\ \vdots & \ddots & \ddots & \ddots & \alpha_{N-1} \mu_{N-1} \bar{P}_\delta \bar{B} \\ 0 & \cdots & 0 & \alpha_{N-1} \mu_{N-1} \bar{B}' \bar{P}_\delta & -\alpha_N I \end{pmatrix}.$$

Using the above Schur argument, it is not hard to show that if

$$\begin{pmatrix} -\alpha_1 I & \alpha_1 \mu_1 \bar{P}_\delta \bar{B} & 0 & \cdots & 0 \\ \alpha_1 \mu_1 \bar{B}' \bar{P}_\delta & \ddots & \ddots & \ddots & \vdots \\ 0 & \ddots & \ddots & \ddots & 0 \\ \vdots & \ddots & \ddots & \ddots & \alpha_{N-2} \mu_{N-2} \bar{P}_\delta \bar{B} \\ 0 & \cdots & 0 & \alpha_{N-2} \mu_{N-2} \bar{B}' \bar{P}_\delta & -\alpha_{N-1} I \end{pmatrix} \leq -6I,$$

then there exists α_N such that:

$$\begin{pmatrix} -\alpha_1 I & \alpha_1 \mu_1 \bar{P}_\delta \bar{B} & 0 & \cdots & 0 \\ \alpha_1 \mu_1 \bar{B}' \bar{P}_\delta & \ddots & \ddots & \ddots & \vdots \\ 0 & \ddots & \ddots & \ddots & 0 \\ \vdots & \ddots & \ddots & \ddots & \alpha_{N-1} \mu_{N-1} \bar{P}_\delta \bar{B} \\ 0 & \cdots & 0 & \alpha_{N-1} \mu_{N-1} \bar{B}' \bar{P}_\delta & -\alpha_N I \end{pmatrix} \leq -5I.$$

Using a recursive argument, we can then prove there exists $\alpha_1, \ldots, \alpha_N$ such that:

$$P_\delta \tilde{A}_{\delta,t} + \tilde{A}_{\delta,t}' P_\delta \leq -5I.$$

This obviously implies that for ν small enough we have (6.31). If this ν is additionally small enough such that $A_a + \frac{\nu}{2} I$ is asymptotically stable (recall that A_a is asymptotically stable) then we obtain that there also exists P_a satisfying (6.32). ■

Define $V_a = \varepsilon^2 \eta_a' P_a \eta_a$ as a Lyapunov function for the dynamics of η_a in (6.29). Similarly, we define $V_d = \varepsilon \eta_d' P_\delta \eta_d$ as a Lyapunov function for the η_d dynamics in (6.29). Then the derivative of V_a is bounded by:

$$\begin{aligned} \mathrm{d}V_a &= -\nu V_a \mathrm{d}t - \varepsilon^2 \|\eta_a\|^2 \mathrm{d}t + 2\varepsilon^2 \mathrm{Re}(\eta_a' P_a \tilde{W}_{ad,t} \eta_d) \mathrm{d}t \\ &\qquad + \varepsilon^2 \operatorname{trace}(P_a G_a Q_0 G_a') \mathrm{d}t + 2\varepsilon^2 \mathrm{Re}(\eta_a' P_a G_a) \mathrm{d}w \\ &\le -\nu V_a \mathrm{d}t + \varepsilon c_3 V_d \mathrm{d}t \\ &\qquad + \varepsilon^2 r_5 \operatorname{trace}(Q_0) \mathrm{d}t + 2\varepsilon^2 \mathrm{Re}(\eta_a' P_a G_a) \mathrm{d}w, \end{aligned} \tag{6.33}$$

where r_5 and c_3 are such that:

$$\operatorname{trace}(P_a G_a Q_0 G_a') \le r_5 \operatorname{trace} Q_0$$

and

$$2\mathrm{Re}(\eta_a' P_a \tilde{W}_{ad,t} \eta_d) \le 2r_4 \|\eta_a\| \|\eta_d\| \le \tfrac{1}{2} \|\eta_a\|^2 + 2r_4^2 \|\eta_d\|^2 \le \tfrac{1}{2} \|\eta_a\|^2 + \varepsilon^{-1} c_3 V_d.$$

Note that we can choose r_4, r_5, and c_3 independent of the network graph but only depending on our bounds on the eigenvalues and condition number of our expand Laplacian $\bar{L}(t)$. Taking the expectation, we get:

$$\mathrm{d}\mathbb{E}V_a \le -\nu \mathbb{E} V_a \mathrm{d}t + \varepsilon c_3 \mathbb{E} V_d \mathrm{d}t + \varepsilon^2 r_5 \operatorname{trace}(Q_0) \mathrm{d}t.$$

Next, the derivative of V_d is bounded by

$$\begin{aligned} \mathrm{d}V_d &= -\nu \varepsilon^{-1} V_d \mathrm{d}t - 4\|\eta_d\|^2 \mathrm{d}t + 2\mathrm{Re}(\eta_d' P_\delta \tilde{W}^{\varepsilon}_{da,t} \eta_a) \mathrm{d}t \\ &\qquad + 2\mathrm{Re}(\eta_d' P_\delta \tilde{W}^{\varepsilon}_{dd,t} \eta_d) \mathrm{d}t + \varepsilon \operatorname{trace}(P_\delta \tilde{G}^{\varepsilon}_{d,t} Q_0 (\tilde{G}^{\varepsilon}_{d,t})') \mathrm{d}t \\ &\qquad + 2\varepsilon \mathrm{Re}(\eta_d' P_\delta \tilde{G}^{\varepsilon}_{d,t}) \mathrm{d}w \\ &\le c_2 V_a \mathrm{d}t - (\nu \varepsilon^{-1} + \nu - \varepsilon^2 \frac{c_2 c_3}{\nu}) V_d \mathrm{d}t - \|\eta_d\|^2 \mathrm{d}t \\ &\qquad + \varepsilon r_3 \operatorname{trace}(Q_0) \mathrm{d}t + 2\varepsilon \mathrm{Re}(\eta_d' P_\delta \tilde{G}^{\varepsilon}_d) \mathrm{d}w, \end{aligned} \tag{6.34}$$

where

$$2\mathrm{Re}(\eta_d' P_\delta \tilde{W}^{\varepsilon}_{dd,t} \eta_d) \le \|\eta_d\|^2$$

for small ε and

$$2\mathrm{Re}(\eta_d' P_\delta \tilde{W}^{\varepsilon}_{da,t} \eta_a) \le \varepsilon r_1 \|\eta_a\| \|\eta_d\| \le \varepsilon^2 r_1^2 \|\eta_a\|^2 + \|\eta_d\|^2 \le c_2 V_a + \|\eta_d\|^2,$$

provided r_1 is such that we have $\varepsilon r_1 \ge \| P_\delta \tilde{W}^{\varepsilon}_{da,t} \|$ and c_2 sufficiently large. Finally

$$\operatorname{trace}(P_\delta \tilde{G}^{\varepsilon}_{d,t} Q_0 (\tilde{G}^{\varepsilon}_{d,t})') \le r_3 \operatorname{trace} Q_0$$

for suitably chosen r_3. Taking the expectation, we get:

$$\mathrm{d}\mathbb{E}V_d \leq c_2\mathbb{E}V_a\mathrm{d}t - (\nu\varepsilon^{-1} + \nu - \varepsilon^2\frac{c_2c_3}{\nu})\mathbb{E}V_d\mathrm{d}t \\ - \mathbb{E}\|\eta_d\|^2\mathrm{d}t + \varepsilon r_3 \operatorname{trace}(Q_0)\mathrm{d}t.$$

We get:

$$\frac{\mathrm{d}}{\mathrm{d}t}\begin{pmatrix}\mathbb{E}V_a\\ \mathbb{E}V_d\end{pmatrix} \leq A_e\begin{pmatrix}\mathbb{E}V_a\\ \mathbb{E}V_d\end{pmatrix} + \begin{pmatrix}\varepsilon^2 r_5 \operatorname{trace}(Q_0)\\ \varepsilon r_3 \operatorname{trace}(Q_0)\end{pmatrix},$$

where

$$A_e = \begin{pmatrix}-\nu & \varepsilon c_3\\ c_2 & -\varepsilon^{-1}\nu - \nu + \varepsilon^2\frac{c_2c_3}{\nu}\end{pmatrix}.$$

Note that the inequality here is componentwise. We find by integration that

$$\begin{pmatrix}\mathbb{E}V_a\\ \mathbb{E}V_d\end{pmatrix}(t_k^-) \leq e^{A_e(t_k-t_{k-1})}\begin{pmatrix}\mathbb{E}V_a\\ \mathbb{E}V_d\end{pmatrix}(t_{k-1}^+) + \int_{t_{k-1}}^{t_k} e^{A_e(t_k-s)}\begin{pmatrix}\varepsilon^2 r_5 \operatorname{trace}(Q_0)\\ \varepsilon r_3 \operatorname{trace}(Q_0)\end{pmatrix}\mathrm{d}s$$

componentwise. In our case:

$$e^{A_e t} = \frac{1}{1+\varepsilon^3\frac{c_2c_3}{\nu^2}}\begin{pmatrix}e^{\lambda_1 t} + \varepsilon^3\frac{c_2c_3}{\nu^2}e^{\lambda_2 t} & \varepsilon^2\frac{c_3}{\nu}\left(e^{\lambda_1 t} - e^{\lambda_2 t}\right)\\ \varepsilon\frac{c_2}{\nu}\left(e^{\lambda_1 t} - e^{\lambda_2 t}\right) & e^{\lambda_2 t} + \varepsilon^3\frac{c_2c_3}{\nu^2}e^{\lambda_1 t}\end{pmatrix},$$

where $\lambda_1 = -\nu + \varepsilon^2\frac{c_2c_3}{\nu}$ and $\lambda_2 = -\varepsilon^{-1}\nu - \nu$. We have a potential jump at time t_{k-1} in V_d. However, there exists m such that

$$V_d(t_{k-1}^+) \leq mV_d(t_{k-1}^-),$$

while V_a is continuous. Using our explicit expression for $e^{A_e t}$ and the fact that $t_k - t_{k-1} > \tau$ we find:

$$(1\ 1)\,e^{A_e(t_k-t_{k-1})}\begin{pmatrix}\mathbb{E}V_a\\ \mathbb{E}V_d\end{pmatrix}(t_{k-1}^+) \leq e^{\lambda_3(t_k-t_{k-1})}\left[\mathbb{E}V_a(t_{k-1}^-) + \mathbb{E}V_d(t_{k-1}^-)\right],$$

where $\lambda_3 = -\nu/2$. Moreover, it can be easily verified that:

$$(1\ 1)\int_{t_{k-1}}^{t_k} e^{A_e(t_k-s)}\begin{pmatrix}\varepsilon^2 r_5 \operatorname{trace}(Q_0)\\ \varepsilon r_3 \operatorname{trace}(Q_0)\end{pmatrix}\mathrm{d}s \leq r\varepsilon^2 \operatorname{trace}(Q_0),$$

where r is a sufficiently large constant. We find

$$\left[\mathbb{E}V_a(t_k^-) + \mathbb{E}V_d(t_k^-)\right] \leq e^{\lambda_3(t_k-t_{k-1})}\left[\mathbb{E}V_a(t_{k-1}^-) + \mathbb{E}V_d(t_{k-1}^-)\right] + r\varepsilon^2 \operatorname{trace}(Q_0).$$

Combining these time intervals, we get:

$$\left[\mathbb{E}V_a(t_k^-) + \mathbb{E}V_d(t_k^-)\right] \leq e^{\lambda_3 t_k}\left[\mathbb{E}V_a(0) + \mathbb{E}V_d(0)\right] + \frac{r\varepsilon^2}{1-\mu}\operatorname{trace}(Q_0),$$

where $\mu < 1$ is such that

$$e^{\lambda_3(t_k - t_{k-1})} \leq e^{\lambda_3 \tau} \leq \mu$$

for all k. Assume $t_{k+1} > t > t_k$. Since we do not necessarily have that $t - t_k > \tau$ we use the bound:

$$e^{A_e(t-t_k)}\begin{pmatrix}\mathbb{E}V_a \\ \mathbb{E}V_d\end{pmatrix}(t_k^+) \leq 2me^{\lambda_3(t-t_k)}\begin{pmatrix}\mathbb{E}V_a \\ \mathbb{E}V_d\end{pmatrix}(t_k^-),$$

where the factor m is due to the potential discontinuous jump. Combining all together, we get:

$$\left[\mathbb{E}V_a(t) + \mathbb{E}V_d(t)\right] \leq 2me^{\lambda_3 t}\left[\mathbb{E}V_a(0) + \mathbb{E}V_d(0)\right] + (2m+1)\frac{r\varepsilon^2}{1-\mu}\operatorname{trace}(Q_0).$$

This implies:

$$\lim_{t\to\infty} \mathbb{E}[\eta_d'(t)\eta_d(t)] \leq \frac{2m+1}{\sigma_{\min}(P_\delta)}\frac{r\varepsilon}{1-\mu}\operatorname{trace}(Q_0).$$

Following the proof above, we find that

$$\begin{aligned}
\mathbf{e} &= (I_N \otimes C_d)(I_N \otimes S_\varepsilon^{-1})(U_t J_t^{-1} \otimes I_{p\rho})\begin{pmatrix}I_{Np\rho} & 0\end{pmatrix} N_d^{-1}\eta_d \\
&= (U_t J_t^{-1} \otimes C_d)\begin{pmatrix}I_{Np\rho} & 0\end{pmatrix} N_d^{-1}\eta_d \\
&= \Theta_t \eta_d,
\end{aligned}$$

for suitably chosen matrix Θ_t. Although Θ_t is time-varying, it is uniformly bounded, because for graphs in $\mathbb{G}_{\alpha,\beta,\pi}^{\varphi,N}$ the matrices U_t and J_t are bounded. The fact that we can make the asymptotic variance of η_d arbitrarily small then immediately implies that the asymptotic variance of $\mathbf{e}$ can be made arbitrarily small. Because all agents and protocols are linear it is obvious that the expectation of $\mathbf{e}$ is equal to zero. ■

Step 3: Combining the precompensator (6.16) and the controller (6.25) in Step 2, we obtain the protocol in the form of (6.15) in Theorem 6.17 (or if we replaced (6.13) by (6.14) we find the protocol for Theorem 6.15) as:

$$\begin{aligned}
\mathscr{A}_i &= \begin{pmatrix}A_d - K_\varepsilon C_d & 0 \\ B_{ip}F_{\delta\varepsilon} & A_{ip}\end{pmatrix}, & \mathscr{B}_i &= \begin{pmatrix}K_\varepsilon & K_\varepsilon \\ 0 & 0\end{pmatrix}, \\
\mathscr{C}_i &= \begin{pmatrix}0 & C_{ip}\end{pmatrix}, & \mathscr{D}_i &= \begin{pmatrix}0 & 0\end{pmatrix}.
\end{aligned} \tag{6.35}$$

6.5 Examples

In this section, we will present two examples. The first example is connected to Theorem 6.15 (without disturbances with known frequencies). The second example is connected to Theorem 6.17 (with disturbances with known frequencies).

6.5.1 Example 1

We illustrate the result in this section on a network of 10 nonidentical agents, which are of the form (6.1) with

$$A_{i_1} = \begin{pmatrix} -1 & 1 & 0 \\ 0 & 0 & 1 \\ 0.1 & 0 & 0.1 \end{pmatrix}, B_{i_1} = \begin{pmatrix} 0 \\ 0 \\ 1 \end{pmatrix}, C'_{i_1} = \begin{pmatrix} 0 \\ 1 \\ 0 \end{pmatrix}, G_{i_1} = \begin{pmatrix} 1 \\ 0 \\ 1.5 \end{pmatrix},$$

$$A_{i_2} = \begin{pmatrix} -3 & 1 & 0 \\ 0 & 0 & 1 \\ 1 & 0.5 & 1 \end{pmatrix}, B_{i_2} = \begin{pmatrix} 0 \\ 0 \\ 1 \end{pmatrix}, C'_{i_2} = \begin{pmatrix} 0 \\ 1 \\ 0 \end{pmatrix}, G_{i_2} = \begin{pmatrix} 0.5 \\ 1 \\ 1 \end{pmatrix},$$

$$A_{i_3} = \begin{pmatrix} -2 & 1 & 0 \\ 0 & 0 & 1 \\ 2 & 2 & 2 \end{pmatrix}, B_{i_3} = \begin{pmatrix} 0 \\ 0 \\ 1 \end{pmatrix}, C'_{i_3} = \begin{pmatrix} 0 \\ 1 \\ 0 \end{pmatrix}, G_{i_3} = \begin{pmatrix} 2 \\ 1 \\ 2 \end{pmatrix},$$

and $i_1 \in \{1, 2, 3\}$, $i_2 \in \{4, 5, 6\}$, $i_3 \in \{7, 8, 9, 10\}$, which will also be used as indices for the following precompensators and interconnection systems. The degree of the infinite zeros for each of the agent is equal to 2.

Assume the reference system as $y_0 = 1$, which is in the form of (6.6) with $S_r = 0$, $R_r = 1$, $x_r(0) = 1$. By using the method given in Sect. 6.4, precompensators are designed of the form (6.16) as

$$A_{i_1 p} = 0, \quad B_{i_1 p} = 10, \quad C_{i_1 p} = -0.1,$$

$$A_{i_2 p} = 0, \quad B_{i_2 p} = -1.2, \quad C_{i_2 p} = -\frac{5}{6},$$

$$A_{i_3 p} = 0, \quad B_{i_3 p} = -\frac{1}{3}, \quad C_{i_3 p} = -3.$$

The interconnection of the above precompensators and agents have the degree of the infinite zeros equal to 3, and can be written in SCB form:

$$A_{i_1} = \begin{pmatrix} -1 & 1.4142 & 0 & 0 \\ 0 & 0 & 1 & 0 \\ 0 & 0 & 0 & 1 \\ -0.0707 & 0.1 & 0 & 0.1 \end{pmatrix}, B_{i_1} = \begin{pmatrix} 0 \\ 0 \\ 0 \\ 1 \end{pmatrix}, C'_{i_1} = \begin{pmatrix} 0 \\ 1 \\ 0 \\ 0 \end{pmatrix}, G_{i_1} = \begin{pmatrix} 1.4142 \\ 0 \\ 1.5 \\ 0.25 \end{pmatrix},$$

$$A_{i_2} = \begin{pmatrix} -3 & 1.562 & 0 & 0 \\ 0 & 0 & 1 & 0 \\ 0 & 0 & 0 & 1 \\ -1.9206 & 1 & 0.5 & 1 \end{pmatrix},\ B_{i_2} = \begin{pmatrix} 0 \\ 0 \\ 0 \\ 1 \end{pmatrix},\ C'_{i_2} = \begin{pmatrix} 0 \\ 1 \\ 0 \\ 0 \end{pmatrix},\ G_{i_2} = \begin{pmatrix} 0.781 \\ 1 \\ 1 \\ 2 \end{pmatrix},$$

$$A_{i_3} = \begin{pmatrix} -2 & 1.2019 & 0 & 0 \\ 0 & 0 & 1 & 0 \\ 0 & 0 & 0 & 1 \\ -3.3282 & 2 & 2 & 2 \end{pmatrix},\ B_{i_3} = \begin{pmatrix} 0 \\ 0 \\ 0 \\ 1 \end{pmatrix},\ C'_{i_3} = \begin{pmatrix} 0 \\ 1 \\ 0 \\ 0 \end{pmatrix},\ G_{i_3} = \begin{pmatrix} 2.4037 \\ 1 \\ 2 \\ 10 \end{pmatrix}.$$

We select $K = (3\ 7\ 3)'$ such that eigenvalues of $(A_d - KC_d)$ are given by $(-0.5265,\ -1.2367 \pm j2.0416)$, and then choose $\delta = 10^{-10}$, $\varepsilon = 0.01$ such that

$$F_{\delta\varepsilon} = \begin{pmatrix} -0.0018 & -0.0021 & -0.0012 \end{pmatrix},\ K_{\varepsilon} = \begin{pmatrix} 300 \\ 70000 \\ 3000000 \end{pmatrix}.$$

Together with A_d, C_d with $\rho = 3$, we get the controller of the form (6.25) for each interconnection system.

As stated in Theorem 6.15, the time-varying network topology switches in a set of network graph $\mathbb{G}^{\varphi,N}_{\alpha,\beta,\pi}$ with minimum dwell time τ, and a priori given α, β, π, φ, N. In this example, we assume a graph set consists of three directed graphs $\mathscr{G}_1$, $\mathscr{G}_2$, $\mathscr{G}_3$, with $N = 10, \alpha = 10, \beta = 0.3$, π only contains node of agent 2, and φ can be any bounded real number for this set is finite (with only 3 graphs). These graphs are shown in Fig. 6.1. The reference system is connected to agent 2, which is in the root set.

Figure 6.2 shows the outputs of 10 agents with reference system $y_0 = 1$ with $\varepsilon = 0.01, \delta = 10^{-10}$. When tuning parameter ε to 0.001, regulated output synchronization errors are squeezed to small and outputs of agents are much closer to the reference trajectory, shown in Fig. 6.3.

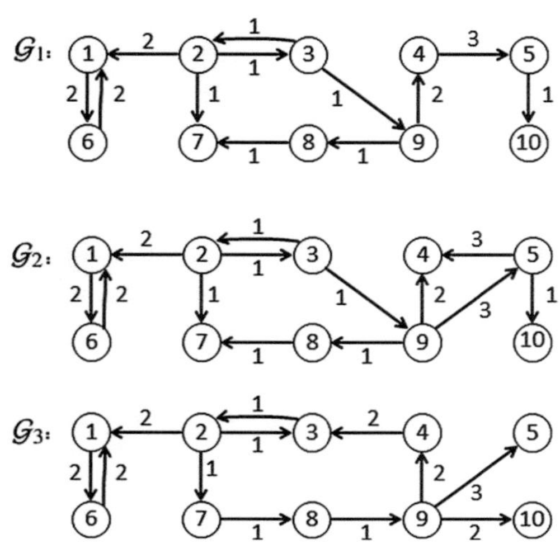

Fig. 6.1 The network topologies

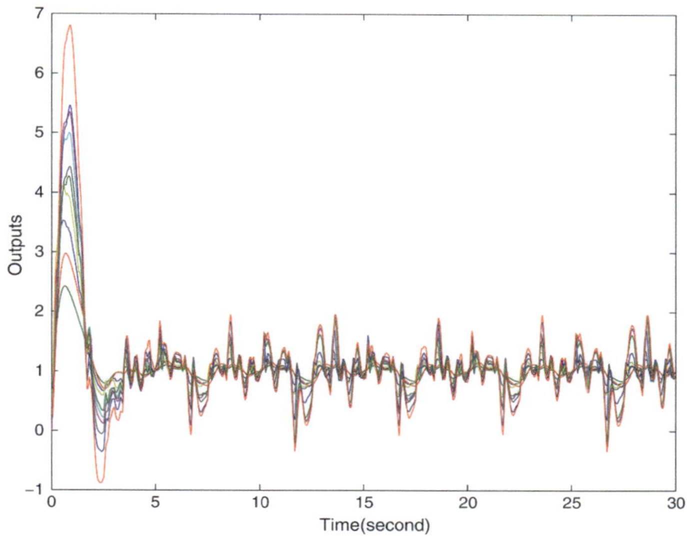

Fig. 6.2 Low- and high-gain parameters $\varepsilon = 0.01$, $\delta = 10^{-10}$

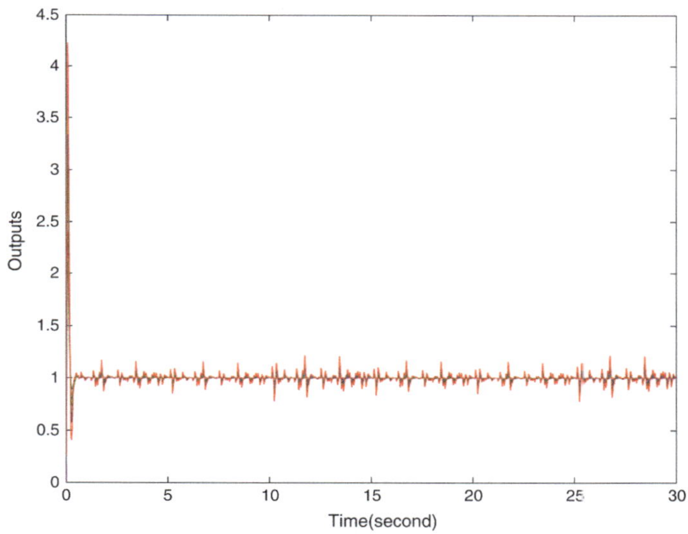

Fig. 6.3 Low- and high-gain parameters $\varepsilon = 0.001$, $\delta = 10^{-10}$

6.5.2 Example 2

In this section, we will modify Sect. 6.5.1 by adding disturbances with known frequencies. The $\tilde{H}_i^1$, $\tilde{H}_i^2$, S_{id}, and R_{id} for agent i are given by:

$$\tilde{H}_{i_1}^1 = \begin{pmatrix} 0 & 1 \\ 0 & 0 \\ 0 & 1.5 \end{pmatrix},\ \tilde{H}_{i_1}^2 = \begin{pmatrix} 1 & 0 \end{pmatrix},\ S_{i_1d} = \begin{pmatrix} 0 & 0 & 0 \\ 0 & 0 & 9 \\ 0 & -9 & 0 \end{pmatrix},\ R_{i_1d} = \begin{pmatrix} 1 & 0 & 0 \\ 0 & 0 & 1 \end{pmatrix},$$

$$\tilde{H}_{i_2}^1 = \begin{pmatrix} 0 & 0.5 \\ 0 & 1 \\ 0 & 1 \end{pmatrix},\ \tilde{H}_{i_2}^2 = \begin{pmatrix} 1 & 0 \end{pmatrix},\ S_{i_2d} = \begin{pmatrix} 0 & 0 & 0 \\ 0 & 0 & 3 \\ 0 & -3 & 0 \end{pmatrix},\ R_{i_2d} = \begin{pmatrix} 1 & 0 & 0 \\ 0 & 1 & 0 \end{pmatrix},$$

$$\tilde{H}_{i_3}^1 = \begin{pmatrix} 0 & 2 \\ 0 & 1 \\ 0 & 2 \end{pmatrix},\ \tilde{H}_{i_3}^2 = \begin{pmatrix} 1 & 0 \end{pmatrix},\ S_{i_3d} = \begin{pmatrix} 0 & 0 & 0 \\ 0 & 0 & 5 \\ 0 & -5 & 0 \end{pmatrix},\ R_{i_3d} = \begin{pmatrix} 1 & 0 & 0 \\ 0 & 0 & 1 \end{pmatrix},$$

where from S_{id} we find that the disturbances with known frequencies are constant and sinusoid waves. Please note that $i_1 \in \{1, 2, 3\}$, $i_2 \in \{4, 5, 6\}$, $i_3 \in \{7, 8, 9, 10\}$. Assume we also have the constant reference trajectory $y_0 = 1$. By applying the method given in Sect. 6.4, we get precompensators

$$A_{i_1p} = \begin{pmatrix} 0 & -0.8441 & 0 \\ 0.8441 & 0 & -8.9603 \\ 0 & 8.9603 & 0 \end{pmatrix},\ B_{i_1p} = \begin{pmatrix} 0.7779 \\ 0.5959 \\ 0.6631 \end{pmatrix},\ C'_{i_1p} = \begin{pmatrix} 0 \\ 0 \\ 1.5079 \end{pmatrix},$$

$$A_{i_2p} = \begin{pmatrix} 0 & -1.1235 & 0 \\ 1.1235 & 0 & -2.7817 \\ 0 & 2.7817 & 0 \end{pmatrix},\ B_{i_2p} = \begin{pmatrix} 0.2106 \\ 0.2285 \\ 0.3177 \end{pmatrix},\ C'_{i_2p} = \begin{pmatrix} 0 \\ 0 \\ 3.1469 \end{pmatrix},$$

$$A_{i_3p} = \begin{pmatrix} 0 & -3.5038 & 0 \\ 3.5038 & 0 & -3.5670 \\ 0 & 3.5670 & 0 \end{pmatrix},\ B_{i_3p} = \begin{pmatrix} 0.0353 \\ 0.1059 \\ 0.1652 \end{pmatrix},\ C'_{i_3p} = \begin{pmatrix} 0 \\ 0 \\ 6.0544 \end{pmatrix}.$$

We also use the same parameters as those in Sect. 6.5.1, i.e., $K = (3 \quad 7 \quad 3)'$, $\delta = 10^{-10}$, $\varepsilon = 0.01$. Then, we have

$$F_{\delta\varepsilon} = \begin{pmatrix} -18.2574 & -20.7160 & -11.7519 \end{pmatrix},\ K_\varepsilon = \begin{pmatrix} 300 \\ 70000 \\ 3000000 \end{pmatrix}.$$

For A_d, C_d with $\rho = 3$ given, we can get the controller of the form (6.25) for each interconnection system.

The network topology also switches among the set of graph shown in Fig. 6.1 in the same way. Figure 6.4 shows the outputs of 10 agents with reference system $y_0 = 1$ with $\varepsilon = 0.01, \delta = 10^{-10}$. When tuning parameter ε to 0.001, regulated output synchronization errors are squeezed to small and outputs of agents are much closer to the reference trajectory, shown in Fig. 6.5. We can find that even agents are affected by any constant and any sinusoid wave with known frequencies, stochastic almost regulated output synchronization is still obtained.

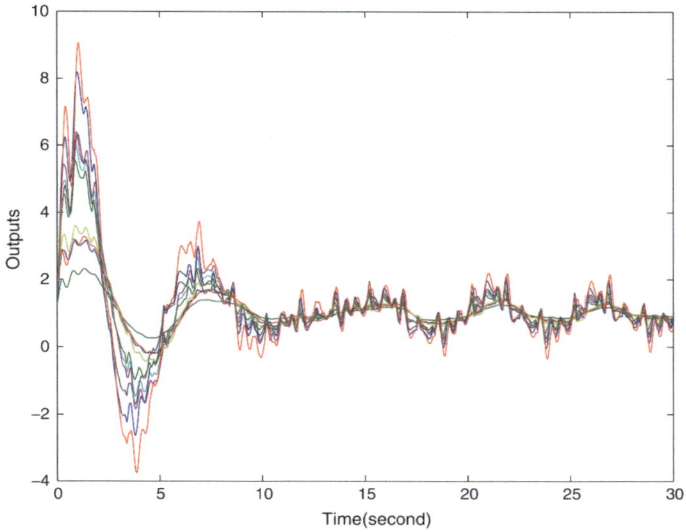

Fig. 6.4 Low- and high-gain parameters $\varepsilon = 0.01$, $\delta = 10^{-10}$

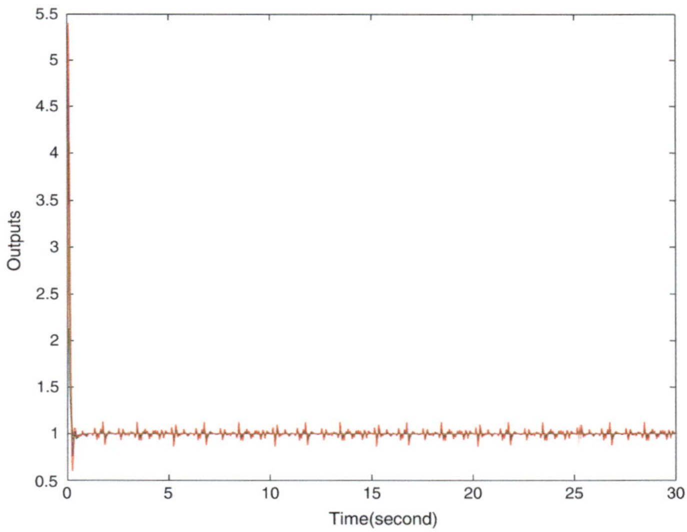

Fig. 6.5 Low- and high-gain parameters $\varepsilon = 0.001$, $\delta = 10^{-10}$

References

1. Bai, H., Arcak, M., Wen, J.: Cooperative Control Design: A Systematic, Passivity-Based Approach. Communications and Control Engineering. Springer, New York (2011)
2. Boyd, S., Barratt, C.: Linear Controller Design: Limits of Performance. Information and System Sciences. Prentice Hall, Englewood Cliffs (1991)
3. Chopra, N., Spong, W.: Output synchronization of nonlinear systems with relative degree one. In: Blondel, V., Boyd, S., Kimura H. (eds.) Recent Advances in Learning and Control. Lecture Notes in Control and Information Sciences, vol. 371. Springer, London, pp. 51–64 (2008)
4. Grip, H., Saberi, A., Stoorvogel, A.: Synchronization in networks of minimum-phase, non-introspective agents without exchange of controller states: homogeneous, heterogeneous, and nonlinear. Automatica **54**, 246–255 (2015)
5. Grip, H., Yang, T., Saberi, A., Stoorvogel, A.: Output synchronization for heterogeneous networks of non-introspective agents. Automatica **48**(10), 2444–2453 (2012)
6. Kim, H., Shim, H., Seo, J.: Output consensus of heterogeneous uncertain linear multi-agent systems. IEEE Trans. Autom. Control **56**(1), 200–206 (2011)
7. Li, Z., Duan, Z., Chen, G., Huang, L.: Consensus of multi-agent systems and synchronization of complex networks: a unified viewpoint. IEEE Trans. Circuit. Syst. I Regul. Pap. **57**(1), 213–224 (2010)
8. Liberzon, D., Morse, A.: Basic problem in stability and design of switched systems. IEEE Control Syst. Mag. **19**(5), 59–70 (1999)
9. Liu, X., Chen, B., Lin, Z.: On the problem of general structure assignments of linear systems through sensor/actuator selection. Automatica **39**(2), 233–241 (2003)
10. Lunze, J.: An internal-model principle for the synchronisation of autonomous agents with individual dynamics. In: Proceedings of Joint 50th CDC and ECC, pp. 2106–2111. Orlando, FL (2011)
11. Meng, Z., Yang, T., Dimarogonas, D. V., and Johansson, K. H. Coordinated output regulation of multiple heterogeneous linear systems. In Proc. 52nd CDC (Florence, Italy, 2013), pp. 2175–2180
12. Mesbahi, M., Egerstedt, M.: Graph Theoretic Methods in Multiagent Networks. Princeton University Press, Princeton (2010)
13. Øksendal, B.: Stochastic Differential Equations: An Introduction with Applications. Universitext, 6th edn. Springer, Berlin (2003)
14. Olfati-Saber, R., Murray, R.: Agreement problems in networks with direct graphs and switching topology. In Proceedings of 42nd CDC, pp. 4126–4132. Maui, Hawaii (2003)
15. Olfati-Saber, R., Murray, R.: Consensus problems in networks of agents with switching topology and time-delays. IEEE Trans. Autom Control **49**(9), 1520–1533 (2004)
16. Peymani, E., Grip, H., Saberi, A.: Homogeneous networks of non-introspective agents under external disturbances—H_∞ almost synchronization. Automatica **52**, 363–372 (2015)
17. Peymani, E., Grip, H., Saberi, A., Wang, X., Fossen, T.: H_∞ almost output synchronization for heterogeneous networks of introspective agents under external disturbances. Automatica **50**(4), 1026–1036 (2014)
18. Ren, W., Cao, Y.: Distributed Coordination of Multi-agent Networks. Communications and Control Engineering. Springer, London (2011)
19. Saberi, A., Stoorvogel, A., Sannuti, P.: Control of Linear Systems with Regulation and Input Constraints. Communication and Control Engineering Series. Springer, Berlin (2000)
20. Sannuti, P., Saberi, A., Zhang, M.: Squaring down of general MIMO systems to invertible uniform rank systems via pre- and/or post-compensators. Automatica **50**(8), 2136–2141 (2014)
21. Seo, J., Shim, H., Back, J.: Consensus of high-order linear systems using dynamic output feedback compensator: low gain approach. Automatica **45**(11), 2659–2664 (2009)
22. Shi, G., Johansson, K.: Robust consensus for continuous-time multi-agent dynamics. SIAM J. Control Optim. **51**(5), 3673–3691 (2013)
23. Su, Y., Huang, J.: Stability of a class of linear switching systems with applications to two consensus problem. IEEE Trans. Autom. Control **57**(6), 1420–1430 (2012)

24. Tuna, S.: LQR-based coupling gain for synchronization of linear systems. arXiv:0801.3390v1 (2008)
25. Tuna, S.: Synchronizing linear systems via partial-state coupling. Automatica **44**(8), 2179–2184 (2008)
26. Wieland, P., Kim, J., Allgöwer, F.: On topology and dynamics of consensus among linear high-order agents. Int. J. Syst. Sci. **42**(10), 1831–1842 (2011)
27. Wieland, P., Sepulchre, R., Allgöwer, F.: An internal model principle is necessary and sufficient for linear output synchronization. Automatica **47**(5), 1068–1074 (2011)
28. Wu, C.: Synchronization in Complex Networks of Nonlinear Dynamical Systems. World Scientific Publishing Company, Singapore (2007)
29. Yang, T., Roy, S., Wan, Y., Saberi, A.: Constructing consensus controllers for networks with identical general linear agents. Int. J. Robust Nonlinear Control **21**(11), 1237–1256 (2011)
30. Yang, T., Saberi, A., Stoorvogel, A., Grip, H.: Output synchronization for heterogeneous networks of introspective right-invertible agents. Int. J. Robust Nonlinear Control **24**(13), 1821–1844 (2014)
31. Zhang, M., Saberi, A., Grip, H.F., Stoorvogel, A.A.: $\mathscr{H}_\infty$ almost output synchronization for heterogeneous networks without exchange of controller states. IEEE Trans. Control of Network Systems (2015). doi:10.1109/TCNS.2015.2426754
32. Zhao, J., Hill, D.J., Liu, T.: Synchronization of complex dynamical networks with switching topology: a switched system point of view. Automatica **45**(11), 2502–2511 (2009)

Chapter 7
A Characterization of Solutions of the ARE and ARI

A. Sanand Amita Dilip and Harish K. Pillai

Abstract This article is about a characterization of the solution set of algebraic Riccati equation (ARE) and the algebraic Riccati inequality (ARI) over the reals, for both controllable and uncontrollable systems. We characterize these solutions using simple linear algebraic arguments. It turns out that solutions of ARE of maximal rank have lower rank solutions encoded within it. We demonstrate how these lower rank solutions are encoded within the full rank solution and how one can retrieve the lower rank solutions from the maximal rank solution. We also obtain a parametrization for solutions of certain specific ARIs. We generalize Willems' result $K_{min} \leq K \leq K_{max}$ for ARI arising out of controllable systems to some specific kind of uncontrollable systems.

7.1 Introduction and Preliminaries

About 15 years ago, Harry Trentelman and the second author occupied adjacent offices for nearly two years. Those were very exciting days for me in IWI, Groningen. The lunches were the most important part of the day when debates on just about everything under the sun took place. More often than not, Harry and I ended up taking diametrically opposite viewpoints, especially when it came to world politics. There were also several topics on which we agreed—sports, mountaineering, and mathematics being some of them. Unfortunately, Harry and I have not written a paper together but perhaps there is still time.

A topic on which Harry has worked extensively is the topic of Algebraic Riccati equations (ARE) and Algebraic Riccati Inequality (ARI). Some of Harry's best works have been closely related to this topic and I find his paper on Quadratic Differential Forms [25], one of the best papers I have read. Moreover, a lot of my understanding

A. Sanand Amita Dilip · H.K. Pillai (✉)
Department of Electrical Engineering, IIT Bombay, Powai, Mumbai, India
e-mail: hp@ee.iitb.ac.in

A. Sanand Amita Dilip
e-mail: asanand@ee.iitb.ac.in

M.N. Belur et al. (eds.), *Mathematical Control Theory II*,
Lecture Notes in Control and Information Sciences 462,
DOI 10.1007/978-3-319-21003-2_7

about ARE and ARI started from my interactions with Harry. I, therefore, think it appropriate to write something about ARE and ARI. This article is based on work done with my student Sanand.

Algebraic Riccati equation occurs naturally in control theory, filtering, numerical analysis, and many other engineering applications. In optimal control, algebraic Riccati equation (ARE) arises in infinite horizon continuous time LQR problem. ARE is also related to power method, QR factorization in matrix computations [1, 9], spectral factorization [5, 6, 16]. Riccati equation shows up in Kalman filters too [11]. In [16], solutions of ARE are used in the study of acausal realizations of stationary processes. Further it is shown how AREs are involved in spectral factorization and in balancing algorithm (related to stochastic balancing) in [16]. In [5, 6] solutions ARE are used for parametrization of minimal spectral factors. In [4], solutions of ARE are used to obtain parametrization of minimal stochastic realizations. For a treatment on discrete-time ARE, refer [7, 26].

Algebraic Riccati inequality (ARI) arises in H_∞ control [18, 19, 22] and also in formulation of storage functions for dissipative systems in behavioral theory of systems [25]. The solution set of the ARI (which is a spectrahedron) characterizes the set of all possible storage functions [25]. This ARI comes from an LMI arising from the dissipation inequality [25]. Study of symmetric solutions of ARI has also appeared in [3, 8, 12–15, 17, 21, 24] and some of the references therein.

The ARE and ARI originate from the Differential Riccati equation (DRE) given by $\dot{K} = -A^T K - KA - Q + KBB^T K$. This DRE defines a flow on the set of symmetric matrices. The equilibrium points of DRE are the solutions of the corresponding algebraic Riccati equation $-A^T K - KA - Q + KBB^T K = 0$.

In this article, we concentrate on the ARE of the form $-A^T K - KA - Q + KBB^T K = 0$ and ARIs of the form $-A^T K - KA - Q + KBB^T K \leq 0$ where A, B, Q are real constant matrices having dimensions $n \times n, n \times m$ and $n \times n$, respectively, with Q being symmetric. (Note that in the characterization of storage functions, the ARI takes the form $Q - A^T K - KA - KBB^T K \geq 0$ which one obtains from ARI: $-A^T K - KA - Q + KBB^T K \leq 0$ by replacing K by $-K$.)

We begin our analysis by considering the scalar DRE: $\frac{dk}{dt} = \dot{k} = -q - 2ak + b^2k^2$ with $k \in \mathbb{R}$. If the discriminant of the polynomial $-q - 2ak + b^2k^2$ is greater than or equal to zero, then the polynomial has real roots and there are two real equilibrium points. If the discriminant is strictly greater than zero, then there are two distinct roots of $-q - 2ak + b^2k^2 = 0$. Let k_{min} and k_{max} be the two solutions (equilibrium points of the DRE) such that $k_{max} > k_{min}$. Clearly, $\dot{k} \rightarrow +\infty$ when $k \rightarrow \pm\infty$. In the region $k_{min} \leq k \leq k_{max}$, $\dot{k} \leq 0$ and hence direction of flow is toward k_{min}. Outside the line segment joining k_{min} and k_{max}, i.e., for $k < k_{min}$ and $k > k_{max}$, $\dot{k} > 0$. For $k < k_{min}$, $\dot{k} > 0$, thereby implying that the direction of flow is toward k_{min} for all $k < k_{min}$. Therefore k_{min} is a stable equilibrium point of the DRE. For $k > k_{max}$, $\dot{k} > 0$ which implies k_{max} is an unstable equilibrium point. The line segment joining k_{min} and k_{max} are precisely those values of k that satisfy the ARI.

If the discriminant of the polynomial $-q - 2ak + b^2k^2$ is equal to zero, then the polynomial has a double root and therefore $k_{min} = k_{max}$ in this case. Thus, the region satisfying the strict ARI is absent in this case. Note further that in this case, the equilibrium point is unstable.

If the discriminant of the polynomial $-q - 2ak + b^2k^2$ is negative, then the polynomial has no real solutions and therefore there are no real solutions for the corresponding ARE and ARI.

We want to investigate the situation for the matrix case. One can expect some sort of similarity between the scalar and matrix cases. From the literature, we know that for matrix case, an equilibrium point exists if there exists an n-dimensional Lagrangian H-invariant subspace. We assume that this is the case and fix an arbitrary solution K_0 of the ARE. Let $K = K_0 + X$ where X can be thought of as a perturbation from K_0. We can then rewrite $-A^T K - KA - Q + KBB^T K$ as

$$\begin{aligned}
&= -A^T(K_0 + X) - (K_0 + X)A - Q + (K_0 + X)BB^T(K_0 + X) \\
&= -A^T K_0 - K_0A - Q + K_0BB^T K_0 - A^T X - XA + K_0BB^T X \\
&\quad + XBB^T K_0 + XBB^T X \\
&= -(A - BB^T K_0)^T X - X(A - BB^T K_0) + XBB^T X. \qquad (7.1) \\
&\quad (\text{Since } -A^T K_0 - K_0A - Q + K_0BB^T K_0 = 0)
\end{aligned}$$

Let $A_0 = A - BB^T K_0$. The original DRE can now be thought of as $\dot{K} = \dot{X} = -A_0^T X - XA_0 + XBB^T X$. We denote $-A_0^T X - XA_0 + XBB^T X$ by $\text{Ric}(X)$. Then solutions to the equation $\text{Ric}(X) = 0$, would characterize all the equilibrium points of the original DRE, i.e., the solutions of the ARE. Similarly, all solutions to the equation $\text{Ric}(X) \leq 0$ would characterize all the solutions of the original ARI.

7.2 Building Blocks for Solutions of Ric(X) = 0 and Ric(X) ≤ 0

Clearly, $X = 0$ is a solution of $\text{Ric}(X) = -A_0^T X - XA_0 + XBB^T X = 0$. We now look for nonzero X that satisfy $\text{Ric}(X) = 0$. We start with the simplest case of matrices X that have rank one. Since X is symmetric, let $X = \alpha vv^T$ where $\alpha \in \mathbb{R}$ and $v \in \mathbb{R}^n$ with $||v|| = 1$. Further, let $\beta = ||B^T v||$.

Theorem 7.1 *Let $X = \alpha vv^T$, such that $||v|| = 1$. Then*
1. if v is an eigenvector of A_0^T, the rank of Ric(X) is at most one and Ric(X) is definite.
2. for all other v, Ric(X) is indefinite.

Proof Let $X = \alpha vv^T$, where v is a right eigenvector of A_0^T, with λ the corresponding eigenvalue. Clearly, $\text{Ric}(X) = (-2\alpha\lambda + \alpha^2\beta^2)vv^T$ has rank at most one and depending on the sign of $(-2\alpha\lambda + \alpha^2\beta^2)$ it is positive or negative semidefinite.

Let $X = \alpha vv^T$ where v is not an eigenvector of A_0^T. Let $A_0^T v = \gamma u$ with $||u|| = 1$. Therefore,

$$\mathrm{Ric}(X) = \begin{bmatrix} u & v \end{bmatrix} \begin{bmatrix} 0 & -\alpha\gamma \\ -\alpha\gamma & (\alpha\beta)^2 \end{bmatrix} \begin{bmatrix} u^T \\ v^T \end{bmatrix}$$

Now $\begin{bmatrix} 0 & -\alpha\gamma \\ -\alpha\gamma & (\alpha\beta)^2 \end{bmatrix}$ is indefinite which implies $\mathrm{Ric}(X)$ is indefinite. ■

Observe that the zero matrix is a very special case of a definite matrix. From the above theorem it is clear that if v is not an eigenvector of A_0^T, then we are not going to get solutions of $\mathrm{Ric}(X) = 0$ along $X = vv^T$.

Consider the DRE: $\dot{X} = -A_0^T X - XA_0 + XBB^T X$ when $X = \alpha vv^T$, where v is an eigenvector of A_0^T and $\alpha \in \mathbb{R}$. We obtain the differential equation $v\dot{\alpha}v^T = v(-2\alpha\lambda + \alpha^2\beta^2)v^T$ which is actually equivalent to the scalar Riccati differential equation: $\dot{\alpha} = -2\alpha\lambda + \alpha^2\beta^2$. If $\beta \neq 0$, then the equilibrium points are at $\alpha = 0$ and $\alpha = 2\lambda/\beta^2$. Note that $\alpha = 0$ corresponds to the equilibrium point K_0 of the original DRE, whereas $\alpha = 2\lambda/\beta^2$ corresponds to a new equilibrium point of the original DRE. Thus $\alpha \in [0, 2\lambda/\beta]$ corresponds to X that satisfy $\mathrm{Ric}(X) \leq 0$. If $\beta = 0$, then $\alpha = 0$ is the only equilibrium point. Note further that for $X = \alpha vv^T$, $\dot{X}$ remains along the direction vv^T if v is an eigenvector of A_0^T.

Consider the pair of matrices (A_0, B). Using a change of basis, it is possible to write A_0 and B in the following form [10, 27]

$$A_0 = \begin{bmatrix} A_0^{11} & A_0^{12} \\ 0 & A_0^{22} \end{bmatrix}, B = \begin{bmatrix} B_1 \\ 0 \end{bmatrix}$$

where (A_0^{11}, B_1) form a controllable pair. Eigenvalues of A_0^{11} are controllable while eigenvalues of A_0^{22} are uncontrollable. Left eigenvectors corresponding to uncontrollable eigenvalues are of the form $v^T = \begin{bmatrix} 0 \; u^T \end{bmatrix}$. All such vectors v are right eigenvectors of A_0^T such that $B^T v = 0$. All right eigenvectors of A_0^T that belong to kernel of B^T, are called eigenvectors that correspond to uncontrollable modes. If v is an eigenvector of A_0^T associated with a controllable eigenvalue, then $B^T v \neq 0$ and so these eigenvectors correspond to controllable modes.

Theorem 7.2 *If A_0 has an uncontrollable zero eigenvalue, then rank one solutions of $Ric(X) = 0$ and $Ric(X) \leq 0$ form an unbounded set.*

Proof Let v be an eigenvector of A_0^T corresponding to zero eigenvalue which is associated to an uncontrollable mode. Therefore, $A_0^T v = B^T v = 0$. Let $X = \alpha vv^T$ where $\alpha \in \mathbb{R}$. $\mathrm{Ric}(X) = 0$ for all $\alpha \in \mathbb{R}$ and therefore one has an unbounded set of rank one solutions for $\mathrm{Ric}(X) = 0$ and $\mathrm{Ric}(X) \leq 0$. ■

Assuming that there are no uncontrollable modes corresponding to zero eigenvalue, one obtains:

Theorem 7.3 *There is one-to-one correspondence between rank one solutions of $Ric(X) = 0$ and eigenvectors v of nonzero real eigenvalues of A_0^T that correspond to controllable modes.*

Proof Let $X = \alpha vv^T$ where $\alpha \in \mathbb{R}$ and v is an eigenvector of A_0^T with corresponding eigenvalue λ. Therefore, $\mathrm{Ric}(X) = (-2\alpha\lambda + \alpha^2\beta^2)vv^T$. If $A_0^T v = 0$, then $\lambda = 0$ and $\mathrm{Ric}(X) = (\alpha^2\beta^2)vv^T$. Hence $\mathrm{Ric}(X) = (\alpha\beta)^2 vv^T = 0$ has only one solution given by $\alpha = 0$. This solution to $\mathrm{Ric}(X) = 0$ is a rank zero solution and not a rank one solution.

Assume $\lambda \neq 0$ and let $v \in \mathbb{R}^n$ be an associated eigenvector of A_0^T corresponding to a controllable mode. Hence, $\beta = ||B^T v|| \neq 0$. Therefore, $\mathrm{Ric}(X) = 0$ for $\alpha = 2\lambda/\beta^2$, i.e., for $X = (2\lambda/\beta^2)vv^T$. If the eigenvector v corresponds to an uncontrollable mode, then $\beta = 0$ and $\mathrm{Ric}(X) = -2\alpha\lambda vv^T$ which is zero only when $\alpha = 0$. But $\alpha = 0$ implies $X = 0$ which is of rank zero. ∎

Following theorem gives parametrization of all rank one solutions of the ARI $\mathrm{Ric}(X) \leq 0$.

Theorem 7.4 *If v is an eigenvector of A_0^T corresponding to some nonzero real eigenvalue, then*

- *if v is an eigenvector corresponding to a controllable mode, then there is one-to-one correspondence between all rank one solutions of $Ric(X) \leq 0$ along vv^T and a bounded interval.*
- *if v is an eigenvector corresponding to an uncontrollable mode, then there is one-to-one correspondence between all rank one solutions of $Ric(X) \leq 0$ along vv^T and a half line.*

If (A, B) is uncontrollable, then (A_0, B) is also uncontrollable. If A_0 has a real eigenvalue which has an uncontrollable mode, then rank one solutions of $\mathrm{Ric}(X) \leq 0$ are unbounded. Conversely, if rank one solutions of $\mathrm{Ric}(X) \leq 0$ are unbounded, then (A, B) must be uncontrollable.

If all real uncontrollable eigenvalues of A_0 are in $\mathbb{R}_+$, then all rank one solutions of $\mathrm{Ric}(X) \leq 0$ are bounded from below. Similarly, if all real uncontrollable eigenvalues of A_0 are in $\mathbb{R}_-$, then all rank one solutions of $\mathrm{Ric}(X) \leq 0$ are bounded from above. If real uncontrollable eigenvalues lie in both $\mathbb{R}_+$ and $\mathbb{R}_-$, then rank one solutions of $\mathrm{Ric}(X) \leq 0$ are neither bounded above nor below. Thus, we have a characterization of all real rank one solutions of $\mathrm{Ric}(X) \leq 0$.

7.2.1 Rank Two Solutions of $Ric(X) = 0$ and $Ric(X) \leq 0$

Notice that all real rank one solutions of the ARE and ARI are related to eigendirections corresponding to real eigenvalues. The complex eigenvalues of A_0 play no role in these rank one solutions. As the complex eigenvalues come in a conjugate pair, one can expect them to play a role in determining the real rank two solutions of the ARE and ARI. We, therefore, now concentrate on all rank two solutions of the ARE and the ARI. Let $X = L\mathscr{L}L^T$ where $L = \begin{bmatrix} u & v \end{bmatrix}$ and $\mathscr{L} = \begin{bmatrix} \alpha_1 & \alpha_3 \\ \alpha_3 & \alpha_2 \end{bmatrix}$ is a rank two symmetric matrix. Here $u, v \in \mathbb{R}^n$, with $||u|| = ||v|| = 1$.

Theorem 7.5 *If $X = L\mathscr{L}L^T$ (where L is $n \times 2$ and $\mathscr{L}$ is 2×2) such that two columns of L are linearly independent, then $Ric(X)$ is definite only if column span of L is a A_0^T−invariant subspace.*

Proof See Theorem 4 of [2]. ■

This theorem holds for general rank k perturbations also and proof runs along similar lines. From the above theorem, it is clear that to find solutions of $\mathrm{Ric}(X) = 0$ or $\mathrm{Ric}(X) \leq 0$ where $X = L\mathscr{L}L^T$, column span of L must be A_0^T−invariant.

7.2.2 Complex Eigenvalues of A_0

Let $p_{A_0}(x) \in \mathbb{R}[x]$ be the characteristic polynomial of A_0. Every two dimensional A_0^T−invariant subspace has a minimal polynomial given by a degree two polynomial which is a factor of $p_{A_0}(x)$. Consider the case when A_0 has a pair of complex conjugate eigenvalues $\lambda \pm i\mu$. Let $v = v_1 + iv_2$ be the complex eigenvector of A_0^T for $\lambda + i\mu$. Therefore, $A_0^T v_1 = \lambda v_1 - \mu v_2$ and $A_0^T v_2 = \mu v_1 + \lambda v_2$. Thus v_1, v_2 span a two-dimensional A_0^T-invariant subspace whose minimal polynomial is a degree two irreducible factor of $p_{A_0}(x)$. Let $X = L\mathscr{L}L^T$ where L is $n \times 2$ matrix having columns v_1 and v_2 and $\mathscr{L}$ is 2×2 symmetric matrix.

Theorem 7.6 *Let $X = L\mathscr{L}L^T$, where $\mathscr{L}$ is a symmetric (2×2) matrix of rank 2 and the two columns of L are the real and imaginary parts of a complex eigenvector corresponding to an eigenvalue $\lambda + i\mu$. Further assume $\lambda \neq 0$ and the eigenvalues $\lambda \pm i\mu$ are controllable. Then $Ric(X) = 0$ has a unique rank two solution of the given form. Further, this rank two solution is definite.*

Proof See Theorem 5 of [2]. ■

Consider $X = L\mathscr{L}L^T$, where columns of L are obtained from real and imaginary parts of an eigenvector of A_0^T corresponding to a complex eigenvalue $\lambda + i\mu$. Since column span of L does not contain any real one-dimensional A_0^T-invariant subspace, there are no rank one solutions of $\mathrm{Ric}(X) \leq 0$ where X has the structure specified above. For the above case, note that $\mathrm{Ric}\ (X) = L(-D\mathscr{L} - \mathscr{L}D^T + \mathscr{L}M\mathscr{L})L^T$ where $D = \begin{bmatrix} \lambda & \mu \\ -\mu & \lambda \end{bmatrix}$ and $M = L^T BB^T L$. $\mathrm{Ric}(X) \leq 0$ iff $-D\mathscr{L} - \mathscr{L}D^T + \mathscr{L}M\mathscr{L} \leq 0$. We now state a lemma applicable to this specific case where D has the structure given above and $M \geq 0$.

Lemma 7.7 *Let $\lambda > 0$. If $\mathscr{L}_1$ is a rank two solution of ARI: $-D\mathscr{L} - \mathscr{L}D^T + \mathscr{L}M\mathscr{L} \leq 0$ and $\mathscr{L}^*$ is the rank two solution of ARE: $-D\mathscr{L} - \mathscr{L}D^T + \mathscr{L}M\mathscr{L} = 0$, then $0 < \mathscr{L}_1 < \mathscr{L}^*$ and $\mathscr{L}^* - \mathscr{L}_1$ is rank two positive definite matrix.*

In the above lemma, if $M = 0$, the ARE becomes: $-D\mathscr{L} - \mathscr{L}D^T = 0$ which has only one solution, namely $\mathscr{L} = 0$. The corresponding ARI becomes $-D\mathscr{L} - \mathscr{L}D^T \leq 0$. When $\lambda > 0$, for every $C \leq 0$, the linear equation $-D\mathscr{L} - \mathscr{L}D^T = C$ has a solution $\hat{\mathscr{L}} > 0$. Further if $\alpha > 0$, then $\mathscr{L} = \alpha\hat{\mathscr{L}}$ also satisfies $-D\mathscr{L} - \mathscr{L}D^T \leq 0$.

Remark 7.8 If real part of the complex conjugate eigenvalues is strictly less than zero and $M \neq 0$, using arguments similar to those used in Lemma 7.7, one can show that if $\mathcal{L}_1$ is a rank two solution of ARI: $-D\mathcal{L} - \mathcal{L}D^T + \mathcal{L}M\mathcal{L} \leq 0$ and $\mathcal{L}^*$ is the rank two solution of ARE: $-D\mathcal{L} - \mathcal{L}D^T + \mathcal{L}M\mathcal{L} = 0$, then $0 > \mathcal{L}_1 > \mathcal{L}^*$ and $\mathcal{L}^* - \mathcal{L}_1$ is rank two negative definite matrix. Further, if real part of the complex conjugate eigenvalues is strictly less than zero and $M = 0$, then one can show that if $\hat{\mathcal{L}}$ is a rank two solution of ARI: $-D\mathcal{L} - \mathcal{L}D^T \leq 0$, then $0 > \hat{\mathcal{L}}$.

Now we consider the case when A_0 has purely imaginary eigenvalues $\pm i\mu$.

Theorem 7.9 *If $X = L\mathcal{L}L^T$ where columns of L form the two-dimensional A_0^T-invariant subspace associated with a complex conjugate pair of purely imaginary eigenvalues $\pm i\mu$ of A_0^T, then both $Ric(X) = 0$ and $Ric(X) \leq 0$ are not satisfied for any nonzero X of the given form.*

We, therefore, conclude

Theorem 7.10 *If $X = L\mathcal{L}L^T$ ($\mathcal{L}$ is 2×2 matrix of rank 2) where two columns of L form A_0^T-invariant subspace corresponding to a pair of complex conjugate eigenvalues $(\lambda \pm i\mu)$ which are not purely imaginary, then*

- *if the pair of complex conjugate eigenvalues are controllable, then rank two solutions of ARI: $Ric(X) \leq 0$ of the form $X = L\mathcal{L}L^T$ are bounded.*
- *if the pair of complex conjugate eigenvalues are uncontrollable, then rank two solutions of ARI: $Ric(X) \leq 0$ of the form $X = L\mathcal{L}L^T$ are unbounded.*

Note that when complex eigenvalues were controllable and not purely imaginary, rank 2 solutions of $\text{Ric}(X) \leq 0$ along the A_0^T-invariant subspaces corresponding to a pair of complex conjugate eigenvalues lie in the matrix interval $(0, \mathcal{L}^*]$ or $[\mathcal{L}^*, 0)$ depending on the sign of the real part of the complex eigenvalue pair. ($\mathcal{L}^*$ is the rank two solution of the reduced 2×2 algebraic Riccati equation.) As the real part of this complex conjugate pair tends to zero, the corresponding matrix interval of solutions collapse to a single point which is the zero solution and there are no nonzero rank two solutions of $\text{Ric}(X) \leq 0$ along these A_0^T-invariant subspaces corresponding to a pair of purely imaginary eigenvalues. Recall that for real eigenvalues that are controllable, rank one solutions of $\text{Ric}(X) \leq 0$ along the A_0^T-invariant subspace corresponding to the eigenvalue are in one-to-one correspondence with numbers in the interval $(0, (2\lambda/\beta^2)]$ or $[(2\lambda/\beta^2), 0)$. As the real eigenvalue tends to zero, this interval collapses to a single point and there are no nonzero solution in the direction corresponding to the eigenvector of the eigenvalue 0. If a real eigenvalue which is uncontrollable tends to zero, then rank one solutions of $\text{Ric}(X) = 0$ and $\text{Ric}(X) \leq 0$ which were unbounded from one side, becomes unbounded from above and below. However, if the real part of complex conjugate pair of eigenvalues which are uncontrollable tends to zero, then there are no nonzero solutions of $\text{Ric}(X) = 0$ and $\text{Ric}(X) \leq 0$ along the direction of A_0^T–invariant subspace corresponding to this pair of purely imaginary eigenvalues. Thus the set of solutions collapse from an unbounded set to just the zero solution.

We have now characterized all solutions of $\text{Ric}(X) = 0$ and $\text{Ric}(X) \le 0$ that can arise from a two-dimensional A_0^T-invariant subspace associated to a pair of complex conjugate eigenvalues.

7.2.3 Real Eigenvalues of A_0

We now consider a two-dimensional subspace spanned by two independent eigenvectors/generalized eigenvectors of A_0^T, say v_1 and v_2 corresponding to real eigenvalues λ_1 and λ_2, respectively, such that $\lambda_1 + \lambda_2 \neq 0$. Let $X = L\mathscr{L}L^T$ where v_1, v_2 form columns of L and $\mathscr{L}$ is a rank two 2×2 symmetric matrix.

Theorem 7.11 *Let $X = L\mathscr{L}L^T$ ($\mathscr{L}$ is 2×2 with rank 2) where the columns of L span a A_0^T-invariant subspace. Further assume that the two columns of L are linearly independent (generalized) eigenvectors of A_0^T corresponding to a pair of controllable modes associated to nonzero real eigenvalues λ_1, λ_2 such that $\lambda_1 + \lambda_2 \neq 0$. Then $Ric(X) = 0$ has a unique rank two solution of the given form.*

Proof See Theorem 7 of [2]. ■

Next we consider the case when A_0 has eigenvalues λ and $-\lambda$.

Theorem 7.12 *If $X = L\mathscr{L}L^T$ ($\mathscr{L}$ is 2×2 with rank 2) where two columns of L form linearly independent eigenvectors of A_0^T corresponding to a pair of controllable modes associated to nonzero real eigenvalues λ_1, λ_2 such that $\lambda_1 + \lambda_2 = 0$, then $Ric(X) = 0$ has either (a) no rank two solution of the given form or (b) infinite rank two solutions of the given form.*

Proof See Theorem 8 of [2]. ■

We demonstrate the above result with an example.

Example 7.13 Let

$$A_0 = \begin{bmatrix} -1 & 0 \\ 0 & 1 \end{bmatrix}, B = \begin{bmatrix} 1 \\ 1 \end{bmatrix}$$

Clearly $A_0 = A_0^T = D$ has eigenvalues $-1, 1$. $M = BB^T = \begin{bmatrix} 1 & 1 \\ 1 & 1 \end{bmatrix}$ (since we can choose L as the identity matrix). One wants to find rank two solutions for this simplified ARE: $-D\mathscr{L} - \mathscr{L}D^T + \mathscr{L}M\mathscr{L} = 0$. If such a solution exists, then its inverse Y satisfies the Lyapunov equation $YD + D^TY = M$. Let $Y = [y_{ij}]$. Clearly, $(-1+1)y_{12} = m_{12}$. Since $m_{12} \neq 0$, Lyapunov equation $YD + D^TY = M$ has no solution. Thus, there are no rank two solutions of $\text{Ric}(X) = 0$ in this case.

Now consider a system with same A_0 as above but $B = \begin{bmatrix} 1 & 0 \\ 0 & 1 \end{bmatrix}$. Therefore, $M = BB^T$ is also the identity matrix. Since $m_{12} = m_{21} = 0$, $Y = \begin{bmatrix} -1/2 & \alpha \\ \alpha & 1/2 \end{bmatrix}$ (for any $\alpha \in \mathbb{R}$) satisfies the Lyapunov equation $YD + D^T Y = M$ (where $D = A_0$). Thus one obtains several Ys that are invertible and their inverses give infinitely many rank two solutions of $\mathrm{Ric}(X) = 0$.

Next we consider a special case when both real eigenvalues of A_0 are either positive or negative and both of them are controllable. Without loss of generality, we assume that both eigenvalues are positive. If both eigenvalues of A_0 are positive, then rank two solution of $\mathrm{Ric}(X) = 0$ is positive definite (this follows from the solution of Lyapunov equation $-YA_0^T - A_0Y + BB^T = 0$ being positive definite if eigenvalues of A_0 lie in open right half plane [23]). If both the eigenvalues are negative, then the rank two solution is negative definite. We denote by D_J an upper triangular matrix in Jordan canonical form. We now state a couple of lemmas without their proofs.

Lemma 7.14 *Suppose both eigenvalues of D_J are real and positive and $\mathscr{L}^*$ be a rank two solution of simplified ARE $-D_J\mathscr{L} - \mathscr{L}D_J^T + \mathscr{L}M\mathscr{L} = 0$. If $\mathscr{L}_1$ is a rank one solution, of this simplified ARE, then $\mathscr{L}_1 \leq \mathscr{L}^*$.*

Lemma 7.15 *Let D_J be a 2×2 matrix in Jordan canonical form such that both eigenvalues of D_J are positive. Let the unique rank 2 solution of simplified ARE: $-D_J\mathscr{L} - \mathscr{L}D_J^T + \mathscr{L}M\mathscr{L} = 0$ be denoted by $\mathscr{L}^*$. Then every solution $\hat{\mathscr{L}}$ of the simplified ARI: $-D_J\mathscr{L} - \mathscr{L}D_J^T + \mathscr{L}M\mathscr{L} \leq 0$ is such that $0 \leq \hat{\mathscr{L}} \leq \mathscr{L}^*$.*

Consider the case when D has eigenvalues λ_1 and $-\lambda_2$ such that both are controllable ($\lambda_1, \lambda_2 > 0$). Note that we may have $\lambda_1 = \lambda_2$. By Theorem 7.12, there may not be a rank two solution or there could be infinitely many rank two solutions of $\mathrm{Ric}(X) = 0$. Let $\mathscr{L}_r^*$ be a rank one solution of ARE $-D\mathscr{L} - \mathscr{L}D^T + \mathscr{L}M\mathscr{L} = 0$ corresponding eigenvector associated with positive eigenvalue λ_1 of D and $\mathscr{L}_\ell^*$ be a rank one solution corresponding eigenvector associated with negative eigenvalue $-\lambda_2$ of D such that $\mathscr{L}_r^* = \mathrm{diag}(\mathscr{L}_r, 0)$ and $\mathscr{L}_\ell^* = \mathrm{diag}(0, \mathscr{L}_\ell)$. Note that $\mathscr{L}_r = 2\lambda_1/(m_{11})$ and $\mathscr{L}_\ell = -2\lambda_2/(m_{22})$.

Theorem 7.16 *Consider $D = \mathrm{diag}(\lambda_1, -\lambda_2)$ (where $\lambda_1, \lambda_2 > 0$ and they may or may not be distinct). Let $\mathscr{L}_r^*$ and $\mathscr{L}_\ell^*$ be the solutions of ARE as stated above. Then every solution $\hat{\mathscr{L}}$ of the simplified ARI: $-D\mathscr{L} - \mathscr{L}D^T + \mathscr{L}M\mathscr{L} \leq 0$ is such that $\mathscr{L}_\ell^* \leq \hat{\mathscr{L}} \leq \mathscr{L}_r^*$.*

Finally, if either one of the eigenvalues is uncontrollable, then from Theorem 7.4, rank one solutions become unbounded. If both the eigenvalues are uncontrollable, then $M = 0$ and rank two solutions become unbounded. By Theorem 9 of [2], if λ is a repeated eigenvalue of A_0 with nontrivial Jordan form such that (A_0, B) is partially controllable, then $\mathrm{Ric}(X) = 0$ has no nontrivial solution. On the other hand, the set of rank one solutions corresponding to an eigenvector of the uncontrollable eigenvalue become unbounded. Now suppose A_0 has two eigenvalues λ and $-\lambda$ such that one is controllable and other is uncontrollable. By Theorem 10 of [2], $\mathrm{Ric}(X) = 0$ has infinitely many rank two solutions and this forms an unbounded set. Therefore, rank two solutions of $\mathrm{Ric}(X) \leq 0$ also form an unbounded set. Set of rank one solutions corresponding to eigenvector of the uncontrollable eigenvalue also forms an unbounded set.

In this section, we have enumerated nearly all situations that give rise to either rank one or rank two solutions of $\mathrm{Ric}(X) = 0$ and $\mathrm{Ric}(X) \leq 0$. As it turns out, this is all that is required to completely understand all the solutions of $\mathrm{Ric}(X) = 0$ and $\mathrm{Ric}(X) \leq 0$. These rank one and rank two solutions are like building blocks and all the other solutions can be build up from them. We, therefore, now look for higher rank solutions of $\mathrm{Ric}(X) = 0$.

7.3 Solutions of ARE, ARI of General Rank

Any rank k solution of $\mathrm{Ric}(X) = 0$ and $\mathrm{Ric}(X) \leq 0$ can be written as $X = L\mathscr{L}L^T$ where L is $n \times k$ and $\mathscr{L}$ is $k \times k$ symmetric matrix. Using arguments from Theorem 7.5, if columns of L do not form an A_0^T-invariant subspace, then $\mathrm{Ric}(X)$ is indefinite. Therefore, to get rank k solutions of the ARE or the ARI, columns of L must span an A_0^T-invariant subspace. When A_0^T is diagonalizable, without loss of generality we can take columns of L as eigenvectors of A_0^T. When A_0 has complex eigenvalues, then one takes the real and imaginary parts of the complex eigenvector of A_0^T as the columns of L. For the more general case of repeated eigenvalues in a Jordan block, one takes generalized eigenvectors of A_0^T as the columns of L. Using $X = L\mathscr{L}L^T$ (where $\mathscr{L}$ is $k \times k$ matrix), $\mathrm{Ric}(X)$ is reduced to expression $L(-D_J\mathscr{L} - \mathscr{L}D_J^T + \mathscr{L}M\mathscr{L})L^T$ where $M = L^T BB^T L$ and D_J is the Jordan form associated with the A_0^T-invariant subspace.

Theorem 7.17 *If A_0 has a zero eigenvalue which is controllable, then the corresponding eigenvectors/generalized eigenvectors do not correspond to any nonzero solution of $\mathrm{Ric}(X) \leq 0$.*

Corollary 7.18 *If A_0 has a zero eigenvalue which is uncontrollable, then $X = L_{uc}\mathscr{L}L_{uc}^T$ is a solution of $\mathrm{Ric}(X) \leq 0$ for any symmetric $\mathscr{L}$, where columns of L_{uc} are eigenvectors of A_0^T associated with the uncontrollable modes of the zero eigenvalue.*

If A_0 has purely imaginary eigenvalues with nontrivial Jordan structure, then using Theorem 7.9, one can show that the invariant subspace corresponding to these purely imaginary eigenvalues of A_0^T does not support the nonzero solutions of $\mathrm{Ric}(X) \leq 0$. We, therefore, consider the case when A_0 has only nonzero eigenvalues that are not purely imaginary.

7.3.1 Maximal Rank Solutions of $Ric(X) = 0$

Using $X = L\mathscr{L}L^T$ (where columns of L are eigenvectors/generalized eigenvectors of A_0^T corresponding to real eigenvalues and real and imaginary part of the complex eigenvectors/generalized eigenvectors associated with complex eigenvalues), the problem reduces to the simplified ARE: $-D_J\mathscr{L} - \mathscr{L}D_J^T + \mathscr{L}M\mathscr{L} = 0$. By assumption, all eigenvalues of D_J are nonzero and not purely imaginary.

Theorem 7.19 *If* (A_0, B) *is controllable and* A_0 *has eigenvalues* λ_i $(1 \leq i \leq n)$ *such that* $\lambda_i + \lambda_j \neq 0$ *for all* $1 \leq i, j \leq n$ *, then* $Ric(X) = 0$ *has a unique rank* n *solution.*

Proof See Theorem 12 of [2]. ■

Putting together the results so far in this article, one can conclude that a unique full rank solution exists for $\text{Ric}(X) = 0$ whenever all the eigenvalues of A_0 are nonzero, controllable and the sum of any two eigenvalues is never equal to zero. It has been demonstrated in Theorem 7.17 that a controllable zero eigenvalue of A_0 results in rank deficient solutions of $\text{Ric}(X) = 0$. On the other hand, Corollary 7.18 shows that an uncontrollable zero eigenvalue of A_0 does not hinder the existence of full rank solutions for $\text{Ric}(X) = 0$, but then uniqueness is lost. Theorem 7.19 demonstrates the conditions for existence of a unique full rank solution for $\text{Ric}(X) = 0$ when all the eigenvalues are controllable. Finally, for the case when the sum of two nonzero eigenvalues of A_0 add up to zero, a full rank solution may or may not exist. If a full rank solution does exist, then there are an infinite number of such full rank solutions.

One can isolate several special cases where all the conditions listed above are satisfied. For example, if one considers all the eigenvalues of A_0 to be controllable and lying in the open right half complex plane (i.e., having real parts that are strictly positive), then $\text{Ric}(X) = 0$ has a unique full rank solution. In this case, one can in fact show that this full rank solution of $\text{Ric}(X) = 0$ is positive definite.

On similar lines, one can also conclude that if all eigenvalues of A_0 are controllable and lies in the open left half plane, then $\text{Ric}(X) = 0$ has a full rank solution which is negative definite.

On the other hand, if (A_0, B) is not controllable, then one can divide the eigenvalues into two sets: those eigenvalues which are controllable (denoted by $\text{Spec}(A_0)_c$) and those eigenvalues which are not controllable (denoted by $\text{Spec}(A_0)_{uc}$). If the controllable subspace is k-dimensional, then one can guarantee a rank k solution, provided $\lambda_i + \lambda_j \neq 0$ for all $\lambda_i, \lambda_j \in \text{Spec}(A_0)_c$. If this condition is not satisfied, then either no rank k solution exists or infinitely many rank k solutions exist. As for the uncontrollable part, if all the eigenvalues are nonzero, then no nontrivial solution of $\text{Ric}(X) = 0$ comes from the A_0^T-invariant subspace corresponding to the uncontrollable eigenvalues in general. The only exception to this rule arises out of a very special situation—if an eigenvalue $\lambda \in \text{Spec}(A_0)_c$ and an eigenvalue $-\lambda \in \text{Spec}(A_0)_{uc}$, then one gets solutions whose rank is greater than k (see [2]). Finally, if $0 \in \text{Spec}(A_0)_{uc}$, then again one gets solutions whose rank is greater than k. In all these cases, where the uncontrollable eigenspaces contribute nontrivially to solutions of $\mathrm{Ric}(X) = 0$, the uniqueness of maximal rank solution is lost.

7.3.2 Information Content in a Maximal Rank Solution of $Ric(X) = 0$

Assume (A_0, B) is controllable and a unique full rank solution of $\mathrm{Ric}(X) = 0$ exists. From our earlier discussions, it is clear that this full rank solution of $\mathrm{Ric}(X) = 0$ has the form $X = L\mathscr{L}^* L^T$ where L is a real $n \times n$ matrix whose columns are either the (generalized) eigenvectors of A_0^T corresponding to real eigenvalues or the real and imaginary parts of complex (generalized) eigenvectors of A_0^T corresponding to complex eigenvalues. The matrix $\mathscr{L}^*$ is the solution of the simplified ARE $-D_J\mathscr{L} - \mathscr{L}D_J^T + \mathscr{L}M\mathscr{L} = 0$. The matrix $M = L^T BB^T L$ and the matrix D_J is a block diagonal Jordan matrix. Note that the eigenvalues of A_0 are such that $\lambda_i + \lambda_j \neq 0$ for all eigenvalues $\lambda_i, \lambda_j \in \mathrm{Spec}(A_0)$.

In order to find a rank k solution of $\mathrm{Ric}(X) = 0$ (where $k < n$), we take $X = L_k\mathscr{L}L_k^T$ where columns of L_k are either (generalized) eigenvectors of A_0^T corresponding to real eigenvalues or the real and complex parts of complex (generalized) eigenvectors of A_0^T corresponding to complex eigenvalues. In other words, L_k is a $n \times k$ submatrix of L. Therefore, $A_0^T L_k = L_k D_k$ where D_k is the corresponding $k \times k$ submatrix of D_J. Then the ARE: $-A_0^T X - XA_0 + XBB^T X = 0$ becomes

$$\begin{aligned} -A_0^T X - XA_0 + XBB^T X &= -A_0^T L_k\mathscr{L}L_k^T - L_k\mathscr{L}L_k^T A_0 + L_k\mathscr{L}L_k^T BB^T L_k\mathscr{L}L_k^T \\ &= -L_k D_k\mathscr{L}L_k^T - L_k\mathscr{L}D_k^T L^T + L_k\mathscr{L}M_k\mathscr{L}L_k^T \\ &= L_k(-D_k\mathscr{L} - \mathscr{L}D_k^T + \mathscr{L}M_k\mathscr{L})L_k^T \\ &\quad \text{where } M_k = L_k^T BB^T L_k \end{aligned}$$

Note that M_k is the appropriate $k \times k$ submatrix of the original matrix M. Further observe that the rank k solution is obtained by solving the simplified ARE: $-D_k\mathscr{L} - \mathscr{L}D_k^T + \mathscr{L}M_k\mathscr{L} = 0$ which is a chopped up version of the original full rank simplified ARE $-D_J\mathscr{L} - \mathscr{L}D_J^T + \mathscr{L}M\mathscr{L} = 0$. Assuming that $\mathscr{L}$ has rank k, we can pre- and post-multiply the chopped up version of the ARE by $Y = \mathscr{L}^{-1}$ thereby obtaining the linear equation

$$-YD_k - D_k^T Y + M_k = 0 \tag{7.2}$$

This Lyapunov equation has a unique solution Y_k (note that our assumption in this section of controllability and $\lambda_i + \lambda_j \neq 0$ where $\lambda_i, \lambda_j \in \mathrm{Spec}(A_0)$ is necessary for this conclusion). From Y_k, one obtains a rank k solution $\mathscr{L}_k = (Y_k)^{-1}$ of the chopped up ARE: $-D_k\mathscr{L} - \mathscr{L}D_k^T + \mathscr{L}M_k\mathscr{L} = 0$. One can show that such a Y_k is invertible. But clearly, this rank k solution of $\mathrm{Ric}(X) = 0$ is related to the full rank solution of $\mathrm{Ric}(X) = 0$ as the governing equation of the former is a chopped up version of the latter. We now bring out this relationship between the various solutions of $\mathrm{Ric}(X) = 0$.

Theorem 7.20 *Let (A_0, B) be controllable. If A_0 has real and distinct eigenvalues $\lambda_1, \ldots, \lambda_n$ such that $\lambda_i + \lambda_j \neq 0$ (for $1 \leq i, j \leq n$), then the lower rank $2^n - 2$ nonzero solutions of simplified ARE: $-D\mathscr{L} - \mathscr{L}D + \mathscr{L}M\mathscr{L} = 0$ are obtained from Schur complements of all the $2^n - 2$ strict principal submatrices of $\mathscr{L}^*$, the unique full rank solution of the simplified ARE.*

Proof See [2]. ■

Note that from the above theorem, if one obtains $\mathscr{L}$ of rank k that satisfies simplified ARE: $-D\mathscr{L} - \mathscr{L}D + \mathscr{L}M\mathscr{L} = 0$, then $X = L\mathscr{L}L^T$ gives a rank k solution of $\mathrm{Ric}(X) = 0$. Thus, the above theorem states that the full rank solution of $-D\mathscr{L} - \mathscr{L}D + \mathscr{L}M\mathscr{L} = 0$ has all the other solutions encoded within it. It is enough to find the full rank solution and all other solutions can be read off from this solution.

Example 7.21

$$A = \begin{bmatrix} 1 & 1 & 1 & 1 \\ 2 & 4 & 5 & 1 \\ 0 & 2 & 4 & 0 \\ 0 & 0 & 5 & 4 \end{bmatrix}, \quad B = \begin{bmatrix} 1 & -1 \\ 1 & 1 \\ 1 & 1 \\ 1 & 2 \end{bmatrix}, \quad Q = \begin{bmatrix} 4 & 1 & 1 & 1 \\ 1 & 5 & 1 & 1 \\ 1 & 1 & 2 & 1 \\ 1 & 1 & 1 & 5 \end{bmatrix},$$

Fixing K_0 (a solution of the ARE: $-A^TK - KA - Q + KBB^TK = 0$), we get $A_0 = A - BB^TK_0$ with eigenvalues $\{10.4939, 3.8178, 2.6097, 1.5784\}$. Forming the matrix L using eigenvectors of A_0^T, the Riccati equation gets modified into diagonal form $-D\mathscr{L} - \mathscr{L}D + \mathscr{L}M\mathscr{L} = 0$ with

$$M = L^TBB^TL = \begin{bmatrix} 10.7996 & 6.8110 & 5.8111 & 1.3180 \\ 6.8110 & 26.5759 & 39.5777 & -19.1530 \\ 5.8111 & 39.5777 & 61.0133 & -31.5026 \\ 1.3180 & -19.1530 & -31.5026 & 18.0857 \end{bmatrix}$$

Let Y^* be the solution of Lyapunov equation $-Y^*D - DY^* + M = 0$. Full rank solution of $-D\mathscr{L} - \mathscr{L}D + \mathscr{L}M\mathscr{L} = 0$ is given by $\mathscr{L}^* = (Y^*)^{-1}$

$$\mathscr{L}^* = \begin{bmatrix} 4.0715 & -5.3083 & 3.0651 & 0.6581 \\ -5.3083 & 28.9839 & -22.2105 & -11.1036 \\ 3.0651 & -22.2105 & 17.8956 & 9.6775 \\ 0.6581 & -11.1036 & 9.6775 & 5.9889 \end{bmatrix},$$

For the (1, 1) principal submatrix of $\mathcal{L}^*$, one obtains a rank one solution by taking appropriate Schur complement to get $\mathcal{L}_1 = \begin{bmatrix} 1.9434 & 0 & 0 & 0 \\ 0 & 0 & 0 & 0 \\ 0 & 0 & 0 & 0 \\ 0 & 0 & 0 & 0 \end{bmatrix}$. Similarly, for the leading 2×2 principal submatrix, taking the appropriate Schur complement one gets $\mathcal{L}_{12} = \begin{bmatrix} 2.2247 & -0.3042 & 0 & 0 \\ -0.3042 & 0.3289 & 0 & 0 \\ 0 & 0 & 0 & 0 \\ 0 & 0 & 0 & 0 \end{bmatrix}$. A rank one solution of $\text{Ric}(X) = 0$ is given by $X_1 = L\mathcal{L}_1 L^T$. Similarly a rank two solution is given by $X_{12} = L\mathcal{L}_{12}L^T$. Full rank solution is given by $X^* = L\mathcal{L}^* L^T$. Thus, all the 14 lower rank nonzero solutions can be obtained by taking appropriate Schur complements of the full rank solution $\mathcal{L}^*$. Finally, all 16 solutions of the ARE are obtained as $K_0 + X$, where X is a solution of $\text{Ric}(X) = 0$.

Theorem 7.20 can be generalized to include A_0 that have complex eigenvalues. In this case too, the maximal rank solution encodes all the lower rank solutions within it. The only difference from Theorem 7.20 is that the Schur complements of all principal submatrices are not admissible in this case. We now give a corollary and an example that conveys the general layout of these results.

Corollary 7.22 *Let (A_0, B) be controllable. Let A_0 have nonzero distinct eigenvalues $\lambda_1, \ldots, \lambda_n$ such that $\lambda_i + \lambda_j \neq 0$ for all $1 \leq i, j \leq n$. Let $\lambda \pm i\mu$ be a complex conjugate pair of eigenvalues of A_0. Then the Schur complement of the principal submatrix of $\mathcal{L}^*$ associated with all the other eigenvalues of A_0, determines a rank two solution of the simplified ARE $-D\mathcal{L} - \mathcal{L}D^T + \mathcal{L}M\mathcal{L} = 0$.*

Example 7.23

$$A = \begin{bmatrix} 1 & 1 & 0 \\ 0 & 1 & 1 \\ 0 & 0 & 1 \end{bmatrix}, B = \begin{bmatrix} 0 \\ 0 \\ 1 \end{bmatrix}, Q = \begin{bmatrix} 1 & 0 & 0 \\ 0 & 1 & 0 \\ 0 & 0 & 1 \end{bmatrix},$$

Fixing K_0 (a solution of the ARE: $-A^T K - KA - Q + KBB^T K = 0$) one gets $A_0 = A - BB^T K_0$ with eigenvalues $\{1.4142, 1.0987 + i0.4551, 1.0987 - i0.4551\}$. Let L be a matrix whose first column is an eigenvector of A_0^T corresponding to the eigenvalue 1.4142 whereas the second and third columns of L are the real and imaginary parts respectively of the complex eigenvector of A_0^T corresponding to the eigenvalue $1.0987 + i0.4551$.

$$M = L^T BB^T L = \begin{bmatrix} 14.6489 & 15.7125 & 10.8940 \\ 15.7125 & 16.8533 & 11.6850 \\ 10.8940 & 11.6850 & 8.1016 \end{bmatrix}$$

Let Y^* be the solution of Lyapunov equation $-YD - D^T Y + M = 0$. Rank 3 solution $\mathscr{L}^*$ is given by $\mathscr{L}^* = (Y^*)^{-1}$.

$$\mathscr{L}^* = \begin{bmatrix} 87.3255 & -60.2755 & -24.9669 \\ -60.2755 & 42.8452 & 16.1228 \\ -24.9669 & 16.1228 & 8.3025 \end{bmatrix}$$

From the Schur complement of lower 2×2 principal submatrix of $\mathscr{L}^*$, one obtains $\mathscr{L}_1 = \begin{bmatrix} 0.1931 & 0 & 0 \\ 0 & 0 & 0 \\ 0 & 0 & 0 \end{bmatrix}$, and from the Schur complement of the $(1, 1)$ principal submatrix, one obtains

$\mathscr{L}_{23} = \begin{bmatrix} 0 & 0 & 0 \\ 0 & 1.2407 & -1.1103 \\ 0 & -1.1103 & 1.1643 \end{bmatrix}$. From these solutions, one constructs the other three solutions of the original Riccati equation.

Now assume that A_0 has repeated eigenvalues whose algebraic multiplicity equals its geometric multiplicity. We continue to impose the condition (A_0, B) is controllable and $\lambda_i + \lambda_j \neq 0$ for $\lambda_i, \lambda_j \in \text{Spec}(A_0)$. Let columns of L be real eigenvectors of A_0^T or real/imaginary parts of complex eigenvectors of A_0^T associated with complex eigenvalues. Due to the repeated eigenvalues, the matrix L is far from unique. One can show in this case that the full rank solution of $\text{Ric}(X) = 0$ is unique. For each choice of L, one gets a unique rank n solution $\mathscr{L}^*$ and the Schur complements of appropriate $(n-k) \times (n-k)$ principal submatrices $(k < n)$ give rank k solutions of simplified ARE: $-D\mathscr{L} - \mathscr{L}D^T + \mathscr{L}M\mathscr{L} = 0$. Note that for different choices of L, the matrix M changes and therefore one gets different $\mathscr{L}^*$s. Thus, the Schur complements of the principal submatrices of the various $\mathscr{L}^*$ need not give the same solutions as the choice of columns of L were different. As a result, the number of rank k solutions, for $k < n$ need not be finite.

Corollary 7.24 *Let (A_0, B) be controllable. If A_0 has nonzero repeated eigenvalues with trivial Jordan structure such that λ and $-\lambda$ do not coexist in Spec(A_0), then the Schur complements of appropriate principal submatrices of $\mathscr{L}^*$ give lower rank nonzero solutions of $-D\mathscr{L} - \mathscr{L}D^T + \mathscr{L}M\mathscr{L} = 0$. Here $\mathscr{L}^*$ is the full rank solution of $-D\mathscr{L} - \mathscr{L}D^T + \mathscr{L}M\mathscr{L} = 0$.*

Next we consider the case of nontrivial Jordan blocks. In this case, the Schur complement of every principal submatrix of the maximal solution $\mathscr{L}^*$ need not give a solution of the equation $-D_J\mathscr{L} - \mathscr{L}D_J^T + \mathscr{L}M\mathscr{L} = 0$. If D_J has a Jordan block of size k corresponding to a real eigenvalue λ, then there are precisely $(k+1)$ choices as far as taking Schur complements are concerned. We demonstrate with an example.

Example 7.25 Consider the ARE with

$$A = \begin{bmatrix} 1 & 1 & 0 \\ 1 & 1 & 0 \\ 0 & 0 & 1 \end{bmatrix}, B = \begin{bmatrix} 2 & 0 \\ 0 & 0 \\ 0 & 1 \end{bmatrix}, Q = \begin{bmatrix} 1 & 1 & 0 \\ 1 & 5 & 0 \\ 0 & 0 & 1 \end{bmatrix},$$

Choosing a solution $K_0 = \begin{bmatrix} -0.5 & 0.5 & 0 \\ 0.5 & -2.5 & 0 \\ 0 & 0 & -0.4142 \end{bmatrix}$ of the ARE, we can form the matrix $A_0^T = \begin{bmatrix} 3 & 1 & 0 \\ -1 & 1 & 0 \\ 0 & 0 & \sqrt{2} \end{bmatrix}$. Let $X = L\mathscr{L}L^T$ where

$$L = \begin{bmatrix} 1 & 1 & 0 \\ -1 & 0 & 0 \\ 0 & 0 & 1 \end{bmatrix}$$

contains generalized eigenvectors of A_0^T. $\mathrm{Ric}(X) = 0$ is reduced to $L(-D_J\mathscr{L} - \mathscr{L}D_J^T + \mathscr{L}M\mathscr{L})L^T = 0$ where $D_J = \begin{bmatrix} 2 & 1 & 0 \\ 0 & 2 & 0 \\ 0 & 0 & \sqrt{2} \end{bmatrix}$. Observe that D_J has a Jordan block of size 2 corresponding to the eigenvalue 2 and a block of size 1 corresponding to eigenvalue $\sqrt{2}$. Thus there are $(2 + 1)(1 + 1) = 6$ solutions to $\mathrm{Ric}(X) = 0$. Observe that A_0^T has two eigenvectors which generate two rank one solutions of $\mathrm{Ric}(X) = 0$. There are two subspaces which are two-dimensional and A_0^T-invariant that generate rank two solutions. The full space generates a rank three solution and zero subspace generates the zero solution. So there are six solutions of $\mathrm{Ric}(X) = 0$.

We obtain the full rank solution from $\mathscr{L}^* = \begin{bmatrix} 10 & -12 & 0 \\ -12 & 16 & 0 \\ 0 & 0 & 2.8284 \end{bmatrix}$. Taking the Schur complements of the principal submatrix involving row and column indices 2 and 3 give us a rank one solution. Similarly, taking the Schur complement of the principal matrix involving row and column indices 1 and 2 gives another rank one solution. Similarly, the Schur complements of the 1×1 submatrices $\mathscr{L}^*(2, 2)$ and $\mathscr{L}^*(3, 3)$ gives the two rank two solutions. We give below the solution obtained by taking the Schur complement of the principal submatrix involving row and column indices 2 and 3, which turns out to be a rank one solution.

$$X_1 = L\mathscr{L}_1L^T = \begin{bmatrix} 1 & -1 & 0 \\ -1 & 1 & 0 \\ 0 & 0 & 0 \end{bmatrix}$$

The rank three solution involves all of $\mathscr{L}^*$ whereas the last solution is $X = 0$.

Remark 7.26 Note that we began with the assumption that a unique full rank solution exists for the case under consideration. There are, however, cases where λ and $-\lambda$ simultaneously belong to $\mathrm{Spec}(A_0)$, where an infinite number of full rank solutions exist. In these cases too, taking Schur complements of any full rank solution $\mathscr{L}^*$ of the equation $-D_J\mathscr{L} - \mathscr{L}D_J^T + \mathscr{L}M\mathscr{L} = 0$, one arrives at lower rank solutions of the equation.

Earlier, we had enumerated various special cases, where there is no possibility of a full rank solution of $\mathrm{Ric}(X) = 0$. For these special cases where full rank solution of the simplified ARE $-D_J\mathscr{L} - \mathscr{L}D_J^T + \mathscr{L}M\mathscr{L} = 0$ do not exist, one can still obtain useful information from those solutions (of the simplified ARE) which have the largest rank.

Throughout this subsection, we had assumed that (A_0, B) is controllable. If we relax this condition, then one needs to consider the controllable subspace. If k is the dimension of the controllable subspace, then the maximum rank of a solution of $\mathrm{Ric}(X) = 0$ one can generically expect is k, provided there are no uncontrollable modes corresponding to the zero eigenvalue and the situation of an eigenvalue being controllable and its negative being another eigenvalue which is not controllable does not arise.

In this section, a complete characterization of all solutions of $\mathrm{Ric}(X) = 0$ was obtained in terms of the unique maximal rank solution of $\mathrm{Ric}(X) = 0$. In case of multiple maximal rank solutions too, this characterization holds—but now the Schur complements of each one of the maximal solutions yield solutions of $\mathrm{Ric}(X) = 0$.

7.3.3 Higher Rank Solutions of $Ric(X) \leq 0$

The rich structure displayed by the solutions of the ARE $\mathrm{Ric}(X) = 0$ makes one expect some similar rich structure in the solutions of the ARI $\mathrm{Ric}(X) \leq 0$. As it turns out, this is indeed true. Recall that the solutions of the ARE were the endpoints of the interval that satisfied the ARI in the scalar Riccati equation. For the matrix case too, one could think of the solutions of the ARE $\mathrm{Ric}(X) = 0$ as some sort of endpoints of an interval that satisfies the ARI. Of course, this interval is a matrix interval, where matrices have to be thought of as being partially ordered by positive definiteness. By this we mean that a matrix $X_1 \geq X_2$ is defined as the matrix $X_1 - X_2$ being positive (semi)-definite. We now state some results that bring out this fact.

Lemma 7.27 *Let D_{J_k} be a $k \times k$ $(1 \leq k \leq n)$ matrix in Jordan canonical form such that $\mathrm{Spec}(D_{J_k})$ lies in the open right half complex plane. Let the unique rank k solution of simplified ARE: $-D_{J_k}\mathscr{L} - \mathscr{L}D_{J_k}^T + \mathscr{L}M_k\mathscr{L} = 0$ be denoted by $\mathscr{L}^*$. Then every solution $\hat{\mathscr{L}}$ of the simplified ARI: $-D_{J_k}\mathscr{L} - \mathscr{L}D_{J_k}^T + \mathscr{L}M_k\mathscr{L} \leq 0$ is such that $0 \leq \hat{\mathscr{L}} \leq \mathscr{L}^*$.*

Observe that if $\mathrm{Spec}(D_{J_k})$ lies in the open left half plane, then one can conclude that $0 \geq \hat{\mathscr{L}} \geq \mathscr{L}^*$, where $\mathscr{L}^*$ is the unique rank k solution of the corresponding ARE. Note that the result above implies that solutions of $\mathrm{Ric}(X) \leq 0$ satisfy $0 \leq X \leq X^*$ when A_0 has all eigenvalues in the open right half complex plane where X^* is the full rank solution of ARE. Therefore, one can obtain Willems' result $K_{min} \leq K \leq K_{max}$ [17, 24] for ARI by using $K = K_{min} + X$.

Lemma 7.28 *Let D_{J_k} be a $k \times k$ $(1 \leq k \leq n)$ matrix in Jordan form such that $Spec(D_{J_k})$ lies in the open right half complex plane. Then, all rank k solutions of simplified strict ARI: $-D_{J_k}\mathscr{L} - \mathscr{L}D_{J_k}^T + \mathscr{L}M_k\mathscr{L} < 0$ are parametrized by all $k \times k$ positive definite matrices.*

One can now combine earlier results to obtain results about the most general case of the simplified ARI: $-D_J\mathscr{L} - \mathscr{L}D_J^T + \mathscr{L}M\mathscr{L} \leq 0$. Assume that D_J has nonzero eigenvalues which are not purely imaginary. Without loss of generality, assume that $D_J = \text{diag}(D_{J_r}, D_{J_\ell})$ where D_{J_r} contains all the eigenvalues in the open right half plane and D_{J_ℓ} contains all the eigenvalues in the open left half plane. Let $\mathscr{L}^*$ be a maximal rank solution simplified ARE: $-D_J\mathscr{L} - \mathscr{L}D_J^T + \mathscr{L}M\mathscr{L} = 0$. From the earlier results, one knows that Schur complements of $\mathscr{L}^*$ with respect to modes corresponding to the left/right half complex planes, namely $\mathscr{L}_r$ and $\mathscr{L}_\ell$ respectively, are also solutions of the ARE. To be more specific, two special solutions of the ARE are $\mathscr{L}_r^* = \text{diag}(\mathscr{L}_r, 0)$ and $\mathscr{L}_\ell^* = \text{diag}(0, \mathscr{L}_\ell)$. This brings us to an important theorem.

Theorem 7.29 *Suppose (A_0, B) is controllable. Consider $D_J = diag(D_{J_r}, D_{J_\ell})$ which is a Jordan form of A_0^T. Let $\mathscr{L}^*$ be a maximal rank solution of the simplified ARE and let $\mathscr{L}_r^*$ and $\mathscr{L}_\ell^*$ be the solutions of ARE obtained by Schur complements, as stated above. Then every solution $\hat{\mathscr{L}}$ of the simplified ARI: $-D_J\mathscr{L} - \mathscr{L}D_J^T + \mathscr{L}M\mathscr{L} \leq 0$ is such that $\mathscr{L}_\ell^* \leq \hat{\mathscr{L}} \leq \mathscr{L}_r^*$.*

Example 7.30

$$A = \begin{bmatrix} 1 & 0 & 0 \\ 0 & 2 & 0 \\ 0 & 0 & -4 \end{bmatrix}, B = \begin{bmatrix} 1 \\ 1 \\ 1 \end{bmatrix}, Q = 0$$

Since A is diagonal, we can write $A = A_0 = D_J$ and $M = BB^T$. D_J has two eigenvalues in RHP and one eigenvalue in LHP. Let D_2 be the leading 2×2 principal submatrix of D_J and M_{11} be corresponding principal submatrix of M. Restricting to 2×2 case for eigenvalues in RHP and solving $-D_2\mathscr{L} - \mathscr{L}D_2^T + \mathscr{L}M_{11}\mathscr{L} = 0$ for 2×2 case, we get rank two solution $\mathscr{L}_r$ as discussed in the above theorem from which we obtain

$$\mathscr{L}_r^* = \begin{bmatrix} 18 & -24 & 0 \\ -24 & 36 & 0 \\ 0 & 0 & 0 \end{bmatrix}$$

Now similarly, corresponding to eigenvalue in LHP, we obtain rank one solution given by

$$\mathscr{L}_l^* = \begin{bmatrix} 0 & 0 & 0 \\ 0 & 0 & 0 \\ 0 & 0 & -8 \end{bmatrix}$$

Let $\mathscr{L}_1 = \begin{bmatrix} 0.72 & 0 & -1.92 \\ 0 & 0 & 0 \\ -1.92 & 0 & -2.88 \end{bmatrix}$. $\mathscr{L}_1$ satisfies $-D_J \mathscr{L}_1 - \mathscr{L}_1 D_J^T + \mathscr{L}_1 M \mathscr{L}_1 = 0$. It turns out that $\mathscr{L}_l^* \leq \mathscr{L}_1 \leq \mathscr{L}_r^*$. This is true for all the solutions of simplified ARE.

Let $\hat{\mathscr{L}} = \begin{bmatrix} 9.216 & -12 & -0.5760 \\ -12 & 18 & 0 \\ -0.5760 & 0 & -2.4640 \end{bmatrix}$ which satisfies ARI $-D_J \hat{\mathscr{L}} - \hat{\mathscr{L}} D_J^T + \hat{\mathscr{L}} M \hat{\mathscr{L}} \leq 0$. $\hat{\mathscr{L}}$ also satisfies the inequality $\mathscr{L}_l^* \leq \hat{\mathscr{L}} \leq \mathscr{L}_r^*$.

When one goes back to the equation $\mathrm{Ric}(X) \leq 0$, the above theorem translates to the existence of a maximal and a minimal solution. We now utilize this observation to generalize a well-known result from the literature. Willems [24] proved that when (A, B) is controllable, all the solutions K of ARI: $-A^T K - KA - Q + KBB^T K \leq 0$ satisfy inequality $K_{min} \leq K \leq K_{max}$. If one starts with the assumption (A, B) is controllable, fix some solution K_0 of the ARE, obtain $A_0 = A - BB^T K_0$, consider the equation $\mathrm{Ric}(X) \leq 0$ obtained using this data, then one reaches a situation where Theorem 7.29 is applicable. Thus K_{min} and K_{max} in Willems' result really comes from $\mathscr{L}_\ell^*$ and $\mathscr{L}_r^*$, respectively.

Observe that Theorem 7.29 is applicable for cases where a maximal rank solution exists for the simplified ARE. Moreover, the assumption that there are no purely imaginary eigenvalues of D_J has been imposed. If we relax this condition, then one needs to consider a block structure of D_J of the form $\mathrm{diag}(D_{J_0}, D_{J_r}, D_{J_\ell})$, where the submatrix D_{J_0} contains all the purely imaginary eigenvalues of D_J. If the purely imaginary eigenvalues are controllable, then this translates to the following: the submatrix of M corresponding to the submatrix D_{J_0} of D_J is positive semidefinite. From the earlier results (specifically Theorem 7.9), one can therefore conclude that the corresponding block of $\mathscr{L}$ (a solution of the ARI) must be zero. Thus the existence of the maximal and minimal solution for the ARI holds for the controllable case.

Willems' result is about boundedness of the solutions of ARI. We now generalize this result. First we state a condition for the solutions K being unbounded.

Theorem 7.31 *If (A, B) is not a controllable pair and A has uncontrollable eigenvalues which are not purely imaginary, then the set of solutions K of the ARI: $-A^T K - KA - Q + KBB^T K \leq 0$ is unbounded.*

Now we give a complete characterization as to when the solutions K of the ARI $-A^T K - KA - Q + KBB^T K \leq 0$ is bounded. This theorem is a generalization of Willems' result [24].

Theorem 7.32 *Consider the pair (A, B) whose uncontrollable modes correspond to nonzero and purely imaginary eigenvalues. Then the solutions K of the ARI $-A^T K - KA - Q + KBB^T K \leq 0$ satisfy the inequality $K_{min} \leq K \leq K_{max}$.*

Study of existence of solutions of ARE and ARI with uncontrollable modes on the imaginary axis has appeared in [20]. We end this section with a theorem that gives further boundedness properties of ARI for uncontrollable systems. Variants of this result has appeared in [14, 17].

Theorem 7.33 *If all uncontrollable eigenvalues lie in the open right half plane, then the solution set of* $Ric(X) \leq 0$ *is bounded from below. If all uncontrollable eigenvalues lie in open left half plane, then the solution set of* $Ric(X) \leq 0$ *is bounded from above. If uncontrollable eigenvalues lie in both half planes, then solution set of* $Ric(X) \leq 0$ *is neither bounded below nor bounded above.*

7.4 Conclusions

In this article, we have done yet another characterization of all solutions of the algebraic Riccati equation. Importantly, we have only used simple linear algebraic arguments to obtain this characterization. We homogenized the ARE/ARI into an equivalent ARE/ARI problem $\text{Ric}(X) = 0$ / $\text{Ric}(X) \leq 0$. The matrix X may then be viewed as a perturbation matrix from a particular solution of the original ARE. We then characterized all the solutions of $\text{Ric}(X) = 0$, ordering them by their rank. We demonstrated how all the solutions are in some sense build up from rank one and rank two solutions, which may be associated to the real and complex eigenvalues of a matrix related to the ARE. We obtain the characterization for both controllable and uncontrollable situations.

We provided conditions under which a unique full rank solution exists for the equation $\text{Ric}(X) = 0$. Special situations that prevent the existence of a unique full rank solution were enumerated and demonstrated. It was then demonstrated how this unique full rank solution encodes within it all lower rank solutions (under some special conditions). Even when these special conditions are not satisfied, the full rank solution does encode several low rank solutions. Further, it was shown that for the cases when a unique full rank solution of $\text{Ric}(X) = 0$ does not exist, all the maximal rank solutions encode within it information about the lower rank solutions.

In parallel, we also characterized the solutions of the ARI. We obtained a generalization of a well-known result from the literature and gave conditions on boundedness and unboundedness of solutions of the ARI.

Wishing Harry a very Happy Birthday and lots of fun in the years to come.

References

1. Bittanti, S., Laub, A.J., Willems, J.C.: The Riccati Equation. Springer, New York (1991)
2. A.S.A. Dilip, H.K. Pillai: Yet another characterization of solutions of the algebraic Riccati equation. To appear in Linear Algebra and its Applications, (April 2015). doi:10.1016/j.laa.2015.04.026
3. P. Faurre: Realisations markoviennes de processus stationnaires. Technical report 13, INRIA(LABORIA), Le Chesnay, France (1973)
4. Ferrante, A.: A parametrization of minimal stochastic realizations. IEEE Trans. Autom. Control AC **39**, 2122–2126 (1994)
5. Ferrante, A.: A homeomorphic characterization of minimal spectral factors. SIAM J. Control Optim. **35**, 1508–1523 (1997)

6. Ferrante, A., Michaletzky, G., Pavon, M.: Parametrization of all minimal square spectral factorss. Syst. Control Lett. **21**, 249–254 (1993)
7. Ferrante, A., Ntogramatzidis, L.: The generalized discrete algebraic riccati equation in linear-quadratic optimal control. Automatica **49**, 471–478 (2013)
8. Ferrante, A., Pavon, M.: The algebraic riccati inequality: parametrization of solutions, tightest local frames and general feedback matrices. Linear Algebra Appl. **292**, 187–206 (1999)
9. Helmke, U., Moore, J.: Optimization and Dynamical Systems. Springer, Berlin (1994)
10. Kailath, T.: Linear Systems. Prentice Hall Inc, New Jersey (1980)
11. Lancaster, P., Rodman, L.: Algebraic Riccati Equation. Clerendon press, Oxford (1995)
12. Lindquist, A., Michaletzky, G., Picci, G.: Zeros of spectral factors, the geometry of splitting subspaces, and the algebraic riccati inequality. SIAM J. Control Optim. **33**, 365–401 (1995)
13. Lindquist, A., Picci, G.: A geometric approach to modelling and estimation of linear stochastic systems. J. Math. Syst. Estim. Control **1**, 241–333 (1991)
14. Pal, D., Belur, M.N.: Dissipativity Of Uncontrollable Systems, Storage Functions and Lyapunov Functions. SIAM Journal on Control and Optimization **47**, 2930–2966 (2008)
15. Pavon, M.: On the parametrization of nonsquare spectral factors, In: Helmke, U., Mennicken, R., Saurer, J. (eds.) Systems and Networks: Mathematical Theory and Application, Proceedings of the International Symposium on MTNS'93, vol. 2, pp. 413–416 (1993)
16. Picci, G., Pinzoni, S.: Acausal models and balanced realizations of stationary processes. Linear Algebra Appl. **205–206**, 997–1043 (1994)
17. Scherer, C.: Solution set of the algebraic riccati equation and the algebraic riccati inequality. Linear Algebra Appl. **153**, 99–122 (1991)
18. Scherer, C.: $H_\infty-$control by state-feedback for plants with zeros on the imaginary axis. SIAM J. Control Optim. **30**, 123–142 (1992)
19. Scherer, C.: The state feedback $H_\infty-$ problem at optimality. Automatica **30**, 293–305 (1994)
20. Scherer, C.: The algebraic riccati equation and the algebraic riccati inequality for systems with uncontrollable modes on the imaginary axis. SIAM J. Matrix Anal. Appl. **16**, 1308–1327 (1995a)
21. Scherer, C.: The general nonstrict algebraic riccati inequality. Linear Algebra Appl. **219**, 1–33 (1995b)
22. Scherer, C.: The Riccati Inequality and State-space $H_\infty-$Optimal Control. Ph.D. thesis, University of Wurzburg (1995c)
23. Snyders, J., Zakai, M.: On nonnegative solutions of the equation $AD + DA' = -C^*$. SIAM J. Appl. Math. **18**, 704–714 (1970)
24. Willems, J.C.: Least squares stationary optimal control and the algebraic riccati equation. IEEE Trans. Autom. Control AC **16**, 621–634 (1971)
25. Willems, J.C., Trentelman, H.L.: On quadratic differential forms. SIAM J. Control Optim. **36**, 1703–1749 (1998)
26. Wimmer, H.K.: Unmixed Solutions of the Discrete-Time Algebraic Riccati Equation. SIAM Journal of Control and Optimization **30**, 867–878 (1992)
27. Wonham, W.: Linear Multivariable Control. Springer, Berlin (1984)

Chapter 8
Implementation of Behavioral Systems

Diego Napp and Paula Rocha

Abstract In this chapter, we study control by interconnection of a given linear differential system (the plant behavior) with a suitable controller. The problem formulations and their solutions are completely representation free, and specified only in terms of the system dynamics. A controller is a system that constrains the plant behavior through a certain set of variables. In this context, there are two main situations to be considered: either all the system variables are available for control, i.e., are control variables (full control) or only some of the variables are control variables (partial control). For systems evolving over a time domain (1D) the problems of implementability by partial (regular) interconnection are well understood. In this chapter, we study why similar results are not valid in the multidimensional (nD) case. Finally, we study two important classes of controllers, namely, canonical controllers and regular controllers.

8.1 Introduction

It is a pleasure to contribute an article in honor of Harry L. Trentelman on the occasion of his 60th birthday. The first author had the privilege of being one of his Ph.D. students and of developing a fruitful research collaboration with him over the last decade. Although she never directly collaborated with Harry, the second author has always appreciated his work, by which she was inspired in several occasions.

D. Napp (✉)
Department of Mathematics, CIDMA – Center for Research and Development in Mathematics and Applications, University of Aveiro, Campus Universitario de Santiago, 3810-193 Aveiro, Portugal
e-mail: diego@ua.pt

P. Rocha
Faculty of Engineering, Department of Electrical and Computer Engineering, SYSTEC - Research Center for Systems and Technologies, University of Porto, Porto, Portugal
e-mail: mprocha@fe.up.pt

M.N. Belur et al. (eds.), *Mathematical Control Theory II*,
Lecture Notes in Control and Information Sciences 462,
DOI 10.1007/978-3-319-21003-2_8

As the topic of our article, we have chosen an issue which is at the core of systems and control theory, namely control and, in particular, the implementation of systems in the behavioral framework. This topic goes back to the seminal contribution of J.C. Willems in [17] where the fundamental ideas of the problem were established. However, it was Harry who thoroughly investigated this issue and provided many fundamental results in this area. It is our intention to make this article an appropriate tribute to his wide ranging scientific interests and to the influence that his work had in the field of behavioral approach to systems and control theory. For this purpose, we have gathered in this chapter our results that are more connected with Harry's own research, together with some new results and insights. In order to keep the paper self-contained and to give a better idea of the kind of reasoning involved, we have included the proofs of most of those results.

A behavior, denoted by $\mathfrak{B}$, is a set of trajectories that obey certain laws described by a mathematical model. In this context, control is viewed as the ability to impose adequate additional restrictions to the variables of the behavior in order to obtain a desired overall functioning pattern. Hence, the behavioral approach proposes a new perspective to control which is based on interconnection of systems, and where no a priori input/output partition is considered [17]. The act of controlling a system is simply viewed as intersecting its behavior with a controller behavior in order to achieve a desired behavior. Thus, a general control (implementation) problem can be stated as follows: Given, a plant behavior $\mathfrak{B}$ and a control objective corresponding to a desired behavior that we want to implement $\mathscr{K}$, find a controller behavior $\mathscr{C}$, within a certain controller class, such that the behavior resulting from the interconnection of $\mathfrak{B}$ and $\mathscr{C}$, $\mathfrak{B} \cap \mathscr{C}$, coincides with $\mathscr{K}$.

Most of the literature on behavioral control is concerned with the situation in which all variables of $\mathfrak{B}$ are available for control, i.e., it is allowed to impose extra restrictions on all the variables of $\mathfrak{B}$. We refer to this situation as *full* control or *full* interconnection [10, 11, 17]. Another important case considered in the literature is when the system variables are divided into two sets: the variables that we are interested to control (called *to-be-controlled variables*) and the variables on which we are allowed to enforce restrictions (called *control variables*). This situation is known as *partial* control or *partial* interconnection [1, 4, 12, 15, 18]. In this more involved situation, although we cannot act directly upon the to-be-controlled variables, we can nevertheless influence their dynamics by imposing restrictions on the control variables.

Of particular interest is the kind of interconnection that is called *regular interconnection*. In such interconnection, the restrictions imposed on the plant by the controller are independent of the restrictions already present in the plant. These type of interconnections are closely related to the notion of feedback control in the classical state-space systems since only system inputs are restricted, as in a feedback loop [14, 17].

The first results on implementability of full control problems were obtained in [18, 20] for linear systems evolving over a time domain (1D behaviors) and in [16] for a very general class of systems. Later, results for 1D behaviors were generalized to regular partial interconnections in [1] (see also [2, 12, 19]). In the context of multidimensional systems (nD behaviors) full regular interconnections were first investigated in [14, 24] and results on the partial interconnection counterpart were first presented in [13, 15]. The case of nD behaviors constituted by compactly supported functions was investigated in [8].

The problem of implementability by regular interconnections is well understood and fully characterized for 1D behaviors in both contexts of full and partial control, see for instance [1, 10, 11, 18]. In fact, $\mathscr{K} \subset \mathfrak{B}$ is implementable by regular full interconnection if and only if $\mathfrak{B} = \mathfrak{B}^c + \mathscr{K}$ where $\mathfrak{B}^c$ is the controllable part of $\mathfrak{B}$. Moreover, in [1] the solvability of a 1D partial control problem was related to the solvability of a suitable associated full control problem involving only the to-be-controlled variables and in terms of the controllable and autonomous parts of the behavior. The situation in the nD case is somewhat more involved, and a direct characterization in terms of implementation of the to-be-controlled variables seems to be impossible. In this chapter, our aim is to reinvestigate the problem of implementability by full and partial regular interconnections of nD behaviors. More concretely, we study the role of the so-called hidden behavior and also of the controllable-autonomous decomposition.

This chapter is organized as follows: we begin by introducing some necessary background from the field of nD behaviors, centering around concepts such as controllability, autonomy, orthogonal module, etc. We conclude this section with a subsection on behaviors with two types of variables. Section 8.3 is devoted to the study of the problem of implementation by regular interconnection. We first analyze the implementation by full control to conclude the chapter by treating the more general case of implementation by partial interconnections.

8.2 Preliminaries

In order to state more precisely the questions to be considered we introduce in this section the necessary material and notation on behavioral theory for nD systems. The last subsection is concerned with the theory of behaviors with two different types of variables (the to-be-controlled variables and the control variables).

8.2.1 nD (kernel) Behaviors

In the behavioral approach to nD systems, a system or behavior is defined by a triple $(\mathscr{U}, q, \mathfrak{B})$, where $\mathscr{U}$ is the signal space or trajectory universe, $q \in \mathbb{Z}^+$ is the number of components of the system variable vector, and $\mathfrak{B} \subset \mathscr{U}^q$ is the behavior. In this chapter, we assume $\mathscr{U} = (\mathbb{C})^{\mathbb{Z}^n}$.

Since the theory of continuous linear time-invariant systems as discussed in [21] is completely analogous to that of the present chapter, the same tools and conclusions will apply in the continuous case, where $\mathscr{U}$ is the space of all infinitely often differentiable functions from $\mathbb{R}^n$ to $\mathbb{R}$, or all $\mathbb{R}$-valued distributions on $\mathbb{R}^n$. For the sake of simplicity we will however focus on the discrete case.

We call $\mathfrak{B}$ a *linear difference nD behavior* or simply *nD behavior* if it is the solution set of a system of linear, constant-coefficient partial difference equations, more precisely, if $\mathfrak{B}$ is the subset of $\mathscr{U}^q$ given by:

$$\mathfrak{B} = \ker\ R(\underline{\sigma}, \underline{\sigma}^{-1}) := \{w \in \mathscr{U}^q \mid R(\underline{\sigma}, \underline{\sigma}^{-1})w \equiv 0\}, \tag{8.1}$$

$\underline{\sigma} = (\sigma_1, \ldots, \sigma_n)$, $\underline{\sigma}^{-1} = (\sigma_1^{-1}, \ldots, \sigma_n^{-1})$, the σ_i's are the elementary nD shift operators (defined by $\sigma_i w(\underline{k}) = w(\underline{k} + e_i)$, for $\underline{k} \in \mathbb{Z}^n$, where e_i is the ith element of the canonical basis of $\mathbb{C}^n$) and $R(\underline{s}, \underline{s}^{-1}) \in \mathbb{R}^{p\times q}[\underline{s}, \underline{s}^{-1}]$ is an nD Laurent-polynomial matrix known as *representation* of $\mathfrak{B}$. If no confusion arises, given an nD Laurent-polynomial matrix $A(\underline{\sigma}, \underline{\sigma}^{-1})$, we sometimes write A instead of $A(\underline{\sigma}, \underline{\sigma}^{-1})$ and $A(\underline{s}, \underline{s}^{-1})$.

Instead of characterizing $\mathfrak{B}$ by means of a representation matrix R, it is also possible to characterize it by means of its *orthogonal module* $\mathrm{Mod}(\mathfrak{B})$, which consists of all the nD Laurent-polynomial rows $r(\underline{s}, \underline{s}^{-1}) \in \mathbb{C}^q[\underline{s}, \underline{s}^{-1}]$ such that $\mathfrak{B} \subset \ker\ r(\underline{\sigma}, \underline{\sigma}^{-1})$, and can be shown to coincide with the $\mathbb{C}[\underline{s}, \underline{s}^{-1}]$-module $\mathrm{RM}(R)$ generated by the rows of R, i.e., $\mathrm{Mod}(\mathfrak{B}) = \mathrm{RM}(R(\underline{s}, \underline{s}^{-1}))$ [21]. Note that this corresponds to the set of all (linear constant-coefficient difference) equations that are satisfied by all the elements (trajectories) of $\mathfrak{B}$.

It turns out that sums, intersections, and inclusions of behaviors can be formulated in terms of the corresponding modules.

Theorem 8.1 ([24, p. 1074]) *Let $\mathfrak{B}^1$ and $\mathfrak{B}^2$ be two behaviors. Then, $\mathfrak{B}^1 + \mathfrak{B}^2$ and $\mathfrak{B}^1 \cap \mathfrak{B}_2$ are also behaviors and*

1. $\mathrm{Mod}(\mathfrak{B}^1 + \mathfrak{B}^2) = \mathrm{Mod}(\mathfrak{B}^1) \cap \mathrm{Mod}(\mathfrak{B}^2)$.
2. $\mathrm{Mod}(\mathfrak{B}^1 \cap \mathfrak{B}^2) = \mathrm{Mod}(\mathfrak{B}^1) + \mathrm{Mod}(\mathfrak{B}^2)$.
3. $\mathfrak{B}^1 \subset \mathfrak{B}^2 \Leftrightarrow \mathrm{Mod}(\mathfrak{B}^2) \subset \mathrm{Mod}(\mathfrak{B}^1)$.

Note that part 3 in Theorem 8.1 implies that if $\mathfrak{B}^1 = \ker\ R_1 \subset \mathfrak{B}^2 = \ker\ R_2$, then there exists an L-polynomial matrix S such that $R_2 = SR_1$.

For a full column rank L-polynomial matrix $R \in \mathbb{R}^{p\times q}[\underline{s}, \underline{s}^{-1}]$ define its Laurent variety (or zeros) as

$$\mathscr{V}(R) = \{(\lambda_1, \lambda_2) \in \mathbb{C}^2 \mid \mathrm{rank}(R(\lambda_1, \lambda_2)) < \mathrm{rank}(R),\ \lambda_1\lambda_2 \neq 0\},$$

where the first rank is taken over $\mathbb{C}$ and the second one over $\mathbb{R}[\underline{s}, \underline{s}^{-1}]$. Note that $\mathscr{V}(R)$ is equal to the set of common zeros of the $q \times q$ minors of R.

Definition 8.2 A full column rank L-polynomial matrix $R \in \mathbb{R}^{p\times q}[\underline{s}, \underline{s}^{-1}]$ is said to be *right minor prime* (rMP) if $\mathscr{V}(R)$ is finite and *right zero prime* (rZP) if $\mathscr{V}(R)$ is empty. A full row rank L-polynomial matrix $R \in \mathbb{R}^{p\times q}[\underline{s}, \underline{s}^{-1}]$ is said to be left minor/zero prime (ℓMP/ℓZP) if R^T is right minor/zero prime, respectively. An

L-polynomial matrix L is called a *minimal left annihilator* (MLA) of R if $LR = 0$, and for any other L-polynomial matrix S such that $SR = 0$ we have that $S = AL$ for some L-polynomial matrix A. We define minimal right annihilators in a similar way, with the obvious adaptations.

We next review the notions of controllability and autonomy in the context of the behavioral approach.

Definition 8.3 A behavior $\mathfrak{B} \subset (\mathbb{R}^q)^{\mathbb{Z}^n}$ is said to be *controllable* if for all $z_1, z_2 \in \mathfrak{B}$ there exists $\delta > 0$ such that for all subsets $U_1, U_2 \subset \mathbb{Z}^n$ with $d(U_1, U_2) > \delta$, there exists a $z \in \mathfrak{B}$ such that $z\,|_{U_1} = z_1\,|_{U_1}$ and $z\,|_{U_2} = z_2\,|_{U_2}$.

In the above definition, $d(\cdot, \cdot)$ denotes the Euclidean metric on $\mathbb{Z}^n$ and $z\,|_U$, for some $U \subset \mathbb{Z}^n$, denotes the trajectory z restricted to the domain U.

In contrast with the one dimensional case, nD behaviors admit a stronger notion of controllability called rectifiability (also known in the literature as strong controllability). Whereas controllable behaviors are the ones that can be represented by an MLA of some L-polynomial matrix or in others words $\mathbb{C}^q[\underline{s}, \underline{s}^{-1}]/\mathrm{Mod}(\mathfrak{B})$ is torsion free, rectifiable behaviors are the ones that can be represented by ℓZP matrices, i.e., the $\mathbb{R}[\underline{s}, \underline{s}^{-1}]$-module $\mathbb{C}^q[\underline{s}, \underline{s}^{-1}]/\mathrm{Mod}(\mathfrak{B}^c)$ is free.

On the other hand, we shall say that a behavior is autonomous if it has no free variables, i.e., no "inputs". It can be shown that $\mathfrak{B} = \ker\ R$ is autonomous if and only if R has full column rank. In the 1D case, all autonomous behaviors are finite dimensional vector spaces but in the nD case this is no longer true. Whereas for 1D systems initial conditions are given in a finite number of points, nD autonomous systems are generally infinite dimensional. But even in this case the amount of information (initial conditions) necessary to generate the trajectories of an autonomous nD system may vary. Hence, given an autonomous behavior, a natural question to ask is how much information is necessary in order to fully determine the system trajectories, i.e., how large is the initial condition set. This question has been analyzed in [5, 22] by introducing the notion of autonomy degrees for behaviors.

Definition 8.4 Let $\mathfrak{B}$ be a non-zero autonomous behavior and $R \in \mathbb{R}^{p\times q}[\underline{s}, \underline{s}^{-1}]$ be an nD Laurent-polynomial matrix with full column rank such that $\mathfrak{B} = \ker\ R$. We define $autodeg(\mathfrak{B}) = n - dim\mathscr{V}(R)$ to be the *autonomy degree of* $\mathfrak{B}$. The autonomy degree of the zero behavior is defined to be ∞.

It turns out that the larger the autonomy degree, the smaller is the freedom to assign initial conditions, see [5].

Every nD behavior $\mathfrak{B}$ can be decomposed into the sum $\mathfrak{B} = \mathfrak{B}^c + \mathfrak{B}^a$, where $\mathfrak{B}^c$ is the *controllable part* of $\mathfrak{B}$ (defined as the largest controllable sub-behavior of $\mathfrak{B}$) and $\mathfrak{B}^a$ is a (non-unique) autonomous sub-behavior. This sum can be chosen to be direct for 1D behaviors, but this is not always possible for multidimensional behaviors, see [23].

8.2.2 Behaviors with Two Types of Variables

Since in this chapter we are interested in considering different types of variables in a behavior (the to-be-controlled variables and the control variables), we introduce the notation $\mathfrak{B}_{(w,c)}$ for a behavior whose variable z is partitioned into two sub-variables w and c. Partitioning the corresponding representation matrix as $[R\ M]$, we can write

$$\mathfrak{B}_{(w,c)} = \{(w, c) \in \mathcal{U}^{\mathtt{w}+\mathtt{c}} \mid R(\underline{\sigma}, \underline{\sigma}^{-1})w + M(\underline{\sigma}, \underline{\sigma}^{-1})c = 0\} = \ker\ [R\ M].$$

In the case one is only interested in analyzing the evolution of one of the sub-variables, say, w, it is useful to eliminate the other one (c) and consider the projection of the behavior $\mathfrak{B}_{(w,c)}$ into $\mathcal{U}^{\mathtt{w}}$, defined as

$$\pi_w(\mathfrak{B}_{(w,c)}) = \{w \mid \exists\, c \text{ such that } (w, c) \in \mathfrak{B}_{(w,c)}\}.$$

The elimination theorem [9] guarantees that $\pi_w(\mathfrak{B}_{(w,c)})$ is also a (kernel) behavior, for which a representation can be constructed as follows: take a minimal left annihilator (MLA) E of M. Then $\pi_w(\mathfrak{B}_{(w,c)}) = \ker\ (ER)$, see [9, Corollary 2.38].

On the other hand given a behavior $\mathfrak{B} = \ker\ R \subset \mathcal{U}^{\mathtt{w}}$ we define the lifting of $\mathfrak{B}$ into $\mathcal{U}^{\mathtt{w}+\mathtt{c}}$ as

$$\mathfrak{B}^*_{(w,c)} := \{(w, c) \in \mathcal{U}^{\mathtt{w}+\mathtt{c}} \mid\ c \text{ is free and } w \in \mathfrak{B}\}. \tag{8.2}$$

Obviously $\mathfrak{B}^*_{(w,c)} = \ker\ [R\ 0]$. Analogous definitions can be given if the roles of w and c are interchanged. For the sake of brevity, if no confusion arises, we identify $\mathfrak{B}$ and $\mathfrak{B}^*_{(w,c)}$ and denote $\mathfrak{B}_w := \pi_w(\mathfrak{B}_{(w,c)})$ and $\mathfrak{B}_c := \pi_c(\mathfrak{B}_{(w,c)})$.

Definition 8.5 Given a behavior $\mathfrak{B}_{(w,c)} \subset \mathcal{U}^{\mathtt{w}+\mathtt{c}}$ we say that c is *observable* from w if $(w, c_1), (w, c_2) \in \mathfrak{B}_{(w,c)}$ implies $c_1 = c_2$.

Usually, in control problems involving behaviors with two types of variables it is important to consider the set of variables that are not observable or *hidden* from the remaining set of variables, see [15–17]. Hence, given a behavior $\mathfrak{B}_{(w,c)}$ we shall define

$$\mathfrak{B}_{(0,c)} := \{c \in \mathcal{U}^{\mathtt{c}} \mid (0, c) \in \mathfrak{B}_{(w,c)}\},$$

as the behavior of the variables c that are not observable or *hidden* from w. Clearly, if $\mathfrak{B}_{(w,c)} = \ker\ [R\ M]$ then $\mathfrak{B}_{(0,c)} = \ker\ M$. Similarly, we define $\mathfrak{B}_{(w,0)}$ as the set of w variables that are hidden from the variables c. Taking into account that we are dealing with linear behaviors, it is not difficult to verify that c is observable from w if and only if $\mathfrak{B}_{(0,c)}$ is the zero behavior. Similarly, w is observable from c if and only if $\mathfrak{B}_{(w,0)}$ is the zero behavior.

8.3 Implementation

The behavioral approach to control rests on the basic idea that to control a system is to impose appropriate additional restrictions to its variables in order to obtain a new desired behavior. These additional restrictions are achieved by interconnecting the given system with another system called the controller. From the mathematical point of view, system interconnection corresponds to the intersection of the behavior to be controlled with the controller behavior.

Two situations have been considered in the literature. The first one is known as *full interconnection* and corresponds to the case where the controller is allowed to impose restrictions on all the system variables. The second, called *partial interconnection*, considers interconnections where one is only allowed to use some of the system variables for the purpose of interconnection.

8.3.1 Control by Regular Full Interconnection

The full interconnection of a behavior to be controlled, $\mathfrak{B} \subset \mathscr{U}^{\mathtt{W}}$, with a controller behavior, $\mathscr{C} \subset \mathscr{U}^{\mathtt{W}}$, yields a controlled behavior given by

$$\mathscr{K} = \mathfrak{B} \cap \mathscr{C}, \tag{8.3}$$

or alternatively, in module terms, by $\mathrm{Mod}(\mathscr{K}) = \mathrm{Mod}(\mathfrak{B}) + \mathrm{Mod}(\mathscr{C})$. If (8.3) holds, we say that $\mathscr{K}$ is *implementable* by full interconnection from $\mathfrak{B}$.

A particular interesting type of interconnection corresponds to the case where the restrictions imposed by the controller do not overlap with the restrictions already active for the behavior to be controlled. Recalling that the elements of the modules associated with a behavior represent the corresponding equations (or restrictions), this means, in terms of the corresponding modules that

$$\mathrm{Mod}(\mathfrak{B}) \cap \mathrm{Mod}(\mathscr{C}) = \{0\},$$

(or, equivalently, that $\mathfrak{B} + \mathscr{C} = \mathscr{U}^{\mathtt{W}}$) and therefore

$$\mathrm{Mod}(\mathscr{K}) = \mathrm{Mod}(\mathfrak{B}) \oplus \mathrm{Mod}(\mathscr{C}).$$

In this case, we say that the interconnection of $\mathfrak{B}$ and $\mathscr{C}$ is a *regular interconnection* and denote it by $\mathfrak{B} \cap_{reg} \mathscr{C}$. For a 1D behaviors, we know from the work of Willems [18] that controllability is equivalent to implementation of any sub-behavior by means of regular interconnection. Again the situation for nD behaviors is more involved. The following necessary (and not necessarily sufficient) condition for implementation of nD behaviors by regular interconnection has been derived in [14, Theorem 4.5, p. 124].

Theorem 8.6 *Let $\mathfrak{B}$ and $\mathscr{K}$ be two nD behaviors and $\mathfrak{B}^c$ the controllable part of $\mathfrak{B}$. Then if $\mathscr{K}$ is implementable by regular interconnection from $\mathfrak{B}$ then $\mathfrak{B} = \mathfrak{B}^c + \mathscr{K}$.*

This result can be intuitively explained by the fact that an autonomous part of a behavior may be somehow considered as obstructions to the (regular) control of that behavior, as happens for instance with the noncontrollable modes in the context of pole-placement for classical state-space systems. Using this result it is possible to show the next useful Lemma.

Lemma 8.7 ([4, Lemma 6]) *Let $\mathfrak{B}$ and $\mathscr{C}$ be two nD behaviors. If the interconnection of $\mathfrak{B}$ and $\mathscr{C}$ is regular then so is the interconnection between $\mathfrak{B}^c$ and $\mathscr{C}$.*

Proof Let $\mathfrak{B} \cap \mathscr{C} = \mathscr{K}$ with regular interconnection, i.e., $\text{Mod}(\mathfrak{B}) \oplus \text{Mod}(\mathscr{C}) = \text{Mod}(\mathscr{K})$. Using Theorem 8.6 we have that $\mathfrak{B} = \mathfrak{B}^c + \mathscr{K}$ or equivalently $\text{Mod}(\mathfrak{B}) = \text{Mod}(\mathfrak{B}^c) \cap \text{Mod}(\mathscr{K}) = \text{Mod}(\mathfrak{B}^c) \cap (\text{Mod}(\mathfrak{B}) \oplus \text{Mod}(\mathscr{C}))$. Using that $\text{Mod}(\mathfrak{B}) \subset \text{Mod}(\mathfrak{B}^c)$ one easily show that $\text{Mod}(\mathfrak{B}^c) \cap (\text{Mod}(\mathfrak{B}) \oplus \text{Mod}(\mathscr{C})) = (\text{Mod}(\mathfrak{B}^c) \cap \text{Mod}(\mathscr{C})) \oplus \text{Mod}(\mathfrak{B})$. Since $\text{Mod}(\mathfrak{B}) \cap \text{Mod}(\mathscr{C}) = \{0\}$ we have that $\text{Mod}(\mathfrak{B}^c) \cap \text{Mod}(\mathscr{C}) = \{0\}$. ∎

Lemma 8.7 shows that the controllable part of a behavior plays an important role in the context of regular interconnections. Indeed, a controller which does not interconnect with $\mathfrak{B}^c$ in a regular way, can not interconnect with $\mathfrak{B}$ regularly.

Next we present a more surprising result, proven in [5, Theorem 18], that shows that the possibility of implementing autonomous sub-behaviors of $\mathfrak{B}$ by regular interconnection may also impose conditions in the controllable part of $\mathfrak{B}$, depending on the autonomy degree of such sub-behaviors. We shall include its short proof for the sake of completeness.

Theorem 8.8 *Let $\mathfrak{B}$ be a behavior. If $\mathscr{K} \subset \mathfrak{B}$ is regularly implementable from $\mathfrak{B}$ and has autonomy degree larger than 1 then $\mathfrak{B}^c$ (the controllable part of $\mathfrak{B}$) is rectifiable.*

Proof In order to prove the result we will make use of the duality between $\mathfrak{B}$ and $\text{Mod}(\mathfrak{B})$. Obviously, $\mathfrak{B} \cap_{reg} \mathscr{C} = \mathscr{K}$ if and only if $\text{Mod}(\mathfrak{B}) \oplus \text{Mod}(\mathscr{C}) = \text{Mod}(\mathscr{K})$. The assumption that $\mathscr{K}$ has autonomy degree ≥ 2 amounts to saying that the height of the annihilator of $\mathbb{C}^q[\underline{s}, \underline{s}^{-1}]/(\text{Mod}(\mathfrak{B}) \oplus \text{Mod}(\mathscr{C}))$ is ≥ 2, see [22, Lemma 4.7, p. 54]. Equivalently, the annihilator of $\mathbb{C}^q[\underline{s}, \underline{s}^{-1}]/(\text{Mod}(\mathfrak{B}) \oplus \text{Mod}(\mathscr{C}))$ contains at least two coprime elements, see [22, Lemma 3.6].

Further, the interconnection $\mathfrak{B} \cap \mathscr{C}$ is regular if and only if $\mathfrak{B}^c \cap \mathscr{C}^c$ is regular, where $\mathfrak{B}^c$ and $\mathscr{C}^c$ denote the corresponding controllable parts, see [6, Lemma 12]. Obviously $\mathfrak{B}^c \cap \mathscr{C}^c \subset \mathfrak{B} \cap \mathscr{C}$ and therefore $autodeg(\mathfrak{B}^c \cap \mathscr{C}^c) \geq autodeg(\mathfrak{B} \cap \mathscr{C})$.

Thus we have, by assumption, that the annihilator of $\mathbb{C}^q[\underline{s}, \underline{s}^{-1}]/(\text{Mod}(\mathfrak{B}^c) \oplus \text{Mod}(\mathscr{C}^c))$ contains at least two coprime elements, say d_1, d_2. Note that since $\mathfrak{B}^c$ and $\mathscr{C}^c$ are controllable, $\mathbb{C}^q[\underline{s}, \underline{s}^{-1}]/\text{Mod}(\mathfrak{B}^c)$ and $\mathbb{C}^q[\underline{s}, \underline{s}^{-1}]/\text{Mod}(\mathscr{C}^c)$ are torsion free.

We prove that $\mathfrak{B}^c$ is rectifiable by showing that $\mathbb{C}^q[\underline{s}, \underline{s}^{-1}]/\text{Mod}(\mathfrak{B}^c)$ is free as a $\mathbb{R}[\underline{s}, \underline{s}^{-1}]$-module.

Consider, an element $\xi \in \mathbb{C}^q[\underline{s}, \underline{s}^{-1}]$. There are coprime elements d_1, d_2 with $d_1\xi = a_1 + b_1$, $d_2\xi = a_2 + b_2$ with $a_1, a_2 \in \mathrm{Mod}(\mathfrak{B}^c)$, $b_1, b_2 \in \mathrm{Mod}(\mathscr{C}^c)$. The element $\tau_1 = \frac{a_1}{d_1} = \frac{a_2}{d_2} \in \mathbb{C}^q(\underline{s}, \underline{s}^{-1})$ has the property $d_1\tau_1, d_2\tau_1 \in \mathbb{C}^q[\underline{s}, \underline{s}^{-1}]$, where $\mathbb{C}^q(\underline{s}, \underline{s}^{-1})$ stands for the field of rational Laurent polynomials. Since d_1, d_2 are coprime, this implies that $\tau_1 \in \mathbb{C}^q[\underline{s}, \underline{s}^{-1}]$. Since $\mathbb{C}^q[\underline{s}, \underline{s}^{-1}]/\mathrm{Mod}(\mathfrak{B}^c)$ has no torsion, one obtains $\tau_1 \in \mathrm{Mod}(\mathfrak{B}^c)$.

The same argument shows that $\tau_2 = \frac{b_1}{d_1} = \frac{b_2}{d_2}$ belongs to $\mathrm{Mod}(\mathscr{C}^c)$. Hence $\xi = \tau_1 + \tau_2 \in \mathrm{Mod}(\mathfrak{B}^c) \oplus \mathrm{Mod}(\mathscr{C}^c)$ and $\mathbb{C}^q[\underline{s}, \underline{s}^{-1}] = \mathrm{Mod}(\mathfrak{B}^c) \oplus \mathrm{Mod}(\mathscr{C}^c)$. Then $\mathrm{Mod}(\mathfrak{B}^c)$ and $\mathrm{Mod}(\mathbb{C}^c)$ are projective modules and therefore free. Finally, since $\mathrm{Mod}(\mathbb{C}^c) \approx \mathbb{C}^q[\underline{s}, \underline{s}^{-1}]/\mathrm{Mod}(\mathfrak{B}^c)$ one obtains that $\mathbb{C}^q[\underline{s}, \underline{s}^{-1}]/\mathrm{Mod}(\mathfrak{B}^c)$ is free. This concludes the proof. ∎

One can conclude from Theorem 8.8 that, in contrast to the 1D case, regular implementability is a very restrictive property in the context of nD behaviors (with $n \geq 2$).

When the controllable part of $\mathfrak{B}$ is rectifiable, it is possible to further exploit the simplified form of the rectified behavior in order to derive the following result on the autonomous-controllable decomposition of $\mathfrak{B}$.

Theorem 8.9 ([5, Prop. 4]) *Let $\mathfrak{B}$ be a behavior with rectifiable controllable part. Then, there always exists an autonomous sub-behavior $\mathfrak{B}^a$ of $\mathfrak{B}$ such that $\mathfrak{B} = \mathfrak{B}^c \oplus \mathfrak{B}^a$.*

8.3.2 Control by Regular Partial Interconnection

In the case of partial interconnection, one starts from a full behavior $\mathfrak{B}_{(w,c)}$, where w is the variable to be controlled and c is the control variable. The goal is to find a control variable behavior $\mathscr{C}$ whose interconnection with $\mathfrak{B}_{(w,c)}$ yields a desired behavior, $\mathscr{K}$ for the variable w. This can be formulated as finding $\mathscr{C}$ such that:

$$\mathscr{K} = \pi_w(\mathfrak{B}_{(w,c)} \cap \mathscr{C}^*_{(w,c)}).$$

For simplicity of notation we shall write $\mathscr{C}_c$ or instead of $\mathscr{C}^*_{(w,c)}$; moreover we shall skip the subscript with the indication of the variable (and write, for instance, $\mathscr{C}$ and $\mathfrak{B}$ instead of $\mathscr{C}_c$ and $\mathfrak{B}_w$, respectively) if no confusion arises.

Also in this context regularity plays an important role. Given two behaviors $\mathfrak{B}_{(w,c)} \subset \mathscr{U}^{\mathtt{w}+\mathtt{c}}$ and $\mathscr{C} \subset \mathscr{U}^{\mathtt{c}}$, we say that the interconnection $\mathfrak{B}_{(w,c)} \cap \mathscr{C}^*_{(w,c)}$ is *regular* if

$$\mathrm{Mod}(\mathfrak{B}_{(w,c)}) \cap \mathrm{Mod}(\mathscr{C}^*_{(w,c)}) = \{0\},$$

or equivalently if $\mathfrak{B}_{(w,c)} + \mathscr{C}^*_{(w,c)} = \mathscr{U}^{\mathtt{w}+\mathtt{c}}$. In this case, we denote the interconnection by $\mathfrak{B}_{(w,c)} \cap_{reg} \mathscr{C}^*_{(w,c)}$ or (in simplified notation) by $\mathfrak{B}_{(w,c)} \cap_{reg} \mathscr{C}$. Obviously, if $\mathscr{C} = \ker\ C$, $\mathrm{Mod}(\mathscr{C}^*_{(w,c)}) = \mathrm{RM}([0\ \ C])$ and when no confusion arises we write $\mathrm{Mod}(\mathscr{C}^*_{(w,c)}) = \mathrm{Mod}(\mathscr{C})$.

The following lemma presents some interesting results about partial interconnections and hidden behaviors that can be found in [7, Lemma 9] or in [15, Corollary 14].

Lemma 8.10 *Let $\mathfrak{B}_{(w,c)} \subset \mathscr{U}^{\mathrm{w+c}}$ and $\mathscr{C} \subset \mathscr{U}^{\mathrm{c}}$ be two behaviors. Then, the following hold true.*

1. $\pi_w(\mathfrak{B}_{(w,c)} \cap \mathscr{C}) = \pi_w(\mathfrak{B}_{(w,c)} \cap (\mathscr{C} + \mathfrak{B}_{(0,c)}))$.
2. $\mathfrak{B}_{(w,c)} \cap_{reg} \mathscr{C}$ *if and only if* $\mathfrak{B}_{(w,c)} \cap_{reg} (\mathscr{C} + \mathfrak{B}_{(0,c)})$.
3. $\mathfrak{B}_{(w,c)} \cap_{reg} \mathscr{C}$ *if and only if* $\mathfrak{B}_c \cap_{reg} \mathscr{C}$.

Proof Let $\mathfrak{B}_{(w,c)} = \ker\,[R\ \ M]$ and $\mathscr{C} = \ker\ C$. Note that $\mathfrak{B}_{(0,c)} = \ker\ M \subset \mathscr{U}^{\mathrm{c}}$ and since $\mathfrak{B}_{(0,c)} \subset \mathscr{C} + \mathfrak{B}_{(0,c)}$, then $\mathscr{C} + \mathfrak{B}_{(0,c)} = \ker\ KM$ for some L-polynomial matrix K.

1. It is enough to show that $\pi_w(\mathfrak{B}_{(w,c)} \cap (\mathscr{C} + \mathfrak{B}_{(0,c)})) \subset \pi_w(\mathfrak{B}_{(w,c)} \cap \mathscr{C})$ since the other inclusion is trivial. Let $w \in \pi_w(\mathfrak{B}_{(w,c)} \cap (\mathscr{C} + \mathfrak{B}_{(0,c)}))$. Then, by definition of π_w there exists a c such that $(w, c) \in \mathfrak{B}_{(w,c)} \cap (\mathscr{C} + \mathfrak{B}_{(0,c)}) = \ker \begin{bmatrix} R & M \\ 0 & KM \end{bmatrix}$. Clearly, c must satisfy $KMc = 0$, i.e., $c \in \mathscr{C} + \mathfrak{B}_{(0,c)} = \ker\ KM$ and therefore $c = c^* + c^{**}$, where $c^* \in \mathscr{C}$ and $c^{**} \in \mathfrak{B}_{(0,c)} = \ker\ M$. Hence, as $(w, c) \in \ker\ [R\ \ M]$, $(w, c^*) \in \ker\ [R\ \ M]$ which implies that $(w, c^*) \in \ker \begin{bmatrix} R & M \\ 0 & C \end{bmatrix} = \mathfrak{B}_{(w,c)} \cap \mathscr{C}$, and therefore $w \in \pi_w(\mathfrak{B}_{(w,c)} \cap \mathscr{C})$.

2. By Theorem 8.1, the proof of 3 amounts to showing that

$$\mathrm{RM}([R\ \ M]) \cap \mathrm{RM}([0\ \ C]) = \{0\} \text{ if and only if } \mathrm{RM}([R\ \ M]) \cap \mathrm{RM}([0\ \ KM]) = \{0\}.$$

As $\ker\ C = \mathscr{C} \subset \mathscr{C} + \mathfrak{B}_{(0,c)} = \ker\ KM$, $\mathrm{RM}(KM) \subset \mathrm{RM}(C)$ and the "only if" part is obvious. For the converse, let $(0, 0) \neq (r, m) \in \mathrm{RM}([R\ \ M]) \cap \mathrm{RM}([0\ \ C])$. Clearly r must be zero and then there exists an L-polynomial row s such that $s[R\ \ M] = (0, m) \neq (0, 0)$, which implies $sM = m \in \mathrm{RM}(C) \cap \mathrm{RM}(M) = \mathrm{RM}(KM)$. Thus, $(0, m) \in \mathrm{RM}([R\ \ M]) \cap \mathrm{RM}([0\ \ KM])$.

3. In terms of the corresponding modules we need to show that

$$\mathrm{RM}([R\ \ M]) \cap \mathrm{RM}([0\ \ C]) = \{0\} \text{ if and only if } \mathrm{RM}(LM) \cap \mathrm{RM}(C) = \{0\},$$

where L is an MLA of R. In order to prove the "if" part, let $(0, 0) \neq (r, m) \in \mathrm{RM}([R\ \ M]) \cap \mathrm{RM}([0\ \ C])$. It is easy to see that r must be zero and therefore there exists $s \in L$ such that $s[R\ \ M] = (0, m)$. Thus, $0 \neq sM = m \in \mathrm{RM}(LM) \cap \mathrm{RM}(C)$. To prove the converse implication suppose that $0 \neq m \in \mathrm{RM}(LM) \cap \mathrm{RM}(C)$. Then, $m = \alpha LM = \beta C$ for some L-polynomial rows α and β. This implies that $(0, m) = \alpha L[R\ \ M] = \beta[0\ \ C]$ and therefore $(0, 0) \neq (0, m) \in \mathrm{RM}([R\ \ M]) \cap \mathrm{RM}([0\ \ C])$. ■

A behavior $\mathcal{K} \subset \mathcal{U}^{\mathrm{w}}$ is trivially implementable from a given behavior $\mathfrak{B} \subset \mathcal{U}^{\mathrm{w}}$ by full (not necessarily regular) interconnection if and only if $\mathcal{K} \subset \mathfrak{B}$. This condition is however not enough in the partial interconnection case. Indeed, it was proven in [1, 15, 16] that $\mathcal{K}$ is implementable by partial (not necessarily regular) interconnection from $\mathfrak{B}_{(w,c)}$ if and only if

$$\mathfrak{B}_{(w,0)} \subset \mathcal{K} \subset \mathfrak{B}_w = \pi_w(\mathfrak{B}_{(w,c)}).$$

For regular partial interconnections the implementation problem was fully addressed and solved in the 1D context in [1]. In effect, the following necessary and sufficient conditions for the regular implementation of a behavior $\mathcal{K}$ were given:

1. $\mathcal{K}$ is implementable by partial interconnection, i.e., $\mathfrak{B}_{(w,0)} \subset \mathcal{K} \subset \mathfrak{B}_w$,
2. $\mathcal{K} + \mathfrak{B}_w^c = \mathfrak{B}_w$, where $\mathfrak{B}_w^c$ stands for the controllable part of $\mathfrak{B}_w$.

Note that the second condition is equivalent (in the 1D case) to $\mathcal{K}$ being regularly implementable by *full* interconnection from $\mathfrak{B}_w$. It was shown in [13, 15] that these two conditions were neither necessary nor sufficient in the nD case. Next we investigate when similar conditions hold in terms of the associated hidden behaviors. We say that a behavior is *regular* if admits a full row rank representation.

Theorem 8.11 *Let $\mathcal{K} \subset \mathcal{U}^{\mathrm{w}}$ and $\mathfrak{B}_{(w,c)} \subset \mathcal{U}^{\mathrm{w+c}}$ be given. Assume that $\mathcal{K}$ is implementable by partial interconnection and that the hidden behavior $\mathfrak{B}_{(w,0)}$ is regular. If $\mathcal{K}$ is regularly implementable by full interconnection (from $\mathfrak{B}_w$) then it is regularly implementable by partial interconnection.*

Proof Let $[\bar{R} \;\; M]$ be such that $\mathfrak{B}_{(w,c)} = \ker \; [\bar{R} \;\; M]$. Since $\mathfrak{B}_{(w,0)} = \ker \; \bar{R}$ is regular we can assume without loss of generality that $\bar{R} = \begin{bmatrix} R \\ 0 \end{bmatrix}$ with R full row rank and therefore $\mathfrak{B}_{(w,c)} = \ker \begin{bmatrix} R & M_1 \\ 0 & M_2 \end{bmatrix}$, for a suitable partition of M. Then, $\mathfrak{B}_w = \pi_w(\mathfrak{B}_{(w,c)}) = \ker \; XR$, where $[X \;\; Y]$ is an MLA of $\begin{bmatrix} M_1 \\ M_2 \end{bmatrix}$. Let $\bar{\mathcal{C}} = \ker \; C \subset \mathcal{U}^{w}$ be the controller that implements $\mathcal{K}$ by full interconnection. As $\mathcal{K}$ is implementable by partial interconnection, $\mathfrak{B}_{(w,0)} \subset \mathcal{K} \subset \bar{\mathcal{C}}$ it follows that there exists a matrix L such that $C = LR$. Take $\mathcal{C} = \ker \; LM_1 \subset \mathcal{U}^{\mathrm{c}}$. Next we show that $\mathcal{C}$ regularly implements $\mathcal{K}$ by partial interconnection. It is easy to check that $\mathcal{C}$ implements $\mathcal{K}$. To show that the interconnection is regular suppose that the row vector m belongs to $RM([0 \;\; LM_1]) \cap RM(\begin{bmatrix} R & M_1 \\ 0 & M_2 \end{bmatrix})$. This means that there exist row vectors s, and $t = [t_1 \;\; t_2]$ such that $m = s\begin{bmatrix} 0 \; LM_1 \end{bmatrix} = [t_1 \;\; t_2]\begin{bmatrix} R & M_1 \\ 0 & M_2 \end{bmatrix}$. As R is full row rank $t_1 = 0$. This implies that $[sL \;\; -t_2]\begin{bmatrix} M_1 \\ M_2 \end{bmatrix} = 0$, and hence $[sL \;\; -t_2] = v[X \;\; Y]$ for some row vector v. In turn, this implies that $sLR = vXR$. As, by assumption, the interconnection of $\mathfrak{B}_w = \ker \; XR$ and $\bar{\mathcal{C}} = \ker \; LR$ is regular, $sLR = (vXR =) \; 0$, and, since R has full row rank, $sL = 0$. Therefore $m = s[0 \;\; LM_1] = 0$, which concludes the proof. ■

Using part 3 of Lemma 8.10, and applying the same type of reasoning as in the proof of Theorem 8.11, one can derive the following corollary.

Corollary 8.12 *Let $\mathscr{K} \subset \mathscr{U}^{\mathtt{w}}$ and $\mathfrak{B}_{(w,c)} \subset \mathscr{U}^{\mathtt{w}+\mathtt{c}}$ be given. Assume that $\mathscr{K}$ is implementable by partial interconnection and that the hidden behavior $\mathfrak{B}_{(0,c)}$ is regular. If $\mathscr{K}$ is regularly implementable by partial interconnection then it is regularly implementable by full interconnection (from $\mathfrak{B}_w$).*

Remark 8.13 Note that rectifiable behaviors admit a full row rank representation, i.e., are regular, and therefore Theorem 8.11 and Corollary 8.12 are still valid if we assume that $\mathfrak{B}_{(w,0)}$ and $\mathfrak{B}_{(0,c)}$ respectively are rectifiable. Moreover, in the 2D case one can assume controllability instead of rectifiability as controllable behaviors always have a full row rank representation.

8.3.3 Controllers

In this section, we look at a special behavior that has also been introduced in [2, 16, 19] under the name of *canonical controller*. In particular, we study its effectiveness in solving partial control problems—a question which has also been considered in [3, 19] for the 1D case—and generalize the corresponding 1D results to the nD case. We conclude the section by analyzing the performance of regular controllers in this context. The results of this section (except for Theorem 8.20) were first presented in [13] although some can also be found in [15] in a more module-theoretical framework.

It is immediately apparent that the study of partial control problems requires additional tools with respect to full control problems. For this reason, it is desirable to translate partial control problems into full control ones. In the 1D case, it is possible to make this translation in terms of full control problems for behaviors involving only the to-be-controlled variable w. Unfortunately this is no longer true in the higher dimensional (nD) case. Therefore, we shall try to characterize regular implementation (by partial control) in terms of conditions on the control variable behavior, rather than by means of conditions on the behavior of the variables to be controlled. To this end we introduce the notion of *canonical controller* associated to a given control problem. For a given control objective $\mathscr{K} \subset \mathscr{U}^{\mathtt{w}}$, the canonical controller associate with $\mathscr{K}$ is defined as follows:

$$\mathscr{C}^{can}(\mathscr{K}) := \{c \mid \exists\, w \text{ such that } (w, c) \in \mathfrak{B}_{(w,c)} \text{ and } w \in \mathscr{K}\}.$$

For simplicity we use $\mathscr{C}^{can}$ for $\mathscr{C}^{can}(\mathscr{K})$. Thus, the canonical controller consists of all the control variable trajectories compatible with the desired behavior for the variables to be controlled.

We start by relating the implementation of $\mathscr{K}$ from $\mathfrak{B}_{(w,c)}$ (by partial control) with the implementation of the corresponding canonical controller from $\mathfrak{B}_c$. First, we treat the implementation problem and then the regular implementation.

Theorem 8.14 *Given a plant behavior* $\mathfrak{B}_{(w,c)}$ *and an implementable control objective* $\mathcal{K}$*, the following holds.*

1. *If the controller* $\mathcal{C}$ *implements* $\mathcal{C}^{can}$ *from* $\mathfrak{B}_c$ *by full control, then it implements* $\mathcal{K}$ *from* $\mathfrak{B}_{(w,c)}$.
2. *If the controller* $\tilde{\mathcal{C}}$ *implements* $\mathcal{K}$ *from* $\mathfrak{B}_{(w,c)}$*, then the controller* $\tilde{\mathcal{C}} + \mathfrak{B}_{(0,c)}$ *implements* $\mathcal{C}^{can}$ *from* $\mathfrak{B}_c$ *by full control.*

Proof Let $Rw = Mc$ be a representation of $\mathfrak{B}_{(w,c)}$. Then, $\mathfrak{B}_c = \ker\ NM$, where N be an nD polynomial matrix which is an minimal left annihilator (MLA) of R. To prove the first statement assume that the controller $\mathcal{C} = \ker\ K$ implements $\mathcal{C}^{can}$ and apply this controller to the plant. This yields the (w, c)-behavior described by the equations:

$$\begin{cases} Rw = Mc \\ 0 \quad = Kc. \end{cases} \tag{8.4}$$

We next show that the corresponding w-behavior coincides with $\mathcal{K}$, which clearly implies that $\mathcal{C}$ implements $\mathcal{K}$ from $\mathfrak{B}_{(w,c)}$.

Suppose then that w^* belongs to the w-behavior induced by equations (8.4), i.e., that there exists c^* such that the pair (w^*, c^*) satisfies these equations. This implies that $c^* \in \mathfrak{B}_c \cap \mathcal{C} = \mathcal{C}^{can}$ and hence, by the definition of the canonical controller, there exists $\bar{w} \in \mathcal{K}$ such that $(\bar{w}, c) \in \mathfrak{B}_{(w,c)}$. Thus, by linearity, $(w^* - \bar{w}, 0) \in \mathfrak{B}_{(w,c)}$, meaning that $w^* - \bar{w} \in \mathfrak{B}_{(w,0)}$. Since $\mathcal{K}$ is by assumption implementable, $\mathfrak{B}_{(w,0)} \subset \mathcal{K}$ and $w^* - \bar{w} \in \mathcal{K}$. Consequently also $w^* \in \mathcal{K}$ and therefore the w-behavior induced by equations (8.4) is contained in $\mathcal{K}$.

Conversely, suppose that $w^* \in \mathcal{K}$. Then obviously $w^* \in \mathfrak{B}_w$ and hence there exists c^* such that $(w^*, c^*) \in \mathfrak{B}_{(w,c)}$, i.e., such that

$$Rw^* = Mc^*.$$

By the definition of the canonical controller, this means that $c^* \in \mathcal{C}^{can}$. Since $\mathcal{C}^{can}$ is assumed to be implementable by $\mathcal{C}$, $\mathcal{C}^{can} \subset \mathcal{C}$ and therefore $c^* \in \mathcal{C}$, i.e.,

$$Kc^* = 0.$$

Thus, the pair (w^*, c^*) satisfies Eq. (8.4), which means that w^* is in the w-behavior induced by these equations. So, $\mathcal{K}$ is contained in that behavior. As mentioned before, this shows that $\mathcal{C}$ implements $\mathcal{K}$ from $\mathfrak{B}_{(w,c)}$.

As for the second statement assume now that the controller $\tilde{\mathcal{C}} = \ker\ K$ implements $\mathcal{K}$ from $\mathfrak{B}_{(w,c)}$. Let $c^* \in \mathcal{C}^{can}$. This means that there exists w^* such that $(w^*, c^*) \in \mathfrak{B}_{(w,c)}$ and $w^* \in \mathcal{K}$. This last condition implies that there exists $\bar{c} \in \tilde{\mathcal{C}}$ such that $(w^*, \bar{c}) \in \mathfrak{B}_{(w,c)}$. Note that by the linearity of $\mathfrak{B}_{(w,c)}$, $(0, c^* - \bar{c}) \in \mathfrak{B}_{(w,c)}$; hence $c^* - \bar{c} \in \mathfrak{B}_{(0,c)}$ and therefore (taking into account that $\bar{c} \in \tilde{\mathcal{C}}$) we have that $c^* \in \mathfrak{B}_{(0,c)} + \tilde{\mathcal{C}}$. Thus, $\mathcal{C}^{can} \subset \mathfrak{B}_{(0,c)} + \tilde{\mathcal{C}}$ and, since also $\mathcal{C}^{can} \subset \mathfrak{B}_c$, $\mathcal{C}^{can} \subset (\mathfrak{B}_{(0,c)} + \tilde{\mathcal{C}}) \cap \mathfrak{B}_c$.

Conversely, assume that $c^* \in (\mathfrak{B}_{(0,c)} + \tilde{\mathscr{C}}) \cap \mathfrak{B}_c$. Then, there exist w^* and $\bar{c} \in \tilde{\mathscr{C}}$ such that $(w^*, c^*) \in \mathfrak{B}_{(w,c)}$, $\bar{c} \in \tilde{\mathscr{C}}$ and $c^* - \bar{c} \in \mathfrak{B}_{(0,c)}$. This implies that $(w^*, \bar{c}) \in \mathfrak{B}_{(w,c)}$ and, since $\tilde{\mathscr{C}}$ implements $\mathscr{K}$ from $\mathfrak{B}_{(w,c)}$, $w^* \in \mathscr{K}$. Together with the fact that $(w^*, c^*) \in \mathfrak{B}_{(w,c)}$, taking the definition of $\mathscr{C}^{can}$ into account, this allows to conclude that $c^* \in \mathscr{C}^{can}$. Therefore $(\mathfrak{B}_{(0,c)} + \tilde{\mathscr{C}}) \cap \mathfrak{B}_c \subset \mathscr{C}^{can}$. This finally proves that $\mathscr{C}^{can} = (\mathfrak{B}_{(0,c)} + \tilde{\mathscr{C}}) \cap \mathfrak{B}_c$, which amounts to say that $\mathfrak{B}_{(0,c)} + \tilde{\mathscr{C}}$ implements $\mathscr{C}^{can}$ from $\mathfrak{B}_c$ by full control. ■

Note that, as a consequence of this theorem, if the hidden behavior $\mathfrak{B}_{(0,c)} = \{0\}$, then $\mathscr{C}$ implements $\mathscr{K}$ from $\mathfrak{B}_{(w,c)}$ if and only if it implements $\mathscr{C}^{can}$ from $\mathfrak{B}_c$ by full control.

Next we extend Theorem 8.14 for regular interconnections.

Theorem 8.15 *Given a plant behavior $\mathfrak{B}_{(w,c)}$ and an implementable control objective $\mathscr{K}$, the following holds.*

1. *If the controller $\mathscr{C}$ implements $\mathscr{C}^{can}$ from $\mathfrak{B}_c$ by regular full control, then $\mathscr{C}$ implements $\mathscr{K}$ regularly from $\mathfrak{B}_{(w,c)}$.*
2. *If the controller $\tilde{\mathscr{C}}$ implements $\mathscr{K}$ regularly from $\mathfrak{B}_{(w,c)}$, then the controller $\tilde{\mathscr{C}} + \mathfrak{B}_{(0,c)}$ implements $\mathscr{C}^{can}$ from $\mathfrak{B}_c$ by regular full control.*

Proof Since the statements about implementation have already been proven in Theorem 8.14 it now suffices to prove the statements concerning regularity.

To show the first statement let $r = [0 \;\; \bar{r}] \in \mathrm{Mod}(\mathfrak{B}_{(w,c)}) \cap \mathrm{Mod}(\mathscr{C}^*_{(w,c)})$ (note that since w is free in $\mathscr{C}^*_{(w,c)}$, the first components of r must be zero). Then, clearly, $\bar{r} \in \mathrm{Mod}(\mathscr{C})$. Moreover, $\mathfrak{B}_c \subset \ker \bar{r}$, and hence $\bar{r} \in \mathrm{Mod}(\mathfrak{B}_c)$. Therefore $\bar{r} \in \mathrm{Mod}(\mathfrak{B}_c) \cap \mathrm{Mod}(\mathscr{C})$. In this way, if $\mathrm{Mod}(\mathfrak{B}_{(w,c)}) \cap \mathrm{Mod}(\mathscr{C}^*_{(w,c)})$ has a nonzero element $r = [0 \;\; \bar{r}]$ with $\bar{r} \neq 0$ then also $\mathrm{Mod}(\mathfrak{B}_c) \cap \mathrm{Mod}(\mathscr{C})$ has a nonzero element $\bar{r}$, proving the desired implication. Statement 2. can be proved using similar arguments. ■

Again we remark that Theorem 8.15 implies that, in case $\mathfrak{B}_{(0,c)} = \{0\}$, $\mathscr{C}$ regularly implements $\mathscr{K}$ from $\mathfrak{B}_{(w,c)}$ if and only if it implements $\mathscr{C}^{can}$ from $\mathfrak{B}_c$ by regular full control.

Theorem 8.15 yields necessary and sufficient conditions for the problem of regular implementation by partial interconnections.

Corollary 8.16 *Let $\mathfrak{B}_{(w,c)}$ be a given plant behavior and $\mathscr{K}$ a control objective. Assume further that $\mathscr{K}$ is implementable from $\mathfrak{B}_{(w,c)}$. Then $\mathscr{K}$ is regularly implementable from $\mathfrak{B}_{(w,c)}$ if and only if $\mathscr{C}^{can}$ is regularly implementable from $\mathfrak{B}_c$ by full control.*

In the previous considerations, the canonical controller associated to a given control problem has been considered as a control objective itself, whose ability to be implemented provides information on the possibility of implementing the true control objective. We now take a different perspective and consider the canonical controller in its most natural role, i.e., as being itself a controller. In this context, two questions obviously arise: Does the canonical controller implement the control objective? If so, is this implementation regular? The answers to these questions are given below.

Theorem 8.17 *Given a plant behavior* $\mathfrak{B}_{(w,c)}$*, a control objective* $\mathcal{K}$*, let* $\mathcal{C}^{can}$ *be the associated canonical controller. Then,* $\mathcal{C}^{can}$ *implements* $\mathcal{K}$ *if and only of* $\mathcal{K}$ *is implementable.*

Proof The "only if" part of the statement is trivial. As for the "if" part, suppose that $\mathcal{K}$ is implementable, and let $\tilde{\mathcal{C}} = \ker\ K$ be a controller that implements this behavior. Then, by Theorem 8.14, the controller $\tilde{\mathcal{C}} + \mathfrak{B}_{(0,c)}$ implements $\mathcal{C}^{can}$ from $\mathfrak{B}_c$. If $Rw = Mc$ is a representation of $\mathfrak{B}_{(w,c)}$ and N is a MLA of R, $\mathfrak{B}_{(0,c)} = \ker\ M$ and $\mathfrak{B}_c = \ker\ NM$. Therefore, the fact that $\tilde{\mathcal{C}} + \mathfrak{B}_{(0,c)}$ implements $\mathcal{C}^{can}$ from $\mathfrak{B}_c$ means that $\mathcal{C}^{can}$ is the c-behavior induced by the following equations:

$$\begin{cases} NMc = 0 \\ c \quad\ \ = c_1 + c_2 \\ Kc_1 \ \ = 0 \\ Mc_2 \ \ = 0. \end{cases} \tag{8.5}$$

Consequently, applying the canonical controller to the plant $\mathfrak{B}_{(w,c)}$ yields the restrictions:

$$\begin{cases} Rw \quad = Mc \\ NMc = 0 \\ c \quad\ \ = c_1 + c_2 \\ Kc_1 \ \ = 0 \\ Mc_2 \ \ = 0, \end{cases} \tag{8.6}$$

that can easily be shown to have the same w-behavior as

$$\begin{cases} Rw \ \ = Mc_1 \\ Kc_1 = 0. \end{cases} \tag{8.7}$$

But this w-behavior is precisely $\mathcal{K}$, which proves that $\mathcal{C}^{can}$ indeed implements $\mathcal{K}$. ■

Our last results concerns regular implementation by means of the canonical controller.

Theorem 8.18 *Given a plant behavior* $\mathfrak{B}_{(w,c)}$*, a control objective* $\mathcal{K}$*, let* $\mathcal{C}^{can}$ *be the associated canonical controller. Then,* $\mathcal{C}^{can}$ *regularly implements* $\mathcal{K}$ *if and only of* $\mathfrak{B}_c$ *coincides with the whole c-trajectory universe, i.e., if and only if* $\mathrm{Mod}(\mathfrak{B}_c) = \{0\}$.

Proof Assume that $\mathcal{C}^{can}$ regularly implements $\mathcal{K}$. Then, by Corollary 8.16, $\mathcal{C}^{can} + \mathfrak{B}_{(0,c)}$ regularly implements $\mathcal{C}^{can}$ from $\mathfrak{B}_c$. This implies that $\mathrm{Mod}(\mathcal{C}^{can} + \mathfrak{B}_{(0,c)}) \cap \mathrm{Mod}(\mathfrak{B}_c) = \{0\}$. But, $\mathrm{Mod}(\mathcal{C}^{can} + \mathfrak{B}_{(0,c)}) \cap \mathrm{Mod}(\mathfrak{B}_c) = \mathrm{Mod}(\mathcal{C}^{can}) \cap \mathrm{Mod}(\mathfrak{B}_c)$. As $\mathrm{Mod}(\mathfrak{B}_c) \subset \mathrm{Mod}(\mathcal{C}^{can})$ (because $\mathcal{C}^{can} \subset \mathfrak{B}_c$), we obtain that $\mathrm{Mod}(\mathfrak{B}_c) = \{0\}$.

Conversely, if $\mathrm{Mod}(\mathfrak{B}_c) = \{0\}$ then the canonical controller regularly implements itself from $\mathfrak{B}_c$. By Corollary 8.16 this implies that $\mathscr{C}^{can}$ also implements $\mathscr{K}$ regularly. ■

Corollary 8.19 *The canonical controller is regular if and only if every controller is regular.*

Proof The if part is obvious. As for the only if part, we start by noting that, given a controller $\mathscr{C}$, $\mathrm{Mod}(\mathfrak{B}_{(w,c)}) \cap \mathrm{Mod}(\mathscr{C}^*_{(w,c)}) = \{r \mid r = [0\ \ \bar{r}], \bar{r} \in \ \mathrm{mod}\ (\mathscr{C}) \cap \mathrm{Mod}(\mathfrak{B}_c)\}$. Assume now that the canonical controller is regular. Then, by the previous theorem, $\mathrm{Mod}(\mathfrak{B}_c) = \{0\}$ and consequently also $\mathrm{Mod}(\mathfrak{B}_{(w,c)}) \cap \mathrm{Mod}(\mathscr{C}^*_{(w,c)}) = \{0\}$ for any given controller $\mathscr{C}$, which precisely means that the controller $\mathscr{C}$ is regular. This proves the desired result. ■

Theorems 8.17, 8.18 and Corollary 8.19 generalize the corresponding 1D results obtained in [3, 19] to the nD case.

Finally, we study another class of controllers that are of interest in the context of regular partial interconnections, namely, controllers that admit full row rank representations, called *regular controllers*. The regular implementation by means of a regular controller implies the regular implementation by full interconnection (from $\mathfrak{B}_w$).

Theorem 8.20 ([4, Theorem 10]) *Let* $\mathfrak{B}_{(w,c)} = \ker\ [R\ \ M]$ *be a behavior. If a desired behavior* $\mathscr{K}$ *is implementable by regular partial interconnection with a* regular *controller* $\mathscr{C} = \ker\ [0\ \ LM]$ *then* $\mathscr{K} = \mathfrak{B}_w \cap_{reg} \ker\ (LR)$*, i.e.,* $\mathscr{K}$ *can also be implementable by regular* (full) *interconnection from* $\mathfrak{B}_w$.

Proof Without loss of generality we supposed that the matrix LM is full row rank since $\mathscr{C}$ is a regular behavior. Further,

$$\begin{bmatrix} I & 0 \\ L & -I \end{bmatrix} \cdot \begin{bmatrix} R & M \\ 0 & LM \end{bmatrix} = \begin{bmatrix} R & M \\ LR & 0 \end{bmatrix}.$$

Let X be the MLA of M. Hence $\Pi_w(\mathfrak{B}_{(w,c)} \cap \mathscr{C}) = \Pi_w(\ker \begin{bmatrix} R & M \\ LR & 0 \end{bmatrix}) = \ker \begin{bmatrix} XR \\ LR \end{bmatrix} = \mathfrak{B}_{(w,c)} \cap \ker\ LR$. To see that the interconnection between $\mathfrak{B}_{(w,c)}$ and $\ker\ LR$ is regular we prove that the interconnection between $\ker\ [R\ \ M] \cap \ker\ [LR\ \ 0]$ is regular, i.e., $v[R\ \ M] = z[LR\ \ 0]$ for some row vectors v and z, implies $v[R\ \ M] = 0 = z[LR\ \ 0]$. Suppose that $v[R\ \ M] = z[LR\ \ 0]$. Note that $z[LR\ \ 0] = z[0\ \ -LM] + [LR\ \ LM]$ and then $v[R\ M] - z[LR\ LM] = (v - zL)[R\ M] = z[0\ \ -LM]$. By assumption that the interconnection of $\mathfrak{B}_{(w,c)}$ and $\mathscr{C}$ is regular one has that $(v - zL)[R\ \ M] = z[0\ \ -LM] = 0$ and since LM is full row rank one obtains that $z = 0$ and therefore $v[R\ \ M] = z[LR\ \ 0] = 0$ which proves that the interconnection is regular. ■

Acknowledgments The authors are supported by Portuguese funds through the CIDMA - Center for Research and Development in Mathematics and Applications, and the Portuguese Foundation for Science and Technology (FCT-Fundação para a Ciência e a Tecnologia), within project UID/MAT/04106/2013.

References

1. Belur, M.N., Trentelman, H.L.: Stabilization, pole placement, and regular implementability. IEEE Trans. Automat. Control **47**(5), 735–744 (2002)
2. Julius, A.A., Polderman, J.W., van der Schaft, A.: Parametrization of the regular equivalences of the canonical controller. IEEE Trans. Automat. Control **53**(4), 1032–1036 (2008)
3. Julius, A.A., Willems, J.C., Belur, M.N., Trentelman, H.L.: The canonical controllers and regular interconnection. Systems Control Lett. **54**(8), 787–797 (2005)
4. Napp, D., Rocha, P.: Implementation of 2D strongly autonomous behaviors by full and partial interconnections. Lectures Notes in Control and Information Science (Springer) **389**, 369–378 (2009)
5. Napp, D., Rocha, P.: Autonomous multidimensional systems and their implementation by behavioral control. Syst. Control Lett. **59**(3–4), 203–208 (2010)
6. Napp, D., Rocha, P.: Strongly autonomous interconnections and stabilization of 2D behaviors. Asian J. Control **12**(2), 127–135 (2010)
7. Napp, D., Rocha, P.: Stabilization of discrete 2D behaviors by regular partial interconnection. Math. Control Signal Syst. **22**(4), 295–316 (2011)
8. Napp, D., Shankar, S., Trentelman, H.L.: Regular implementation in the space of compactly supported functions. Syst. Control Lett. **57**(10), 851–855 (2008)
9. Oberst, U.: Multidimensional constant linear systems. Acta Appl. Math. **20**(1–2), 1–175 (1990)
10. Polderman, J.W., Mareels, I.: A behavioral approach to adaptive control. In: Polderman, J.W., Trentelman, H.L. (eds.) The Mathematics of Systems and Control: From Intelligent Control to Behavioral Systems. Foundation Systems and Control Groningen, The Netherlands (1999)
11. Willems, J.W., Willems, J.C.: Introduction to Mathematical Systems Theory, a Behavioral Approach. Texts in Applied Mathematics, vol. 26. Springer, New York (1998)
12. Praagman, C., Trentelman, H.L., Zavala Yoe, R.: On the parametrization of all regularly implementing and stabilizing controllers. SIAM J. Control Optim. **45**(6), 2035–2053 (2007). (electronic)
13. P. Rocha. Canonical controllers and regular implementation of nD behaviors. In: Proceedings of the 16th IFAC World Congress (2005)
14. Rocha, P., Wood, J.: Trajectory control and interconnection of 1D and nD systems. SIAM J. Control Optim. **40**(1), 107–134 (2001)
15. Trentelman, H.L., Napp Avelli, D.: On the regular implementability of nD systems. Syst. Control Lett. **56**(4), 265–271 (2007)
16. van der Schaft, A.J.: Achievable behavior of general systems. Syst. Control Lett. **49**(2), 141–149 (2003)
17. Willems, J.C.: Paradigms and puzzles in the theory of dynamical systems. IEEE Trans. Automat. Control **36**(3), 259–294 (1991)
18. Willems, J.C.: On interconnections, control, and feedback. IEEE Trans. Automat. Control **42**(3), 326–339 (1997)
19. J.C. Willems, M.N. Belur, A.A. Julius, H.L. Trentelman. The canonical controller and its regularity. In: Proceedings of the 42nd IEEE Conference on Decision and Control, Hawaii, pp. 1639–1644 (2003)
20. Willems, J.C., Trentelman, H.L.: Synthesis of dissipative systems using quadratic differential forms. I. IEEE Trans. Automat. Control **47**(1), 53–69 (2002)

21. Wood, J.: Modules and behaviours in nD systems theory. Multidimens. Systems Signal Process. **11**(1–2), 11–48 (2000)
22. Wood, J., Rogers, E., Owens, D.H.: A formal theory of matrix primeness. Math. Control Signals Syst. **11**(1), 40–78 (1998)
23. Zerz, E.: Primeness of multivariate polynomial matrices. Syst. Control Lett. **29**(3), 139–145 (1996)
24. Zerz, E., Lomadze, V.: A constructive solution to interconnection and decomposition problems with multidimensional behaviors. SIAM J. Control Optim. **40**(4), 1072–1086 (2001/02)

Chapter 9
Synchronization of Linear Multi-Agent Systems with Input Nonlinearities via Dynamic Protocols

Kiyotsugu Takaba

Abstract This paper is concerned with the local synchronization of linear agents subject to sector-bounded input nonlinearities over an undirected communication graph via dynamic output feedback protocol. We first derive a sufficient condition for achieving the local synchronization for any nonlinearities satisfying a given sector condition with a given dynamic protocol in terms of LMIs. Based on this analysis, we present a sufficient BMI synthesis condition of a dynamic protocol which locally synchronizes the linear agents with arbitrary sector-bounded input nonlinearities. Though the present BMI condition is non-convex, the condition is numerically tractable because it does not depend on the size of the communication graph except for computation of the Laplacian eigenvalues.

9.1 Introduction

It is a great pleasure to contribute this paper to the festschrift of Prof. Trentelman on the occasion of his 60th birthday.

Over the last decade, distributed cooperative control of multi-agent systems has been attracting a great interest in the control theory community (see [1, 2] and the references therein). The key feature of a multi-agent system is that it achieves a certain cooperative task such as synchronization, consensus, and formation, through distributed control of individual agents based on local interactions with their neighboring agents. Trentelman and his co-workers have reported several important results in this research area in recent years [3–6].

Early works on distributed cooperative control of linear multi-agent systems focused mainly on consensus and formation problems with homogeneous agent dynamics without model uncertainties [2, 7–10]. One of the recent research directions

K. Takaba
Department of Electrical and Electronic Engineering,
Ritsumeikan University, 1-1-1 Noji-higashi, Kusatsu, Shiga 525-8577, Japan
e-mail: ktakaba@fc.ritsumei.ac.jp

M.N. Belur et al. (eds.), *Mathematical Control Theory II*,
Lecture Notes in Control and Information Sciences 462,
DOI 10.1007/978-3-319-21003-2_9

of multi-agent control systems has been the robustness to cope with heterogeneity and/or model uncertainties of the agent dynamics [3, 11, 12].

Many control systems are subject to input nonlinearities due to physical characteristics of actuators and/or safety requirements. Of course, this situation is also true for multi-agent systems. In this paper, we will consider the synchronization of linear agents subject to sector-bounded nonlinearities in their input channels. There have been several related works in the literature. Zhang, Trentelman, and Scherpen considered the design of a dynamic protocol that robustly synchronizes a network of Lur'e systems with incrementally passive nonlinearities [5, 6].

As a typical input nonlinearity, several works considered synchronization of linear agents with input saturations [13–18]. Among them, Takaba [17, 18] derived LMI synthesis conditions of relative state feedback laws that achieve the global/local synchronization of linear agents in the presence of input saturations. Although the previous works of [13–18] dealt with only a particular type of nonlinearities, it is important from a robustness viewpoint to guarantee the synchronizability of the multi-agent systems against uncertainties of input nonlinearities, and such an uncertain nonlinearity is often modeled in terms of sector-bounds.

Therefore, we consider the synchronization of linear agents with sector-bounded nonlinearities in their input channels. In line with the philosophy of robustness, we wish to design a dynamic feedback protocol that achieves the local synchronization for arbitrary sector-bounded input nonlinearities.

9.2 Problem Statement

9.2.1 Agent Dynamics

Throughout this article, we consider the synchronization of a multi-agent system consisting of N agents. Since many control systems have input nonlinearities such as saturations and dead-zones due to characteristics of actuators and/or safety requirements, we model the dynamics of the individual agents by

$$\dot{x}_i = Ax_i + B_0\psi_i(u_i, t), \quad y_i = Cx_i, \qquad i = 1, \ldots, N, \tag{9.1}$$

where $x_i : \mathbb{R}_+ \to \mathbb{R}^n$, $u_i : \mathbb{R}_+ \to \mathbb{R}$, and $y_i : \mathbb{R}_+ \to \mathbb{R}$ are the state, input, and output of the ith agent, respectively. The memoryless functions $\psi_i : \mathbb{R} \times \mathbb{R}_+ \to \mathbb{R}$, $i = 1, \ldots, N$ represent the input nonlinearities . We assume that u_i and y_i are scalar-valued variables. The results presented in this paper can be generalized to the case of multi-input multi-output agent dynamics in a straightforward manner.

The nonlinearities ψ_i, $i = 1, \ldots, N$ satisfy the local sector condition

$$\alpha u_i \le \psi_i(u_i, t) \le \beta u_i \ \ \forall u_i \in [-\mu, \ \mu], \ \forall t \ge 0, \tag{9.2}$$

where α, β, and μ are given constants such that $\beta > \alpha \ge 0$ and $\mu > 0$.

For some input nonlinearities such as saturations, it is impossible to globally synchronize or stabilize linear systems with exponentially unstable poles. Therefore, we consider the local synchronization problem under the assumption that the sector condition is satisfied only within the finite interval $[-\mu,\ \mu]$, $\mu < \infty$.

We define $\varphi_i : \mathbb{R} \times \mathbb{R}_+ \to \mathbb{R}$ by

$$\varphi_i(u_i,\ t) = u_i - \frac{1}{\beta}\psi_i(u_i,\ t). \tag{9.3}$$

It is easily verified that φ_i, $i = 1, \ldots, N$ satisfy another sector condition

$$0 \leq \varphi_i(u_i,\ t) \leq \gamma u_i \ \ \forall u_i \in [-\mu,\ \mu],\ \forall t \geq 0,$$

or equivalently,

$$\varphi_i(u_i,\ t)\left[\varphi_i(u_i,\ t) - \gamma u_i\right] \leq 0\ \forall u_i \in [-\mu,\ \mu],\ \ \forall t \geq 0, \tag{9.4}$$

where $\gamma = (\beta - \alpha)/\beta > 0$. The sector-bounded nonlinearities are illustrated in Fig. 9.1.

We can re-write the agent dynamics as

$$\dot{x}_i = Ax_i + Bu_i - B\varphi_i(u_i,\ t), \quad y_i = Cx_i, \quad i = 1, 2, \ldots, N, \tag{9.5}$$

where $B = B_0\beta$. The nonlinearity φ_i can be viewed as sector-bounded uncertainty to the linear time-invariant nominal system. Hereafter, we will discuss the synchronization of the multi-agent system based on the state-space model of (9.5) and the local sector condition (9.4).

Assumption 9.1 *(A, B) is stabilizable, and (C, A) is detectable.*

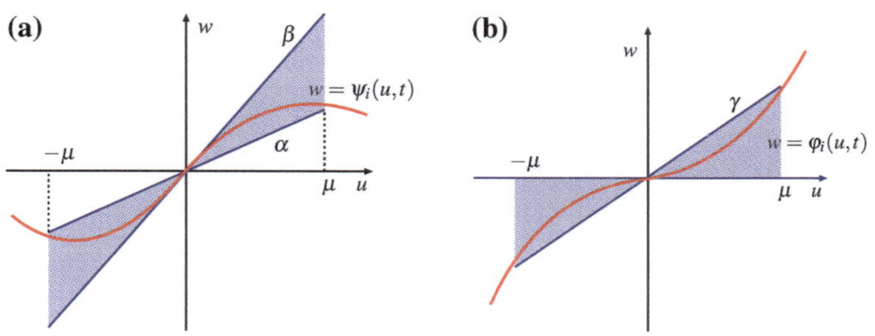

Fig. 9.1 Sector-bounded nonlinearities. **a** sector bound $(\alpha,\ \beta)$. **b** sector bound $(0,\ \gamma)$

9.2.2 Communication Graph

Communications among agents are well described in terms of mathematical graphs [19]. A graph $\mathcal{G}$ is defined as a pair $\mathcal{G} = (\mathcal{V}, \mathcal{E})$, where $\mathcal{V} = \{1, \ldots, N\}$ is the index set of the nodes, $\mathcal{E} \subset \mathcal{V} \times \mathcal{V}$ is the edge set. Each element of $\mathcal{V}$ represents an agent. Also, communication links between two agents are defined by edges of the graph, namely, $(i, j) \in \mathcal{E}$ means that there is a communication link between the agents i and j. The set of the neighbors to the node i is defined by (Fig. 9.2)

$$\mathcal{N}_i = \{j \in \mathcal{V} \mid (i, j) \in \mathcal{E}, \ j \neq i\}.$$

Throughout this paper, we assume that the communication between any two agents is bi-directional, i.e., $(i, j) \in \mathcal{E} \Leftrightarrow (j, i) \in \mathcal{E}$. In this case, the graph $\mathcal{G}$ is identified with an undirected graph. Moreover, if $(i, j) \in \mathcal{E}$, then the agents i and j exchanges their output values y_i and y_j.

Assumption 9.2

(i) The topology of the graph $\mathcal{G} = (\mathcal{V}, \mathcal{E})$ is time-invariant.
(ii) The graph $\mathcal{G}$ is connected, namely, there exists at least one path from any node to another.

The Laplacian $L \in \mathbb{R}^{N \times N}$ of the graph $\mathcal{G}$ is a square matrix defined by

$$L = (\ell_{ij}), \quad \ell_{ij} = \begin{cases} |\mathcal{N}_i|, & \text{if } i = j, \\ -1, & \text{if } (i, j) \in \mathcal{E}, \\ 0, & \text{otherwise} \end{cases}$$

The Laplacian L of an undirected graph is symmetric and nonnegative definite. Moreover, L has a zero eigenvalue whose eigenvector is $\mathbf{1} := [1, 1, \ldots, 1]^\top \in \mathbb{R}^N$.

For later discussion, we define the eigenvalues of L as λ_i, $i = 1, \ldots, N$ in the ascending order:

$$0 = \lambda_1 \leq \lambda_2 \leq \cdots \leq \lambda_{N-1} \leq \lambda_N.$$

It is well known that $\mathcal{G}$ is a connected graph if and only if $\lambda_2 > 0$, or equivalently $\operatorname{rank} L = N - 1$.

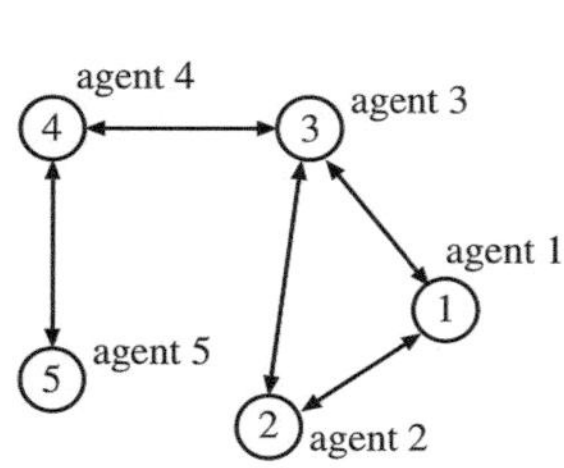

Fig. 9.2 Communication graph

9.2.3 Problem Statement

Since the states x_i, $i = 1, \ldots, N$ of the individual agents cannot be used for synchronization, we employ the following dynamic protocol, i.e., a dynamic feedback law, to achieve synchronization:

$$\dot{x}_{c,i} = A_c x_{c,i} + B_c \sum_{j \in \mathscr{N}_i} (y_i - y_j), \tag{9.6a}$$

$$u_i = C_c x_{c,i} + D_c \sum_{j \in \mathscr{N}_i} (y_i - y_j), \qquad i = 1, 2, \ldots, N, \tag{9.6b}$$

where $x_{c,i} : \mathbb{R}_+ \to \mathbb{R}^{n_c}$ is the state of the protocol for the agent i.

Assumption 9.3 $x_{c,i}(0) = 0$, $i = 1, \ldots, N$.

The local synchronization problem considered in this paper is to design a dynamic protocol of the form (9.6) which satisfies

$$\lim_{t \to \infty} \left\| x_i(t) - x_j(t) \right\| = 0 \ \forall i, \ j \in \mathscr{V} \tag{9.7}$$

$$\lim_{t \to \infty} \left\| x_{c,i}(t) - x_{c,j}(t) \right\| = 0 \ \forall i, \ j \in \mathscr{V} \tag{9.8}$$

for any state trajectories $(x_1, \ldots, x_N)$ starting from the inside of some closed region $\mathscr{R} \in \mathbb{R}^{nN}$, and for any input nonlinearities $\psi_1, \ldots, \psi_N$ satisfying (9.2).

Define

$$x = \begin{bmatrix} x_1 \\ x_2 \\ \vdots \\ x_N \end{bmatrix}, \ u = \begin{bmatrix} u_1 \\ u_2 \\ \vdots \\ u_N \end{bmatrix}, \ y = \begin{bmatrix} y_1 \\ y_2 \\ \vdots \\ y_N \end{bmatrix}, \ x_c = \begin{bmatrix} x_{c1} \\ x_{c2} \\ \vdots \\ x_{cN} \end{bmatrix}, \ \Phi(u, t) = \begin{bmatrix} \varphi_1(u_1, t) \\ \varphi_2(u_2, t) \\ \vdots \\ \varphi_N(u_N, t) \end{bmatrix}.$$

Then, the closed-loop system of (9.5), (9.6) equivalently reduces to

$$\begin{bmatrix} \dot{x} \\ \dot{x}_c \end{bmatrix} = \begin{bmatrix} I_N \otimes A + L \otimes B D_c C & I_N \otimes B C_c \\ L \otimes B_c C & I_N \otimes A_c \end{bmatrix} \begin{bmatrix} x \\ x_c \end{bmatrix} - \begin{bmatrix} I_N \otimes B \\ 0 \end{bmatrix} w, \tag{9.9a}$$

$$u = \begin{bmatrix} L \otimes D_c C & I_N \otimes C_c \end{bmatrix} \begin{bmatrix} x \\ x_c \end{bmatrix}, \tag{9.9b}$$

$$w = \Phi(u, t), \tag{9.9c}$$

where L is the Laplacian of the communication graph, $\otimes$ denotes the Kronecker product, and I_p denotes the $p \times p$ identity matrix.

It follows from (9.4) and the definition of Φ that

$$w^\top (w - \gamma u) \leq 0 \tag{9.10}$$

holds for $w = \Phi(u, t)$, $u \in [-\mu, \mu]^N$, $t \geq 0$. Moreover, we define the diagonal matrix Λ and the orthogonal matrix U by

$$ULU^\top = \Lambda, \quad \Lambda = \mathrm{diag}\{0, \lambda_2, \ldots, \lambda_N\}, \tag{9.11}$$

and perform the change of variables as

$$\xi = \begin{bmatrix} \xi_1 \\ \vdots \\ \xi_N \end{bmatrix} = (U \otimes I_n)x, \qquad \xi_c = \begin{bmatrix} \xi_{c1} \\ \vdots \\ \xi_{c,N} \end{bmatrix} = (U \otimes I_{n_c})x_c. \tag{9.12}$$

Notice the first column of $U^\top$ is the eigenvector associated with the smallest eigenvalue $\lambda_1 = 0$. Hence, the first row of U is equal to $\mathbf{1}^\top/\sqrt{N}$, and we obtain $\xi_1 = \frac{1}{\sqrt{N}} \sum_{i=1}^N x_i$. It then follows from (9.12) that, if $\lim_{t\to\infty} \|\xi_i(t)\| = 0$ is satisfied for $i = 2, \ldots, N$, we get

$$x(t) \to (U^\top \otimes I_n) \begin{bmatrix} \xi_1(t) \\ 0 \\ \vdots \\ 0 \end{bmatrix} = \frac{1}{\sqrt{N}} \mathbf{1} \otimes \xi_1(t) \quad (t \to \infty).$$

The same discussion also applies to x_c and ξ_c. From the above observation, the local synchronization of the states $(x_i, x_{c,i})$, $i = 1, \ldots, N$, reduces to the asymptotic stabilization of $(\xi_i, \xi_{c,i})$, $i = 2, \ldots, N$ to the origin.

9.3 Synchronization Condition

Since U is an orthogonal matrix, application of the coordinate transformation (9.12) to (9.9) yields

$$\begin{bmatrix} \dot{\xi} \\ \dot{\xi}_c \end{bmatrix} = \begin{bmatrix} I_N \otimes A + \Lambda \otimes BD_cC & I_N \otimes BC_c \\ \Lambda \otimes B_cC & I_N \otimes A_c \end{bmatrix} \begin{bmatrix} \xi \\ \xi_c \end{bmatrix} - \begin{bmatrix} I_N \otimes B \\ 0 \end{bmatrix} \tilde{w} \tag{9.13a}$$

$$\tilde{u} = Uu = \begin{bmatrix} \Lambda \otimes D_cC & I_N \otimes C_c \end{bmatrix} \begin{bmatrix} \xi \\ \xi_c \end{bmatrix} \tag{9.13b}$$

$$\tilde{w} = Uw = U\Phi(U^\top \tilde{u}, t) \tag{9.13c}$$

Moreover, since Λ is a diagonal matrix, (9.13) can be equivalently rewritten as

$$\dot{z} = \begin{bmatrix} A_1 & & & \\ & A_2 & & \\ & & \ddots & \\ & & & A_N \end{bmatrix} z - \begin{bmatrix} B_1 & & & \\ & B_2 & & \\ & & \ddots & \\ & & & B_N \end{bmatrix} \tilde{w} \tag{9.14a}$$

$$\tilde{u} = \begin{bmatrix} C_1 & & & \\ & C_2 & & \\ & & \ddots & \\ & & & C_N \end{bmatrix} z \tag{9.14b}$$

$$\tilde{w} = U\Phi(U^\top \tilde{u},\ t) \tag{9.14c}$$

where

$$z = \begin{bmatrix} z_1 \\ \vdots \\ z_N \end{bmatrix},\quad z_i = \begin{bmatrix} \xi_i \\ \xi_{c,i} \end{bmatrix},\quad i = 1, \dots, N$$

and

$$A_i = \begin{bmatrix} A + \lambda_i B D_c C & B C_c \\ \lambda_i B_c C & A_c \end{bmatrix},\quad B_i = \begin{bmatrix} B \\ 0 \end{bmatrix},\quad C_i = \begin{bmatrix} \lambda_i D_c C & C_c \end{bmatrix}.$$

Lemma 9.4 *The inequality*

$$\tilde{w}^\top (\tilde{w} - \gamma \tilde{u}) \le 0$$

holds for $\tilde{w} = U\Phi(U^\top \tilde{u},\ t)$ *and* $t \ge 0$, *where* $\tilde{u} \in \mathbb{R}^N$ *is an arbitrary vector such that* $\tilde{u} = Uu$, $u \in [-\mu,\ \mu]^N$.

Theorem 9.5 follows from Lemma 9.4 and the block diagonal structure of (9.14).

Theorem 9.5 *Under Assumptions 9.1–9.3, let a dynamic protocol of (9.6) be given. Assume that there exists a positive definite matrix* $P \in \mathbb{R}^{(n+n_c)\times(n+n_c)}$ *satisfying*

$$\begin{bmatrix} A_i^\top P + P A_i & \gamma C_i^\top - P B_i \\ \gamma C_i - B_i^\top P & -2 \end{bmatrix} < 0, \tag{9.15}$$

$$\begin{bmatrix} P & C_i^\top \\ C_i & \mu^2 \end{bmatrix} \ge 0 \tag{9.16}$$

for $i = 2, \dots, N$. *Then, the multi-agent system (9.5) with the dynamic protocol achieves the local synchronization for arbitrary nonlinearities* $\varphi_1, \dots, \varphi_N$ *satisfying the sector condition (9.4), or equivalently, the multi-agent system (9.1) with the same dynamic protocol achieves the local synchronization for arbitrary* $\psi_1, \dots, \psi_N$ *satisfying (9.2).*

Proof Based on the discussion in the previous section, we have only to show the convergence of z_i, $i = 2, \dots, N$ to the origin.

Suppose that (9.15) and (9.16) are satisfied. Then,

$$\begin{bmatrix} A_i^\top P + PA_i + \varepsilon I_{n+n_c} & \gamma C_i^\top - PB_i \\ \gamma C_i - B_i^\top P & -2 \end{bmatrix} \le 0, \quad i = 2, \dots, N \tag{9.17}$$

holds for some $\varepsilon > 0$. It follows from (9.14) and (9.17) that

$$\frac{d}{dt}(z_i^\top P z_i) = 2z_i^\top P(A_i z_i - B_i \tilde{w}_i) \le -\varepsilon \|z_i\|^2 + 2\tilde{w}_i(\tilde{w}_i - \gamma \tilde{u}_i)$$

holds for $i = 2, \dots, N$. Thus, by defining

$$V(z) = \sum_{i=2}^{N} z_i^\top P z_i,$$

we obtain

$$\dot{V}(z) + \varepsilon \sum_{i=2}^{N} \|z_i\|^2 \le 2 \sum_{i=2}^{N} \tilde{w}_i(\tilde{w}_i - \gamma \tilde{u}_i). \tag{9.18}$$

On the other hand, we see from $\lambda_1 = 0$ and (9.14) that

$$\dot{\xi}_1 = A\xi_1 + BC_c\xi_{c1} - B\tilde{w}_1, \quad \dot{\xi}_{c1} = A_c\xi_{c1}, \quad \tilde{u}_1 = C_c\xi_{c1}.$$

Since $\xi_{c1}(0) = \frac{1}{\sqrt{N}} \sum_{i=1}^{N} x_{c,i}(0) = 0$ under Assumption 9.3, the above equation imply $\tilde{u}_1(t) = 0 \; \forall t \ge 0$. It thus follows that

$$\tilde{w}_1(\tilde{w}_1 - \gamma \tilde{u}_1) = \tilde{w}_1^2 \ge 0. \tag{9.19}$$

Putting (9.18) and (9.19) together yields

$$\dot{V}(z) + \varepsilon \sum_{i=2}^{N} \|z_i\|^2 \le 2\tilde{w}^\top(\tilde{w} - \gamma \tilde{u}). \tag{9.20}$$

Define the cylindrical region

$$\mathscr{C}(P) = \left\{ z \in \mathbb{R}^{(n+n_c)N} \,\middle|\, V(z) \le 1 \right\}.$$

By the Schur complement formula, (9.16) is equivalent to $P \ge \mu^{-2} C_i^\top C_i$. Since $\tilde{u}_i = C_i z_i$, the following inequality holds for $z \in \mathscr{C}(P)$.

$$\sum_{i=1}^{N}|u_i|^2=\sum_{i=1}^{N}|\tilde{u}_i|^2=\sum_{i=2}^{N}|\tilde{u}_i|^2=\sum_{i=2}^{N}z_i^\top C_i^\top C_i z_i\le\mu^2\sum_{i=2}^{N}z_i^\top P z_i=\mu^2 V(z).$$

Thus, we get $u\in[-\mu,\mu]^N$. As a result, from Lemma 9.4, $z\in\mathscr{C}(P)$ implies

$$\tilde{w}^\top(\tilde{w}-\gamma\tilde{u})\le 0. \tag{9.21}$$

We thus conclude from (9.20) and (9.21) that

$$\dot{V}(z)\le-\varepsilon\sum_{i=2}^{N}\|z_i\|^2\le 0 \tag{9.22}$$

holds for any $z\in\mathscr{C}(P)$. This inequality implies that $\mathscr{C}(P)$ is a positively invariant set for the system of (9.14). Namely, any trajectory of z starting from the inside of $\mathscr{C}(P)$ is confined in $\mathscr{C}(P)$, and satisfies $u_i\in[-\mu,\mu]$, $i=1,\dots,N$ all the time. Moreover, by La Salle's invariant principle, z converges to the maximal invariant subset of $\{z\in\mathbb{R}^{(n+n_c)N}\mid\dot{V}(z)=0\}$. Thus, we have $\|z_i(t)|\to 0$ $(t\to\infty)$, $i=2,\dots,N$ by the Eq. (9.22). This implies that the local synchronization (9.7), (9.8) is achieved. □ ■

The matrix inequalities (9.15), (9.16) are affine with respect to λ_i, and λ_i's are ordered in the ascending order. Therefore, to check the synchronization, it suffice to solve the matrix inequalities in Theorem 9.5 only for $i=2$ and N.

Corollary 9.6 *Under Assumptions 9.1–9.3, let a dynamic protocol of (9.6) be given. Assume that there exists a positive definite matrix $P\in\mathbb{R}^{(n+n_c)\times(n+n_c)}$ satisfying (9.15) and (9.16) for $i=2,\ N$. Then, the multi-agent system (9.5) with the dynamic protocol achieves the local synchronization for arbitrary nonlinearities $\varphi_1,\dots,\varphi_N$ satisfying the sector condition (9.4), or equivalently, the multi-agent system (9.1) with the same protocol achieves the local synchronization for arbitrary $\psi_1,\dots,\psi_N$ satisfying (9.2).*

Remark 9.7 Theorem 9.5 provides an inner approximation of the region of attraction $\mathscr{R}$ to the synchronized states as

$$\Omega(P)=\left\{x\in\mathbb{R}^{nN}\,\middle|\,z\in\mathscr{C}(P),\quad z=T\begin{bmatrix}x\\0\end{bmatrix}\right\},$$

where T is defined by $T=\left[U\otimes\begin{bmatrix}I_n\\0\end{bmatrix}\ \ U\otimes\begin{bmatrix}0\\I_{n_c}\end{bmatrix}\right]$, which maps (x,x_c) to z.

Remark 9.8 In view of the well-known circle criterion, Theorem 9.5 implies that the local synchronization problem is reduced to the stabilization problem of finding an output feedback controller

$$\dot{x}_c = A_c x_c + B_c y, \quad u = C_c x_c + D_c y$$

that asymptotically stabilizes

$$\dot{x} = Ax + Bu - B\varphi(u,t), \quad y = \lambda_i C x$$

for $i = 2, \ldots, N$, and for every $\varphi : \mathbb{R} \times \mathbb{R}_+ \to \mathbb{R}$ satisfying the sector condition

$$\varphi(u,t)[\varphi(u,t) - \gamma u] \leq 0 \ \forall u \in [-\mu, \mu], \ \forall t \geq 0.$$

9.4 Dynamic Protocol Synthesis

On the basis of Theorem 9.5, we shall present a synthesis method of a dynamic protocol which achieves the local synchronization, with the aid of the change of variables technique in [20]. We make the following assumption for simplicity.

Assumption 9.9 *The order of the dynamic protocol (9.6) is equal to that of the agent dynamics (9.5), namely,* $n_c = n$.

We partition P and P^{-1} as

$$P = \begin{bmatrix} Y & V \\ V^\top & * \end{bmatrix}, \quad P^{-1} = \begin{bmatrix} X & W \\ W^\top & * \end{bmatrix}, \tag{9.23}$$

where $*$ denotes irrelevant terms. Note that $WV^\top = I_n - XY$ for the sub-matrices X, Y, W, and V. We also define

$$\Pi_1 = \begin{bmatrix} X & I_n \\ W^\top & 0 \end{bmatrix}, \quad \Pi_2 = \begin{bmatrix} I_n & Y \\ 0 & V^\top \end{bmatrix}. \tag{9.24}$$

Since $P\Pi_1 = \Pi_2$, application of the congruence transform with diag$(\Pi_1, 1)$ to (9.15) and (9.16) yields

$$\mathrm{He} \left[\begin{array}{cc:c} AX + B\hat{C} + \lambda_i B\hat{D}CX & A + \lambda_i B\hat{D}C & -B \\ \hat{A} + \lambda_i \hat{B}CX & YA + \lambda_i \hat{B}C & -YB \\ \hdashline \gamma\hat{C} + \gamma\lambda_i \hat{D}CX & \gamma\lambda_i \hat{D}C & -1 \end{array} \right] < 0 \tag{9.25}$$

$$\left[\begin{array}{cc:c} X & I_n & * \\ I_n & Y & * \\ \hdashline \hat{C} + \lambda_i \hat{D}CX & \lambda_i \hat{D}C & \mu^2 \end{array} \right] \geq 0, \tag{9.26}$$

where we have defined He $(\bullet) := (\bullet) + (\bullet)^\top$, and

$$\hat{A} = VA_cW^\top + YAX + YB_cCX, \quad \hat{B} = VB_c + YBD_c, \quad \hat{C} = C_cW^\top, \quad \hat{D} = D_c. \tag{9.27}$$

Similarly, it follows from $P\Pi_1 = \Pi_2$ that $P > 0$ is equivalent to

$$\begin{bmatrix} X & I_n \\ I_n & Y \end{bmatrix} > 0. \tag{9.28}$$

As a result, a dynamic protocol achieving the synchronization can be designed by solving the matrix inequalities (9.25), (9.26) and (9.28) for $i = 2, N$.

Theorem 9.10 *Under Assumptions 9.1–9.9, assume that there exist symmetric matrices X, Y, and matrices $\hat{A}$, $\hat{B}$, $\hat{C}$, $\hat{D}$ satisfying the matrix inequalities (9.25), (9.26), and (9.28) for $i = 2, N$. Then, there exists a dynamic protocol which locally synchronizes the multi-agent system (9.1) for arbitrary input nonlinearities $\psi_1, \ldots, \psi_N$ satisfying the sector condition (9.2). One of such dynamic protocols is given by*

$$\begin{aligned} A_c &= V^{-1}(\hat{A} - YAX - YB_cCX)W^{-\top}, \\ B_c &= V^{-1}(\hat{B} - YBD_c), \\ C_c &= \hat{C}W^{-\top}, \\ D_c &= \hat{D}, \end{aligned}$$

where W and V are constant matrices such that $WV^\top = I_n - XY$.

According to Theorem 9.10, the design of a dynamic protocol which achieves the synchronization reduces to the mathematical programming problem of finding a solution to (9.25), (9.26), and (9.28) for $i = 2, N$. In particular, it is often desirable to maximize the size of the approximated region of attraction $\Omega(P)$ (see Remark 9.7). In view of the Eq. (9.23), this can be done by solving the following optimization problem.

$$\text{(OP)} \quad \underset{(X,Y,\hat{A},\hat{B},\hat{C},\hat{D})}{\text{minimize}} \quad \text{trace}\, Y \quad \text{subject to (9.25), (9.26), (9.28)}, \quad i = 2, N$$

This problem is non-convex, because (9.25) and (9.26) contain the bilinear terms between $(\hat{B}, \hat{D})$ and X. However, it is possible to find a solution to the optimization problem (OP), since several techniques to efficiently solve bilinear matrix inequalities (BMI) have been reported for the last two decades (see ,e.g., [21, 22]).

Remark 9.11 The size of the matrix inequalities in Corollary 9.6 and Theorem 9.10 are independent of the size of the communication graph, i.e., the number of agents, except for computation of the eigenvalues of L. This implies that the matrix inequality conditions are scalable as long as λ_2 and λ_N are available to the designers.

9.5 Concluding Remarks

We have studied the local state synchronization of linear agents subject to input nonlinearities over a fixed undirected communication graph. We have derived a sufficient condition for achieving the local synchronization for arbitrary nonlinearities satisfying a given sector bounds. Then, we have shown that the synthesis condition of a synchronizing dynamic protocol based on the above analysis becomes a non-convex condition in terms of BMIs. In view of Remark 9.11, the advantage of the present BMI condition is that the condition is scalable as long as the Laplacian eigenvalues of the communication graph are available to the designers. It remains as a future topic to extend the present results to more general situations such as heterogeneous agent dynamics and/or directed communication graphs. We will also need to develop a more efficient method to design a synchronizing dynamic protocol.

References

1. Mesbahi, M., Egerstedt, M.: Graph Theoretic Methods in Multiagent Networks. Princeton University Press, Princeton (2010)
2. O.-Saber, R., Fax, J.A., Murray, R.M.: Consensus and cooperation in networked multi-agent systems. Proc. IEEE **95**(1), 215–233 (2007)
3. Trentelman, H.L., Takaba, K., Monshizadeh, N.: Robust synchronization of uncertain linear multi-agent systems. IEEE Trans. Autom. Control **58**(6), 1511–1523 (2013)
4. Monshizadeh, N., Trentelman, H.L., Camlibel, M.K.: Stability and synchronization preserving model reduction of multi-agent systems. Syst. Control Lett. **62**(1), 1–10 (2013)
5. Zhang, F., Trentelman, H.L., Scherpen, J.M.A.:Output feedback robust synchronization of networked Lur'e systems with incrementally passive nonlinearities. In: Proceedings of the 21st International Symposium on Mathematical Theory of Networks and Systems (MTNS2014), 1960–1965 (2014)
6. Zhang, F., Trentelman, H.L., Scherpen, J.M.A.: Fully distributed robust synchronization of networked Lur'e systems with incremental nonlinearities. Automatica **50**(10), 2515–2526 (2014)
7. Scardovi, L., Sepulchre, R.: Synchronization in networks of identical linear systems. Automatica **45**(11), 2557–2562 (2009)
8. Zhang, H., Lewis, F.L., Das, A.: Optimal design for synchronization of cooperative systems: state feedback, observer and output feedback. IEEE Trans. Autom. Control **56**(8), 1948–1952 (2011)
9. Li, Z., Duan, Z., Chen, G., Lin, H.: Consensus of multiagent systems and synchronization of complex networks: a unified viewpoint. IEEE Trans. Circuits Syst. I **57**(1), 213–224 (2010)
10. Wieland, P., Sepulchre, R., Allgöwer, F.: Internal model principle is necessary and sufficient for linear output synchronization. Automatica **47**, 1068–1074 (2011)
11. Jönsson, U., Kao, C.-Y.: A scalable robust stability criterion for systems with heterogenous LTI components. IEEE Trans. Autom. Control **55**(10), 2210–2234 (2010)
12. Hara, S., Tanaka, H.: $\mathscr{D}$-stability and robust stability conditions for LTI systems with generalized frequency variables. In: Proceedings of the 49th IEEE Conference on Decision and Control, 5738–5743 (2010)
13. Lin, Y., Xiang, J., Wei, W.: Consensus problems for linear time-invariant multi-agent systems with saturation constraints. IET Control Theory Appl. **5**(6), 823–829 (2011)
14. Meng, Z., Zhao, Z., Lin, Z.: On global consensus of linear multi-agent systems subject to input saturation. In: Proceedings of the American Control Conference, 4516–4521 (2012)

15. Yang, T., Meng, Z., Dimarogonas, D.M., Johannnson, K.H.: Global consensus in homogeneous networks of discrete time agents subject to actuator saturation. In: Proceedings of the European Control Conference, 2045–2049 (2013)
16. Yang, T., Stoorvogel, A., Grip, H.J., Saberi, A.: Semi-global regulation of output synchronization for heterogeneous networks of non-introspective, invertible agents subject to input saturation. In: Proceedings of the 51th IEEE Conference Decision and Control, 5298–5303 (2012)
17. Takaba, K.: Synchronization of linear multi-agent systems under input saturation. In: Proceedings of the 21st International Symposium on Mathematical Theory of Networks and Systems (MTNS2014), 1976–1979 (2014)
18. Takaba, K.: Local synchronization of linear multi-agent systems under input saturation. Submitted for publication (2014)
19. Godsil, C., Royle, G.F.: Algbraic Graph Theory. Springer, New York (2001)
20. Scherer, C., Gahinet, P., Chilali, M.: Multiobjective output feedback control via LMI optimization. IEEE Trans. Autom. Control **42**(7), 896–911 (1997)
21. Goh, K.C., Turan, L., Safonov, M.G., Papavassilopoulos, G.P., Ly, J.H.: Biaffine matrix inequality properties and computational methods. In: Proceedings of the American Control Conference, 850–855 (1994)
22. Hassibi, A., How, J., Boyd, S.: A path-following method for solving BMI problems in control. In: Proceedings of the American Control Conference, 1385–1389 (1999)

Chapter 10
Strong Structural Controllability of Networks

Nima Monshizadeh

Abstract In this chapter, we discuss strong structural controllability and strong targeted controllability of networks from a unified viewpoint. The problem of strong structural controllability accounts for controllability of the whole family of matrices carrying the structure of an underlying graph. By looking into a certain infection process identified by a coloring rule, topological characterizations for strong structural properties of the network is provided. In particular, the strong structurally reachable subspace is translated into the derived set of a given leader set. Moreover, the set of leaders rendering the network strongly structurally controllable are characterized by zero forcing sets. Then, the minimum number of leaders to achieve strong structural controllability is given by the zero forcing number of the graph. Motivated by the fact that network controllability is neither always feasible nor necessary, we discuss the problem of (strong) targeted controllability where controllability is only required for a subset of the nodes in the network. We illustrate graph theoretic sufficient conditions guaranteeing strong targeted controllability of the network.

10.1 Preliminaries

For a given simple directed graph G, the vertex set of G is a nonempty set and is denoted by V. The arc set of G, denoted by E, is a subset of $V \times V$, and $(i, i) \notin E$ for all $i \in V$. The cardinality of a given set V is denoted by $|V|$. Also we sometimes use $|G|$ to denote in short the cardinality of V. We call vertex j an out-neighbor of vertex i if $(i, j) \in E$.

For $V = \{1, 2, \dots, n\}$ and $V' = \{v_1, v_2, \dots, v_r\} \subseteq V$, we define the $n \times r$ matrix $P(V; V') = [P_{ij}]$ by:

N. Monshizadeh (✉)
Faculty of Mathematics and Natural Sciences, Engineering and Technology Institute Groningen, University of Groningen, Nijenborg 4, 9747 Ag Groningen, The Netherlands
e-mail: n.monshizadeh@rug.nl

M.N. Belur et al. (eds.), *Mathematical Control Theory II*, Lecture Notes in Control and Information Sciences 462,
DOI 10.1007/978-3-319-21003-2_10

$$P_{ij} = \begin{cases} 1 & \text{if } i = v_j \\ 0 & \text{otherwise.} \end{cases} \tag{10.1}$$

10.2 Problem Formulation

We consider the following dynamics evolving on a directed graph $G = (V, E)$:

$$\dot{x} = Xx + Uu \tag{10.2}$$

where $x \in \mathbb{R}^{|G|}$ is the state, $u \in \mathbb{R}^m$ is the input, and $U = P(V; V_L)$ for some given *leader set* $V_L \subseteq V$. Note that the matrix X represents the *coupling* in accordance with G, and reveals how information is exchanged throughout the network.

Recall that controllability is the ability of an external input to steer a system from any initial state to any other final state in a finite time. Here, we are primarily interested in "strong structural controllability" of systems of the form (10.2). Roughly speaking, by "structural" we refer to properties which are identified by the graph G rather than a particular realization of the system (10.2). To formalize this, we define the *qualitative class* of G as the family of matrices compatible with the structure of G:

$$Q(G) = \{X \in \mathbb{R}^{|G|\times|G|} : \text{ for } i \neq j,\ X_{ij} \neq 0 \Leftrightarrow (j, i) \in E\}$$

Then, for a given leader set V_L, we call the system (10.2) *strongly structurally controllable* if the pair (X, U) is controllable for all $X \in Q(G)$. In that case, we write $G; V_L)$ is controllable. Note that the term "strong" is used to distinguish with the case of "weak structural controllability" which amounts to the existence of a controllable pair (X, U), $X \in Q(G)$.

Another problem of interest is "minimal leader selection" the goal of which is to choose a leader set V_L with minimum cardinality such that the pair (X, U) is controllable. A strong structural variation of this problem is to select a leader set V_L with minimum cardinality such that (X, U) is controllable for all $X \in Q(G)$. We denote the cardinality of such minimal leader set by $\ell_{\min}(G)$, i.e.,

$$\ell_{\min}(G) = \min_{V_L \subseteq V(G)} \{|V_L| : (G; V_L) \text{ is controllable}\}. \tag{10.3}$$

For simplicity, we use calligraphic notation in this chapter to denote the image of a matrix induced by a subset $V' \subseteq V$. More precisely, $\mathcal{V}'$ denotes, in short, the subspace $\operatorname{im} P(V; V')$.

In the next section, we briefly review the state of the art in controllability as well as structural controllability of systems of the form (10.2).

10.3 Review: Controllability of Networks

One main line of research within the context of controllability of networks has been devoted to the case where $X = L$ in (10.2) with L being the Laplacian matrix of an undirected graph G. This line of research has been initiated by [15] and further developed by [14]. Motivated by the fact that algebraic conditions do not provide much insight into network controllability, topological translation of controllability properties in terms of certain graph partitions has been pursued by some authors [8, 17]. These partitions in fact provide a partial characterization of the reachable subspace associated with the pair (L, U). An extension to the controlled invariant subspace has also been reported in [10]. Characterization of controllability of the pair (L, U) in terms of graph automorphisms is provided in [1].

In addition, a (minimum) leader selection for rendering the pair (L, U) controllable has been investigated for particular classes of undirected graphs [12, 13, 17].

Another instance of systems that has been studied in the context of controllability corresponds to the case where $X = A$ in (10.2) with A being the adjacency matrix of an undirected graph, see, e.g., [4].

In complex networks, the weights of the communications are typically unknown or partially known. Hence, it is interesting to investigate a controllability property which depends on the structure of the graph/network rather than a particular realization. In fact, this property, known as structural controllability deals with a family of pairs (X, U) rather than a particular instance and asks whether the family contains a controllable pair (weak structural controllability [7]) or all members of the family are controllable (strong structural controllability [11]). The latter is also the main focus of the present chapter. For a more general look at controllability of structured systems see, e.g., [2, 6, 9].

10.4 Strong Structural Controllability

First, we recap the notion of the reachability subspace in time-invariant linear systems.

10.4.1 Reachability Subspace

Consider, the system

$$\dot{x}(t) = Ax(t) + Bu(t) \tag{10.4}$$

with state space $\mathbb{R}^n$. For a given initial state x_0 and input function u, we denote the resulting state trajectory of the system by $x_u(t, x_0)$. The smallest A-invariant subspace containing the image of the input matrix U is denoted by $\langle A \mid \text{im } B\rangle$. This subspace, called the *reachable subspace*, consists of all points in the state space that can be reached from the origin in finite time by choosing an appropriate input function, i.e., all points $x_1 \in \mathbb{R}^n$ for which there exists $T > 0$ and u such that $x_1 = x_u(T, 0)$. It is well known that the system is controllable if and only if the reachable subspace $\langle A \mid \text{im } B \rangle$ is equal to the entire state space $\mathbb{R}^n$. In turn, this is equivalent to the condition

$$\text{rank}\begin{bmatrix} B & AB & \cdots & A^{n-1}B \end{bmatrix} = n.$$

10.4.2 Strong Structurally Reachable Subspace

Recall that we are interested in structural controllability properties of systems of the form (10.2). To incorporate these structural properties, we consider all the points in the state space which can be reached by applying appropriate input signals to the nodes in the leader set V_L, irrespective of the choice of $X \in Q(G)$. These points constitute a subspace which we call *strong structurally reachable subspace*. By definition, this subspace is equal to

$$\bigcap_{X \in Q(G)} \langle X \mid \mathscr{V}_L \rangle.$$

Moreover, it is easy to observe that the strong structurally reachable subspace provides a geometric characterization for strong structural controllability, i.e.,

$$(G; V_L) \text{ is controllable} \Leftrightarrow \bigcap_{X \in Q(G)} \langle X \mid \mathscr{V}_L \rangle = \mathbb{R}^{|G|} \tag{10.5}$$

Note that the geometric characterization (10.5) by itself does not provide much insight to the strong structural controllability property. In particular, we would like to "visualize" a network enjoying this property in comparison with the one which lacks such a property. Ultimately, we would like to draw some conclusions on possible minimal leader selections guaranteeingsss strong structural controllability. To this end, we use the notion of zero forcing sets.

10.4.3 Zero Forcing Sets

In this section, we recap the notion of zero forcing sets together with the terminology that will be used later in this chapter. For more details see, e.g., [5].

Let G be a graph, and suppose that each vertex is colored either white or black. Consider the following coloring rule:

◐: If u is a black vertex and exactly one out-neighbor v of u is white, then change the color of v to black.

The following terminology will be used when we apply the color-change rule above to a graph G:

- If the color-change rule is applied to $u \in V$ to change the color of $v \in V$, we say u *forces* or *infects* v, and write $u \to v$.
- Given a coloring set $C \subseteq V$, i.e., C indexes the initially black vertices of G, the *derived set* of C is denoted by $D(C)$, and is the set of black vertices obtained by applying the color-change rule until no more changes are possible.
- The set $Z(G) \subseteq V$ is a *zero forcing set* (ZFS) for G if $D(Z(G)) = V$.
- The *zero forcing number* $Z(G)$ is the minimum of $|Z|$ over all zero forcing sets $Z(G) \subseteq V$.

Figure 10.1 illustrates the coloring rule, where vertices 1, 2, and 5 are initially colored black. As vertex 1 has only one white out-neighbor, $1 \to 3$. Similarly, $5 \to 7$. Consequently, $2 \to 4$, and then $3 \to 9$. Therefore, the derived set of $\{1, 2, 5\}$ is equal to $\{1, 2, 3, 4, 5, 7, 9\}$, and clearly $\{1, 2, 5\}$ is not a zero forcing set. It is easy to observe that the set $\{1, 2, 5, 6\}$ constitutes a minimal zero forcing set, and thus the zero forcing number is equal to 4 in this case.

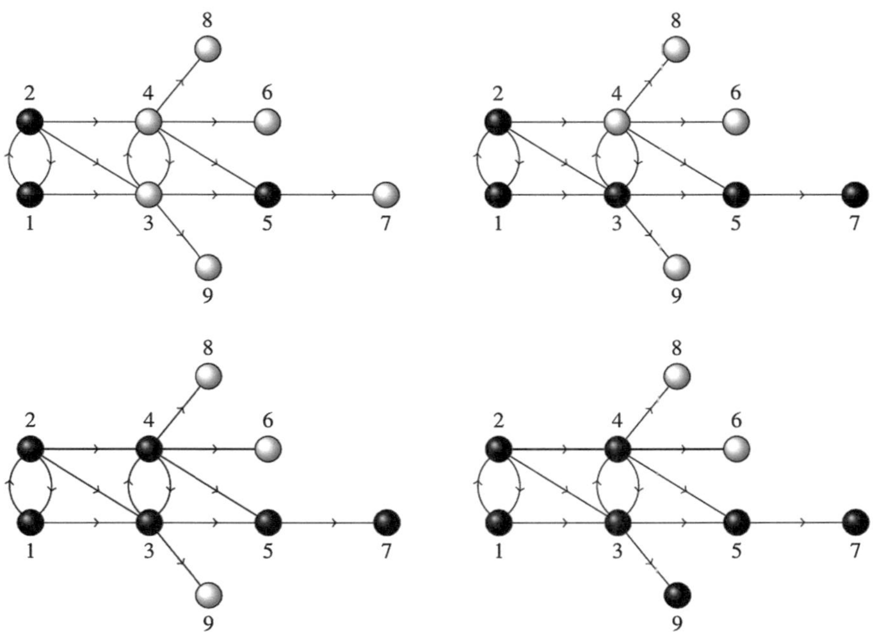

Fig. 10.1 An example of the coloring rule

10.4.4 Topological Characterization

The coloring rule and zero forcing sets discussed in the previous subsection generate a chain of forces/infections in the graph, starting from a leader set V_L. Next, we investigate how controllability properties of the network is affected by this forcing/infection process. More precisely, let $v \in V_L$, $w \notin V_L$ and suppose that $v \to w$. Then, we examine the effect of this infection on the reachable subspace, on the strong structurally reachable subspace, and ultimately we draw conclusions on strong structural controllability of (10.2).

First, we look into the reachability subspace, i.e., $\langle X \mid \mathcal{V}_L \rangle$ for any given $V_L \subseteq V$ and $X \in Q(G)$. We *claim* that

$$\operatorname{im} P(V; V_L') \subseteq \langle X \mid \mathcal{V}_L \rangle. \tag{10.6}$$

where $V_L' = V_L \cup \{w\}$ and P is given by (10.1). Without loss of generality, assume that $V_L = \{1, 2, \ldots, m\}$, $v = m$, and $w = m + 1$. Then, the matrix X can be partitioned as

$$X = \begin{bmatrix} X_{11} & X_{12} & X_{13} & X_{14} \\ X_{21} & X_{22} & X_{23} & X_{24} \\ X_{31} & X_{32} & X_{33} & X_{34} \\ X_{41} & X_{42} & X_{43} & X_{44} \end{bmatrix} \tag{10.7}$$

where the diagonal blocks/elements X_{11}, X_{22}, X_{33}, and X_{44} correspond to the vertices in $V_L \backslash \{v\}$, the vertex v, the vertex w, and the rest of the vertices, respectively.

Let $\xi \in \mathbb{R}^n$ be a vector in $\langle X \mid \mathcal{V}_L \rangle^{\perp}$. Clearly, we have $\xi^T X^{k-1} P(V; V_L) = 0$ for each $k \in \mathbb{N}$. We write $\xi = \operatorname{col}(\xi_1, \xi_2, \xi_3, \xi_4)$ compatible with the partitioning of X. Note that $P(V; V_L')$ now reads as

$$P(V; V_L') = \begin{bmatrix} I_{m-1} & 0 & 0 \\ 0 & 1 & 0 \\ 0 & 0 & 1 \\ 0 & 0 & 0 \end{bmatrix}$$

From the equality $\xi^T P(V; V_L) = 0$, we obtain that $\xi_1 = 0$ and $\xi_2 = 0$. Then, the equality $\xi^T X P(V; V_L) = 0$ yields

$$\begin{bmatrix} \xi_3^T & \xi_4^T \end{bmatrix} \begin{bmatrix} X_{31} & X_{32} & X_{33} \\ X_{41} & X_{42} & X_{43} \end{bmatrix} = 0 \tag{10.8}$$

Observe that, since $v \to w$, the vertex v has exactly one out-neighbor in $V \backslash V_L$, and thus we have $X_{32} \neq 0$ and $X_{42} = 0$. Therefore, by (10.8), we obtain that the scalar ξ_3 is equal to zero. Clearly, $\xi = \operatorname{col}(0, 0, 0, \xi_4)$ is orthogonal to the subspace $\operatorname{im} P(V; V_L')$. Hence, the subspace inclusion (10.6) holds.

Now, by repeating the argument above, we conclude that

$$\mathscr{D}(V_L) \subseteq \langle X \mid \mathscr{V}_L \rangle \tag{10.9}$$

where $\mathscr{D}(V_L) = \text{im}\, P(V; D(V_L))$. This yields

$$\langle X \mid \mathscr{D}(V_L) \rangle \subseteq \langle X \mid \mathscr{V}_L \rangle \tag{10.10}$$

On the other hand, noting that $\mathscr{V}_L \subseteq \mathscr{D}(V_L)$, we obtain that

$$\langle X \mid \mathscr{V}_L \rangle \subseteq \langle X \mid \mathscr{D}(V_L) \rangle.$$

This together with (10.10) yields

$$\langle X \mid \mathscr{V}_L \rangle = \langle X \mid \mathscr{D}(V_L) \rangle \tag{10.11}$$

for any given leader set V_L. The equality above reveals the fact that the reachable subspace is invariant under the infection process. Consequently, the strong structurally reachable subspace enjoys this invariance property as well, i.e.

$$\bigcap_{X \in Q(G)} \langle X \mid \mathscr{V}_L \rangle = \bigcap_{X \in Q(G)} \langle X \mid \mathscr{D}(V_L) \rangle \tag{10.12}$$

The aforementioned invariance properties can be used to provide a topological translation of the strong structurally reachable subspace. In particular, observe that

$$\mathscr{D}(V_L) \subseteq \bigcap_{X \in Q(G)} \langle X \mid \mathscr{D}(V_L) \rangle = \bigcap_{X \in Q(G)} \langle X \mid \mathscr{V}_L \rangle \tag{10.13}$$

Now, we show that $\mathscr{D}(V_L)$ is indeed equal to the strong structurally reachable subspace. Clearly, by (10.13), it remains to show that

$$\bigcap_{X \in Q(G)} \langle X \mid \mathscr{D}(V_L) \rangle \subseteq \mathscr{D}(V_L) \tag{10.14}$$

We define the set Δ as

$$\Delta = \{\delta \in \mathbb{R}^n : \delta_i = 0 \Leftrightarrow i \in D(V_L)\} \tag{10.15}$$

Let δ be a vector in Δ. Without loss of generality, let $D(V_L) = \{1, 2, \ldots, d\}$. Then, δ can be written as $\text{col}(0_d, \delta_2)$ where each element of $\delta_2 \in \mathbb{R}^{n-d}$ is nonzero. Let the matrix X be partitioned accordingly as

$$X = \begin{bmatrix} X_{11} & X_{12} \\ X_{21} & X_{22} \end{bmatrix}$$

Clearly, we have

$$\delta^T X = \delta_2^T \begin{bmatrix} X_{21} & X_{22} \end{bmatrix}$$

Observe that nonzero elements of X_{21} correspond to the arcs from the vertices in $D(V_L)$ to the vertices in $V \setminus D(V_L)$. Hence, by the coloring rule, each column of X_{21} is either identically zero or contains at least two nonzero elements. We choose these nonzero elements, if any, such that $\delta_2^T X_{21} = 0$. Noting that the diagonal elements of X_{22} are free parameters, we conclude that, for any vector $\delta \in \Delta$, there exists a matrix $X \in Q(G)$ such that $\delta^T X = 0$. Therefore, we obtain that

$$\delta \in \langle X \mid \mathscr{D}(V_L) \rangle^{\perp}$$

for some matrix $X \in Q(G)$. Now, let $\xi \in \mathbb{R}^n$ be a vector in $\bigcap_{X \in Q(G)} \langle X \mid \mathscr{D}(V_L) \rangle$. Clearly, we have $\delta^T \xi = 0$ which yields $\delta_2^T \xi_2 = 0$, by writing $\xi = \text{col}(\xi_1, \xi_2)$. As this conclusion holds for an arbitrary choice of $\delta \in \Delta$, we obtain that $\xi_2 = 0$, and thus $\xi \in \mathscr{D}(V_L)$ which proves (10.14). Hence, we conclude that

$$\bigcap_{X \in Q(G)} \langle X \mid \mathscr{V}_L \rangle = \mathscr{D}(V_L). \tag{10.16}$$

By (10.16), the strong structurally reachable subspace is topologically equivalent to the subspace induced by the derived set of V_L, i.e., $\mathscr{D}(V_L)$. An important special case of (10.16) is obtained for $\mathscr{D}(V_L) = \mathbb{R}^{|G|}$, or equivalently $V_L = V$. This results in an exact topological translation of strong structural controllability:

$$(G; V_L) \text{ is controllable} \iff V_L \text{ is a zero forcing set of } G \tag{10.17}$$

In addition, by a simple cardinality argument, we obtain that

$$\ell_{\min}(G) = Z(G). \tag{10.18}$$

Note that the equalities (10.16)–(10.18) provide a topological characterization of structural controllability properties of systems of the form (10.2). In particular, the derived set of V_L determines the strong structurally reachable subspace, there is a one-to-one correspondence between zero forcing sets and sets of leaders rendering the network strongly structurally controllable, and the minimum number of leaders required is given by the zero forcing number of the graph.

10.4.5 Leader Selection

An important question in the context of controllability of dynamical networks is minimal leader selection, i.e., to choose a leader set, with minimum cardinality, such that the network is controllable. The research effort in this direction has been devoted mostly to the case where X is equal to the Laplacian matrix of an undirected graph, in particular undirected path, cycle, complete, and circulant graphs [12, 14, 17]. Minimum leader selections rendering the network weakly structurally controllable is also discussed in [7].

The equality (10.18) reveals the fact that our knowledge about minimal leader selection for strong structural controllability of networks is closely related to the knowledge we have on the zero forcing number of graphs. In fact, for any graph whose zero forcing number is known or can be computed, we immediately obtain the minimum number of leaders for controllability, and, in principle, a minimal leader selection scheme. To illustrate this, we mention next few examples.

Note that an undirected graph can be identified by a corresponding directed graph whose arc set is symmetric, see [11] for more details. Then, clearly, either of the two boundary vertices in an undirected path graph P forms a zero forcing set, and thus $\ell_{\min}(P) = 1$. For an undirected cycle graph C, any two neighboring vertices constitute a minimal zero forcing set and hence $\ell_{\min}(C) = 2$. Similarly, for an undirected complete graph K with n vertices, we have $\ell_{\min}(K) = n - 1$. For a directed cycle graph, merely one vertex suffices for strong structural controllability.

The *path cover number* of G, denoted by $P(G)$, is the minimum number of vertex disjoint paths occurring as induced subgraphs of G that cover all the vertices of G; such a set of paths realizing $P(G)$ is called a *minimal path cover*. For acyclic graphs, it has been shown that the zero forcing number is equal to the path cover number of the graph. Moreover, the initial points of the vertex disjoint paths realizing a minimal path cover form a minimal zero forcing set [5]. Therefore, by selecting those initial points as leaders, we obtain a minimal leader selection scheme for strong structural controllability of acyclic networks.

However, computing a minimal zero forcing set for general graphs with cycles is very difficult. Hence, determining zero forcing number for certain subclasses of graphs, as well as finding suboptimal (non-minimal) zero forcing sets for general directed graphs are interesting problems in the context of strong structural controllability of dynamical networks.

10.5 Strong Targeted Controllability

Network controllability is not always present in complex networks, or it may ask for considerable number of nodes to be directly controlled which is not always feasible. Besides, in certain cases steering the entire network to any arbitrary state may not be necessary, and instead the interest is to drive a subset of the network to a desired state. Following [3], we refer to this problem as *targeted controllability* of networks. By the results discussed in the previous section, we investigate the targeted controllability problem in a strong structural sense. Note that targeted controllability is essentially an "output controllability" problem which is recapped next.

10.5.1 Output Controllability

Consider again the system (10.4) with an additional output equation

$$y(t) = Cx(t) \tag{10.19}$$

where the output $y(t)$ takes its values in the output space $\mathbb{R}^p$. Denote the output trajectory corresponding to the initial state x_0 and input function u by $y_u(t, x_0)$. The system (10.4)–(10.19) is then called *output controllable* if for any $x_0 \in \mathbb{R}^n$ and $y_1 \in \mathbb{R}^p$ there exists an input function u and a $T > 0$ such that $y_u(T, x_0) = y_1$. We also say that the triple (A, B, C) is output controllable meaning that the system (10.4), (10.19) is output controllable. It is well known (see ,e.g., [16, Exc. 3.22]) that (A, B, C) is output controllable if and only if the rank condition

$$\text{rank}\begin{bmatrix} CB & CAB & \cdots & CA^{n-1}B \end{bmatrix} = p.$$

holds. In turn this is equivalent to the condition

$$C\langle A \mid \text{im}\, B\rangle = \mathbb{R}^p,$$

,i.e., the image under C of the reachable subspace is equal to the output space $\mathbb{R}^p$. Obviously, this condition is equivalent to $\ker C + \langle A \mid \text{im}\, B\rangle = \mathbb{R}^n$. Finally, by taking orthogonal complements, the latter holds if and only if

$$\text{im}\, C^\top \cap \langle A \mid \text{im}\, B\rangle^\perp = \{0\}.$$

10.5.2 Topological Characterization

In this subsection, we discuss the "strong targeted controllability" problem for systems of the form (10.2) with an additional output equation:

$$\dot{x} = Xx + Uu \tag{10.20a}$$

$$y = Hx \tag{10.20b}$$

where $X \in \mathcal{Q}(G)$, $U = P(V; V_L)$ for some given leader set $V_L \subseteq V$, and $H = P^T(V; V_T)$ for some given *target set* $V_T \subseteq V$. For a given leader set V_L and a target set V_T, we call the system (10.20) *strongly targeted controllable* if the triple (X, U, H) is output controllable for all $X \in \mathcal{Q}(G)$. In that case, we also write as $(G; V_L; V_T)$ is targeted controllable.

Note that strong targeted controllability is basically a strong structural output controllability property, where the output of the network can be steered to any desired state in $\mathbb{R}^{|V_T|}$, irrespective of the choice of $X \in Q(G)$.

Let $\mathscr{V}_T = \operatorname{im} P(V; V_T)$, $|V_T| = p$, and $|V| = n$. Then, by Sect. 10.5.1, $(G; V_L; V_T)$ is targeted controllable if and only if

$$\operatorname{rank} \begin{bmatrix} HU & HXU & HX^2U & \cdots & HX^{n-1}U \end{bmatrix} = p \quad \text{for all} \quad X \in Q(G),$$

which is equivalent to

$$H \langle X \mid \mathscr{V}_L \rangle = \mathbb{R}^p \quad \text{for all} \quad X \in Q(G),$$

and thus to

$$\mathscr{V}_T \cap \langle X \mid \mathscr{V}_L \rangle^{\perp} = \{0\} \quad \text{for all} \quad X \in Q(G). \tag{10.21}$$

Note that targeted controllability captures the structural controllability as a special case, namely by setting $H = I$, or equivalently $V_T = V$. Hence, investigating targeted controllability of networks is a more general, and thus more challenging problem. Next, by using the results elaborated in the previous section, we discuss topological conditions under which the network is strongly targeted controllable.

Observe that $(G; V_L; V_T)$ is targeted controllable if

$$\mathscr{V}_T \subseteq \bigcap_{X \in Q(G)} \langle X \mid \mathscr{V}_L \rangle \tag{10.22}$$

Note that indeed (10.22) implies (10.21), which results in targeted controllability. Therefore, by (10.16), we conclude that

$$V_T \subseteq D(V_L) \Longrightarrow (G; V_L; V_T) \text{ is targeted controllable.} \tag{10.23}$$

This means that the state components corresponding to the vertices in the derived set of V_L can be steered to any desired point in $\mathbb{R}^d$ with $d = |D(V_L)|$, by applying appropriate inputs to the vertices in V_L.

Consider the graph depicted in Fig. 10.2, and let $V_L = \{1, 2\}$. It is easy to observe that by the color-change rule the derived set of V_L is obtained as $D(V_L) = \{1, 2, 3, 4\}$. By (10.23), we have that $(G; V_L; V_T)$ is targeted controllable for any

$$V_T \subseteq \{1, 2, 3, 4\}. \tag{10.24}$$

However, this is not necessary as one can show that $(G; V_L; V_T)$ is targeted controllable with

$$V_T = \{1, 2, 3, 4, 5, 6, 7\}. \tag{10.25}$$

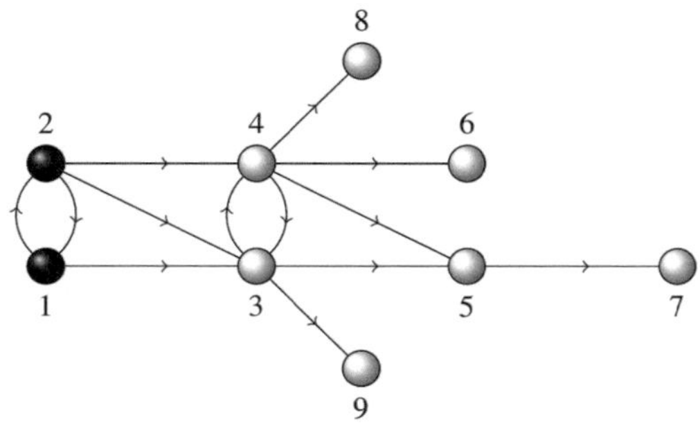

Fig. 10.2 The graph $G = (V, E)$

Next, we discuss how the condition (10.23) can be sharpened. To this end, we define the subgraph $G' = (V, E')$ with

$$E' = \{(i, j) : \ i \in D(V_L) \text{ and } j \in V_T\}. \tag{10.26}$$

Hence, the set E' captures all the arcs from the vertices in $D(V_L)$ to the vertices in V_T. To avoid confusion, let $D(V_L)$ be denoted by V_L'. Then, by $D'(V_L')$ we denote the derived set of V_L' in the graph G'. Note that the set $D'(V_L')$ is in fact constructed as a result of the following steps:

1. Compute the set $D(V_L)$, that is the derived set of V_L in the graph $G = (V, E)$. This means that the vertices in V_L are initially colored black, and we apply the color-change rule based on the arc set E.
2. Construct the subgraph $G' = (V, E')$ from G, with E' given by (10.26)
3. Compute the set $D'(V_L')$, that is derived set of V_L' in G'. This means that the vertices in $V_L' = D(V_L)$ are initially colored black, and we apply the color-change rule based on the arc set E'.

Noting that $D(V_L) \subseteq D'(V_L')$, without loss of generality, assume that $D(V_L) = \{1, 2, \ldots, d\}$ and $D'(V_L') = \{1, 2, \ldots, d, d+1, \ldots, d+e\}$. Consider the condition (10.21). Let $X \in \mathcal{Q}(G)$ and let ξ be a vector in the subspace $\mathscr{V}_T \cap \langle X \mid \mathscr{V}_L\rangle^{\perp}$. Hence, by (10.11),

$$\xi \in \mathscr{V}_T \cap \langle X \mid \mathscr{D}(V_L)\rangle^{\perp}. \tag{10.27}$$

Note that $\xi \in \mathbb{R}^n$ can be written as $\xi = \text{col}(\xi_1, \xi_2, \xi_3)$, where $\xi_1 \in \mathbb{R}^d$, $\xi_2 \in \mathbb{R}^e$, and $\xi_3 \in \mathbb{R}^{n-d-e}$. Compatible with ξ, let the matrix X be partitioned as

$$X = \begin{bmatrix} X_{11} & X_{12} & X_{13} \\ X_{21} & X_{22} & X_{23} \\ X_{31} & X_{32} & X_{33} \end{bmatrix}. \tag{10.28}$$

Now assume that $V_T = D'(V_L')$, and thus

$$\mathscr{V}_T = \mathscr{D}'(V_L')$$

where $\mathscr{D}'(V_L') = \operatorname{im} P(V; D'(V_L'))$. Then, clearly $\xi \in \mathscr{D}'(V_L')$, and hence $\xi_3 = 0$. By (10.27), we have

$$\xi^T X^{k-1} P(V; D(V_L)) = 0 \tag{10.29}$$

for each $k \in \mathbb{N}$. The equality $\xi^T P(V; D(V_L)) = 0$ yields $\xi_1 = 0$. Then, by (10.29) with $k = 2$, we obtain that

$$\xi_2^T X_{21} = 0. \tag{10.30}$$

Now, observe that the matrix

$$X' = \begin{bmatrix} 0 & 0 & 0 \\ X_{21} & 0 & 0 \\ 0 & 0 & 0 \end{bmatrix}$$

belongs to the qualitative class $Q(G')$, where the partitioning is compatible to (10.28). Therefore, by (10.9), we have

$$\mathscr{D}'(V_L') \subseteq \langle X' \mid \mathscr{V}_L' \rangle = \langle X' \mid \mathscr{D}(V_L) \rangle \tag{10.31}$$

where $\mathscr{V}_L' = \operatorname{im} P(V; V_L')$. The subspace inclusion (10.31) yields

$$\mathscr{D}'(V_L') = \operatorname{im} \begin{bmatrix} I & 0 \\ 0 & I \\ 0 & 0 \end{bmatrix} \subseteq \operatorname{im} \begin{bmatrix} I & 0 \\ 0 & X_{21} \\ 0 & 0 \end{bmatrix} = \langle X' \quad \mathscr{D}(V_L) \rangle$$

where the partitioning is again compatible to (10.28), and we have used the fact that $(X')^k = 0$ for $k > 1$. This implies that X_{21} is full row rank. Hence, the equality (10.30) results in $\xi_2 = 0$ which in turn implies targeted controllability of $(G; V_L; V_T)$ by (10.21). Therefore, we conclude that

$$V_T = D'(V_L') \Longrightarrow (G; V_L; V_T) \text{ is targeted controllable,} \tag{10.32}$$

which obviously can be restated as

$$V_T \subseteq D'(V_L') \Longrightarrow (G; V_L; V_T) \text{ is targeted controllable.} \tag{10.33}$$

Noting that $D(V_L) \subseteq D'(V_L')$, the condition (10.23) can be replaced by the sharper condition (10.33) to deduce strong targeted controllability of the network.

As an example consider again the graph depicted in Fig. 10.2 with $V_L = \{1, 2\}$. Recall that the derived set of V_L in G is given by $D(V_L) = \{1, 2, 3, 4\}$ in this case. Let V_T be given by

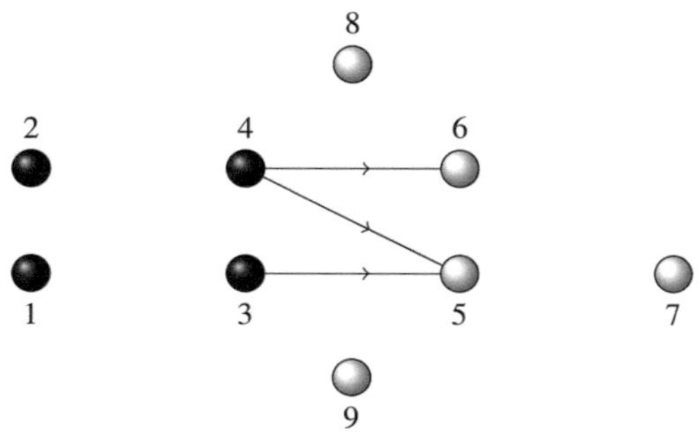

Fig. 10.3 The subgraph $G' = (V, E')$

$$V_T = \{1, 2, 3, 4, 5, 6\}$$

Then, Fig. 10.3 shows the subgraph $G' = (V, E')$ with E' given by (10.26). It is easy to observe that $D'(V'_L)$ is obtained as $D'(V'_L) = \{1, 2, 3, 4, 5, 6\}$. Noting that $V_T = D'(V'_L)$, we conclude that $(G; V_L; V_T)$ is targeted controllable by (10.32). It is worth mentioning that the sufficient condition (10.33) is not tight, as evident by (10.25). In fact, the vertices in $D(V_L)$ do not infect vertex 7 in G'.

10.6 Conclusions and Outlook

We have considered the problem of controllability of the network for a family of matrices carrying the structure of an underlying directed graph. This family of matrices is called the qualitative class, and as observed, there is a one-to-one correspondence between zero forcing sets and the set of leaders rendering the network controllable for all matrices in the qualitative class. As illustrated, this provides a bridge connecting the results available in graph theory for zero forcing sets/number to (minimal) leader selection schemes for strong structural controllability of dynamical networks. As minimal zero forcing sets for general graphs with cycles are very difficult to compute, finding suboptimal (non-minimal) zero forcing sets, or equivalently suboptimal (non-minimal) leader selection schemes for strong structural controllability of networks is an interesting problem for future research.

We have also studied the case where the network is not strongly structurally controllable, yet we are interested in controllability properties in some parts of the network identified by a target set. We have discussed topological sufficient conditions guaranteeing strong targeted controllability of the network. There are still important questions to be addressed in this direction. One notable problem is to establish a tractable topological necessary and sufficient condition verifying strong targeted controllability. Then, given a leader set, a problem of interest is to characterize a target set, with maximum cardinality, such that the network is strongly targeted controllable. Another interesting question is the "dual" problem, i.e., to select a set of leaders, with minimum cardinality, such that the network is strongly targeted controllable for a given target set.

References

1. Chapman, A., Mesbahi, M.: On symmetry and controllability of multi-agent systems. In: Proceedings of the 53rd IEEE Conference on Decision and Control, 625–630 December 2014
2. Dion, J.M., Commault, C., van der Woude, J.: Generic properties and control of linear structured systems: a survey. Automatica **39**(7), 1125–1144 (2003)
3. Gao, J., Liu, Y.Y., D'Souza, R.M., Barabási, A.L.: Target control of complex networks. Nat. Commun. **5**, (2014)
4. Godsil, C.D.: Control by quantum dynamics on graphs. Phys. Rev. A **81**(5), 1–5 (2010). 052316
5. Hogben, L.: Minimum rank problems. Linear Algebra Appl. **432**, 1961–1974 (2010)
6. Lin, C.T.: Structural controllability. IEEE Trans. Autom. Control **19**(3), 201–208 (1974)
7. Liu, Y.Y., Slotine, J.J., Barabasi, A.L.: Controllability of complex networks. Nature **473**, 167–173 (2011)
8. Martini, S., Egerstedt, M., Bicchi, A.: Controllability analysis of multi-agent systems using relaxed equitable partitions. Int. J. Syst. Control Commun. **2**(1/2/3), 100–121 (2010)
9. Mayeda, H., Yamada, T.: Strong structural controllability. SIAM J. Control Optim. **17**(1), 123–138 (1979)
10. Monshizadeh, N., Zhang, S., Camlibel, M.K.: Disturbance decoupling problem for multi-agent systems: a graph topological approach. Syst. Control Lett. **76**, 35–41 (2015)
11. Monshizadeh, N., Zhang, S., Camlibel, M.K.: Zero forcing sets and controllability of dynamical systems defined on graphs. IEEE Trans. Autom. Control **59**(9), 2562–2567 (2014)
12. Nabi-Abdolyousefi, M., Mesbahi, M.: On the controllability properties of circulant networks. IEEE Trans. Autom. Control **58**(12), 3179–3184 (2013)
13. Parlangeli, G., Notarstefano, G.: On the reachability and observability of path and cycle graphs. IEEE Trans. Autom. Control **57**(3), 743–748 (2012)
14. Rahmani, A., Ji, M., Mesbahi, M., Egerstedt, M.: Controllability of multi-agent systems from a graph theoretic perspective. SIAM J. Control Optim. **48**(1), 162–186 (2009)
15. Tanner. H.G.: On the controllability of nearest neighbor interconnections. In: Proceedings of the 43rd IEEE Conference on Decision and Control, 2467–2472 (2004)
16. Trentelman, H.L., Stoorvogel, A.A., Hautus, M.L.J.: Control Theory for Linear Systems. Springer, London (2001)
17. Zhang, S., Cao, M., Camlibel, M.K.: Upper and lower bounds for controllable subspaces of networks of diffusively coupled agents. IEEE Trans. Autom. Control **59**(3), 745–750 (2014)

Chapter 11
Physical Network Systems and Model Reduction

Arjan van der Schaft

Abstract The common structure of a number of physical network systems is identified, based on the incidence structure of the graph, the weights associated to the edges, and the total stored energy. State variables may not only be associated to the vertices, but also to the edges of the graph; in clear contrast with multiagent systems. Structure-preserving model reduction concerns the problem of approximating a complex physical network system by a system of lesser complexity, but within the same class of physical network systems. Two approaches, respectively, based on clustering and on Kron reduction, are explored.

11.1 Introduction

While complexity and large-scale systems have always been important themes in systems and control theory, the current flowering of *network dynamics and control* was not easy to predict. Two main reasons for the enormous research activity are the ubiquity of large-scale networks in a large number of application areas (from power networks to systems biology) and the happy marriage between on the one-hand systems and control theory and algebraic graph theory on the other. Especially, this last aspect can be clearly witnessed in the recent work by my colleague Harry Trentelman.

The research paths of Harry and myself have developed in parallel, at a number of instances tangential, but for some reason never leading to a joint publication. Needless to say that both our scientific developments were heavily influenced and shaped by our joint supervisor and promoter Jan C. Willems, with his unique and inspiring style and taste for doing research.

A. van der Schaft
Jan C. Willems Center for Systems and Control, Johann Bernoulli Institute for Mathematics and Computer Science, University of Groningen, Groningen, The Netherlands
e-mail: a.j.van.der.schaft@rug.nl

M.N. Belur et al. (eds.), *Mathematical Control Theory II*,
Lecture Notes in Control and Information Sciences 462,
DOI 10.1007/978-3-319-21003-2_11

After completing our respective Ph.D. studies with Jan at the Mathematics Institute in Groningen, our ways parted with Harry leaving for Eindhoven (in 1985) and myself for Twente (a few years before in 1982). Nevertheless we remained in close contact, during conferences, activities of the Dutch Institute of Systems and Control, and foremost simply by being part of the "Groningen school on systems theory". Harry returned to his Alma Mater in 1991 as an Associate Professor working closely together with Jan, while in 2005 I could not resist the temptation to return to Groningen. After being colleagues now for 10 years it is a special pleasure to contribute to this Festschrift for Harry's upcoming 60th birthday, and in this way to honor his scientific contributions and simply to herald our pleasant collaboration through all these years. Happy 60th birthday Harry!

11.1.1 Preliminaries from Graph Theory

This paper will be concerned with network dynamics and model reduction, emphasizing *physical network systems*. As preliminaries we recall from [2, 9] the following standard definitions and facts. A *graph* $\mathscr{G}(\mathscr{V}, \mathscr{E})$, is defined by a set $\mathscr{V}$ of *vertices* (nodes) and a set $\mathscr{E}$ of *edges* (links, branches), where $\mathscr{E}$ is identified with a set of unordered pairs $\{i, j\}$ of vertices $i, j \in \mathscr{V}$. We allow for multiple edges between vertices, but not for self-loops $\{i, i\}$. By endowing the graph with an orientation we obtain a *directed graph*. A directed graph with n vertices and k edges is specified by its $n \times k$ *incidence matrix*, denoted by B. Every column of B corresponds to an edge of the graph, and contains exactly one -1 at the row corresponding to its tail vertex and one $+1$ at the row corresponding to its head vertex, while the other elements are 0. In particular, $\mathbb{1}^T B = 0$ where $\mathbb{1}$ is the vector of all ones. Furthermore, $\ker B^T = \text{span}\, \mathbb{1}$ if and only if the graph is *connected* (any point can be reached from any other point by a sequence of, - undirected -, edges). For any diagonal positive semi-definite $k \times k$ matrix R we define the *weighted Laplacian matrix* of the graph as $L := BRB^T$, where the nonnegative diagonal elements $r_1, \ldots, r_k$ of the matrix R are the weights of the edges. It is well known [2] that L is *independent* of the orientation of the graph, and thus is associated with the undirected graph.

11.2 Physical Network Systems

In this section we will discuss a number of examples of *physical network systems* and identify their common structure, based on the incidence structure of the graph, the weights associated to the edges, and a Hamiltonian function given by the total stored energy. It will turn out that a distinguishing feature of physical network systems is the fact that state variables may not only be associated to the vertices of the graph, but also to the edges. This is in sharp contrast with the standard framework of

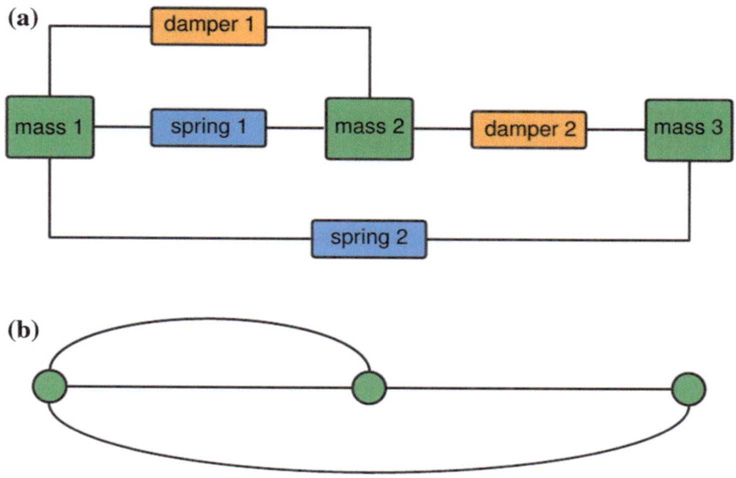

Fig. 11.1 **a** Mass–spring–damper system; **b** the corresponding graph

multiagent systems, where the edges only capture the information exchange structure of the networked system, and is similar to the passivity setup described in [1], where dynamical controllers are associated to the edges.

11.2.1 Mass–Spring–Damper Systems

We will start with the paradigmatic example of *mass–spring–damper* systems. The basic[1] way of modeling mass–spring–damper systems as systems on a graph is to associate the *masses* to the *vertices*, and the *springs* and *dampers* to the *edges* of the graph; see Fig. 11.1.

For clarity of exposition we will start with the separate treatment of mass–damper and mass–spring systems, and then combine the two.

A *mass–damper system* is associated to a graph $\mathcal{G}$ with n vertices (masses), k edges (dampers), and incidence matrix B. Throughout[2] we consider the situation that the masses are located in one-dimensional space $\mathbb{R}$. This leads to the total vector $p \in \mathbb{R}^n$ of the scalar momenta of all n masses. Assuming that the dampers are *linear*, it can be verified that the dynamics is compactly represented as

[1]See for a further discussion [24].

[2]The setup can be easily extended (i.e., by using Kronecker products) to the situation that the scalar variable x_i is replaced by a vector in some higher dimensional physical space, e.g., $\mathbb{R}^3$; see the remarks later on.

$$\begin{aligned} \dot{p} &= -BRB^T M^{-1} p + Eu, \quad p \in \mathbb{R}^n, u \in \mathbb{R}^m \\ y &= E^T M^{-1} p, \end{aligned} \tag{11.1}$$

where R is the $k \times k$ positive semi-definite diagonal matrix of damper coefficients, and M is the positive diagonal matrix of mass parameters. Furthermore, E is an $n \times m$ matrix, with ith column containing exactly one $+1$ element at the row corresponding to the boundary vertex where the input (external force) u_i takes place; all other elements being zero. As outputs we have chosen the corresponding velocities of the boundary vertices. Identifying the kinetic energy $H : \mathbb{R}^n \to \mathbb{R}$ as $H(p) = \frac{1}{2} p^T M^{-1} p$ it follows that

$$\frac{d}{dt} H(p) = -p^T M^{-1} BRB^T M^{-1} p + y^T u \leqslant y^T u,$$

showing *passivity* of the mass–damper system, since the Laplacian matrix $L := BRB^T$ is positive semi-definite.

Remark 11.1 Note that the dynamics of a mass–damper system with *unit masses*, in the absence of external forces, is given by $\dot{p} = -Lp$, which is the standard symmetric continuous-time consensus dynamics on an *undirected* graph.

In case of *mass–spring systems* we associate to the ith spring an elongation $q_i \in \mathbb{R}$, leading to the total vector $q \in \mathbb{R}^k$ of elongations of all k springs. Furthermore, the definition of the Hamiltonian extends to $H : \mathbb{R}^k \times \mathbb{R}^n \to \mathbb{R}$ given as the sum of a potential and kinetic energy

$$H(q, p) = \frac{1}{2} q^T K q + \frac{1}{2} p^T M^{-1} p, \tag{11.2}$$

where the kinetic energy $\frac{1}{2} p^T M^{-1} p$ is defined as before, and the potential energy $\frac{1}{2} q^T K q$ is equal to the sum of the potential energies of the k springs, with K the $k \times k$ diagonal matrix of spring constants. In the absence of inputs and outputs the dynamics of the mass–spring system is then described as the Hamiltonian system

$$\begin{bmatrix} \dot{q} \\ \dot{p} \end{bmatrix} = \begin{bmatrix} 0 & B^T \\ -B & 0 \end{bmatrix} \begin{bmatrix} \frac{\partial H}{\partial q}(q, p) \\ \frac{\partial H}{\partial p}(q, p) \end{bmatrix} \tag{11.3}$$

defined with respect to the Poisson structure on the state space $\mathbb{R}^k \times \mathbb{R}^n$ given by the skew-symmetric matrix

$$J := \begin{bmatrix} 0 & B^T \\ -B & 0 \end{bmatrix}. \tag{11.4}$$

Remark 11.2 Note the fundamental difference with the standard setup of *multiagent systems* on graphs, in which case state variables are only associated to the *vertices* of the graph. In the above formulation of mass–spring systems part of the state variables (namely q) are associated to the edges.

The inclusion of boundary vertices, and thereby of external interaction, can be done in different ways. The first option is to associate *boundary masses* to the boundary vertices. We are then led to the system

$$\begin{aligned} \dot{q} &= B^T \frac{\partial H}{\partial p}(q,p), \\ \dot{p} &= -B\frac{\partial H}{\partial q}(q,p) + Eu, \\ y &= E^T \frac{\partial H}{\partial p}(q,p). \end{aligned} \tag{11.5}$$

Here E is defined as before, while the inputs u are the external *forces* exerted (by the environment) on the boundary masses, and the outputs y are the *velocities* of these boundary masses.

Another possibility is to regard the boundary vertices as being *massless*. In this case we obtain the system (with p now denoting the vector of momenta of the masses associated to all vertices except for the boundary vertices)

$$\begin{aligned} \dot{q} &= B_i^T \frac{\partial H}{\partial p}(q,p) + B_b^T u, \\ \dot{p} &= -B_i\frac{\partial H}{\partial q}(q,p), \\ y &= B_b \frac{\partial H}{\partial q}(q,p), \end{aligned} \tag{11.6}$$

with u the velocities of the massless boundary vertices, and y the forces at the boundary vertices as *experienced* by the environment. Here we have split the incidence matrix as $B = \begin{bmatrix} B_i \\ B_b \end{bmatrix}$, with B_b the rows corresponding to the *boundary* vertices, and B_i the rows corresponding to the remaining *internal* vertices. Note that in this case the velocities u of the boundary vertices can be considered to be *inputs* to the system and the forces y to be *outputs*; in contrast to the previously considered case (boundary vertices corresponding to boundary masses), where the forces are inputs and the velocities the outputs of the system. In both cases we derive the energy balance $\frac{d}{dt}H(p) = y^T u$, showing that the system is *lossless* [19, 22].

Remark 11.3 In the above treatment we have considered springs with *arbitrary* elongation vectors $q \in \mathbb{R}^k$. Often the vector q of elongations is given as $q = B^T q_c$, where $q_c \in \mathbb{R}^n$ denotes the vector of *positions* of the masses at the vertices. Hence in this case $q \in \operatorname{im} B^T \subset \mathbb{R}^k$. Note that the subspace $\operatorname{im} B^T \times \mathbb{R}^n \subset \mathbb{R}^k \times \mathbb{R}^n$ is an invariant subspace, both with regard to the dynamics (11.5) or (11.6). See [24] for the precise connection between these two formulations, in terms of reduction by *symmetry*.

Finally, for a *mass–spring–damper system* the edges will correspond partly to *springs*, and partly to *dampers*. Thus a mass–spring–damper system is described by a graph $\mathcal{G}(\mathcal{V}, \mathcal{E}_s \cup \mathcal{E}_d)$, where the vertices in $\mathcal{V}$ correspond to the *masses*, the edges in $\mathcal{E}_s$ to the *springs*, and the edges in $\mathcal{E}_d$ to the dampers of the system. This corresponds to an incidence matrix $B = \begin{bmatrix} B_s & B_d \end{bmatrix}$, where the columns of B_s reflect the spring edges and the columns of B_d the damper edges. For the case *without* boundary vertices the dynamics of such a mass–spring–damper system with linear dampers takes the form

$$\begin{bmatrix} \dot{q} \\ \dot{p} \end{bmatrix} = \begin{bmatrix} 0 & B_s^T \\ -B_s & -B_d R B_d^T \end{bmatrix} \begin{bmatrix} \frac{\partial H}{\partial q}(q, p) \\ \frac{\partial H}{\partial p}(q, p) \end{bmatrix} \tag{11.7}$$

In the presence of boundary vertices we may distinguish, as above, between *massless* boundary vertices, with inputs u being the boundary velocities and outputs y the boundary (reaction) forces, and *boundary masses*, in which case the inputs u are the external forces and the outputs y the velocities of the boundary masses. In both cases we obtain

$$\frac{d}{dt} H(p) = -\frac{\partial^T H}{\partial p}(q, p) B_d R B_d^T \frac{\partial H}{\partial p}(q, p) + y^T u \leqslant y^T u,$$

showing passivity.

11.2.2 Abstraction and Port-Hamiltonian Formulation

Before moving on to other examples of physical network system we will first introduce some abstractions which directly lead to a general port-Hamiltonian formulation, and are also important for the model reduction approach taken later.

The state space $\mathbb{R}^n$ of a mass–damper system can be more abstractly defined as follows; cf. [24] for further information. It is given by the linear space Λ_0 of all functions from the vertex set $\mathcal{V}$ to $\mathbb{R}$. Obviously, Λ_0 can be identified with $\mathbb{R}^n$. The matrix M^{-1} defines an inner product on Λ_0. As a consequence, any vector $M^{-1}p$, $p \in \Lambda_0$, can be considered to be an element of the dual space of Λ_0, which is denoted by Λ^0. For a mass–damper system, $v := M^{-1}p$ is the vector of velocities of the n masses. It follows that the system (11.1) can be represented in the state vector $v := M^{-1}p \in \Lambda^0$ as

$$\dot{v} = -M^{-1} B R B^T v + M^{-1} E u, \quad v \in \Lambda^0 = \mathbb{R}^n, \tag{11.8}$$

or equivalently in the *gradient system* representation (with M defining an inner product on the space Λ^0)

$$M\dot{v} = -BRB^T v + Eu, \quad v \in \Lambda^0 = \mathbb{R}^n, u \in \mathbb{R}^m \tag{11.9}$$

Furthermore, the edge space $\mathbb{R}^k$ can be defined more abstractly as the linear space Λ_1 of functions from the edge set $\mathscr{E}$ to $\mathbb{R}$, with dual space denoted by Λ^1. It follows that the incidence matrix B defines a linear map (denoted by the same symbol) $B : \Lambda_1 \to \Lambda_0$ with adjoint map $B^T : \Lambda^0 \to \Lambda^1$. Furthermore, R can be considered to define an inner product on Λ^1, or equivalently, a map $R : \Lambda^1 \to \Lambda_1$.

For mass–spring systems we notice that $q \in \Lambda^1$, and that K defines an inner product on Λ^1, or equivalently, a map $K : \Lambda^1 \to \Lambda_1$ mapping the elongation vector $q \in \Lambda^1$ to the vector of spring forces $Kq \in \Lambda_1$

Using these abstractions it is straightforward to extend the mass–spring–damper dynamics to other spatial domains than just the one-dimensional domain $\mathbb{R}$. Indeed, for any linear space $\mathscr{R}$ we can define Λ_0 as the set of functions from $\mathscr{V}$ to $\mathscr{R}$, and Λ_1 as the set of functions from $\mathscr{E}$ to $\mathscr{R}$. In this case we can identify Λ_0 with the tensor product $\mathbb{R}^n \otimes \mathscr{R}$ and Λ_1 with the tensor product $\mathbb{R}^k \otimes \mathscr{R}$. Furthermore, the incidence matrix B defines a linear map $B \otimes I : \Lambda_1 \to \Lambda_0$, where I is the identity map on $\mathscr{R}$. In matrix notation $B \otimes I$ equals the Kronecker product of the incidence matrix B and the identity matrix I. For $\mathscr{R} = \mathbb{R}^3$ this will describe the motion of mass–spring–damper systems in $\mathbb{R}^3$.

An especially interesting generalization are *multibody-systems* in $\mathbb{R}^3$, in which case $\mathscr{R} = se(3)$ (the Lie algebra corresponding to the special Euclidean group in $\mathbb{R}^3$); see [24] for further information.

Furthermore, these abstractions naturally lead to a *port-Hamiltonian formulation*, see, e.g., [19, 22, 23]. Note that in the case of a mass–spring system the Poisson structure

$$J = \begin{bmatrix} 0 & B^T \\ -B & 0 \end{bmatrix}$$

is naturally defined on the state space $\Lambda^1 \times \Lambda_0$. In order to include boundary vertices we may distinguish, as above, between *massless* boundary vertices and *boundary masses*, leading to the definition of two canonical Dirac structures, cf. [24] for details.

11.2.3 Hydraulic Networks

A hydraulic network can be modeled as a directed graph with edges corresponding to pipes; see, e.g., [6]. The vertices may either correspond to *fluid reservoirs* (buffers), or to connection points of the pipes. We concentrate on the first case; the second case being similar to the case of electrical circuits considered later on. Let x_v be the stored fluid at vertex v and let ν_e be the fluid flow through edge e. Collecting all stored fluids x_v into a vector x, and all fluid flows ν_e into a vector ν, the *mass-balance* is summarized as

$$\dot{x} = B\nu, \tag{11.10}$$

with B denoting the incidence matrix of the graph. In the absence of fluid reservoirs this reduces to Kirchhoff's current laws $B\nu = 0$.

For incompressible fluids a standard model of the fluid flow ν_e through pipe e is

$$J_e \dot{\nu}_e = P_i - P_j - \lambda_e(\nu_e), \tag{11.11}$$

where P_i and P_j are the pressures at the tail, respectively, head, vertices of edge e. Note that this captures in fact *two* effects; one corresponding to energy storage and one corresponding to energy dissipation. Defining the energy variable $\varphi_e := J_e \nu_e$ the stored energy in the pipe associated with edge e is given as $\frac{1}{2J_e}\varphi_e^2 = \frac{1}{2}J_e \nu_e^2$. Secondly, $\lambda_e(\nu_e)$ is a damping force corresponding to energy dissipation.

In the case of fluid reservoirs at the vertices the pressures P_v at each vertex v are functions of x_v, and thus, being scalar functions, are always derivable from an energy function $P_v = \frac{\partial H_v}{\partial x_v}(x_v)$, $v \in \mathcal{V}$, for some Hamiltonian $H_v(x_v)$ (e.g., gravitational energy). The resulting dynamics (with state variables x_v and φ_e) is port-Hamiltonian with respect to the Poisson structure (11.4). The setup is immediately extended to boundary vertices (either corresponding to controlled fluid reservoirs or direct in- or outflows).

11.2.4 Detailed-Balanced Chemical Reaction Networks

Consider an isothermal chemical reaction network (under constant pressure) consisting of r reversible reactions involving m chemical species specified by a vector of concentrations $x \in \mathbb{R}^m_+ := \{x \in \mathbb{R}^m \mid x_i > 0, i = 1, \ldots, m\}$. The general form of the dynamics of the chemical reaction network (without inflows and outflows) is

$$\dot{x} = Sv(x),$$

with S the stoichiometric matrix, and $v(x) = \begin{bmatrix} v_1(x) & \cdots & v_r(x) \end{bmatrix}^T \in \mathbb{R}^r$ the vector of *reaction rates*. We assume that $v(x)$ is given by *mass action kinetics*; the most basic way of modeling reaction rates. Following [25] we will show how, under the assumption of existence of a thermodynamic equilibrium, the dynamics of the reaction network can be seen to be very similar to the dynamics of a (nonlinear) mass–damper system.

In order to do so we first need to introduce some concepts and terminology. The collection of all the different left- and right-hand sides of the reactions are called the chemical complexes of the reaction network, or briefly, the *complexes*. Denoting the number of complexes by c, the expression of the complexes in terms of the chemical species concentration vector $x \in \mathbb{R}^m_+$ is formalized by the $m \times c$ *complex composition* matrix Z, whose ρth column captures the expression of the ρth complex in the m chemical species. Note that by definition all elements of the matrix Z are nonnegative integers.

The complexes can be naturally associated with the vertices of a *directed graph*, with edges corresponding to the reversible reactions. The complex on the left-hand side of each reaction is called the *substrate* complex, and the one on the right-hand side the *product* complex. Formally, the reversible reaction $\sigma \rightleftharpoons \pi$ between the substrate σ and the product π defines a directed edge with tail vertex σ and head vertex π. The resulting directed graph is called the *complex graph*, and is characterized by its $c \times r$ incidence matrix B. It is readily verified that the stoichiometric matrix S of the chemical reaction network is given as $S = ZB$.

Mass action kinetics for the reaction rate vector $v(x) \in \mathbb{R}^r$ is defined as follows. Consider first, as an example, the single reaction

$$X_1 + 2X_2 \rightleftharpoons X_3,$$

involving the three chemical species X_1, X_2, X_3 with concentrations x_1, x_2, x_3. It is a combination of the *forward reaction* $X_1 + 2X_2 \rightharpoonup X_3$ with forward rate equation $v_1^+(x_1, x_2) = k^+ x_1 x_2^2$ and the *reverse reaction* $X_1 + 2X_2 \leftharpoondown X_3$, with rate equation $v^-(x_3) = k^- x_3$. The constants k^+, k^- are called, respectively, the *forward* and the *reverse reaction constants*. The net reaction rate is thus

$$v(x_1, x_2, x_3) = v^+(x_1, x_2) - v^-(x_3) = k^+ x_1 x_2^2 - k^- x_3.$$

In general, the mass action reaction rate of the jth reaction of a chemical reaction network, from the substrate complex σ_j to the product complex π_j, is given as

$$v_j(x) = k_j^+ \prod_{i=1}^{m} x_i^{Z_{i\sigma_j}} - k_j^- \prod_{i=1}^{m} x_i^{Z_{i\pi_j}}, \tag{11.12}$$

where $Z_{i\rho}$ is the (i, ρ)th element of the matrix Z, and $k_j^+, k_j^- \geqslant 0$ are the forward/reverse reaction constants of the jth reaction, respectively.

Equation (11.12) can be rewritten in the following way. Let us first introduce some notation. Define the mapping $\mathrm{Ln} : \mathbb{R}_+^n \to \mathbb{R}^n$ as the componentwise natural logarithm. Analogously, define the mapping $\mathrm{Exp} : \mathbb{R}^c \to \mathbb{R}_+^c$ as the componentwise exponential function. Let Z_{σ_j} and Z_{π_j} denote the columns of Z corresponding to the substrate complex σ_j and the product complex π_j of the jth reaction. Then (11.12) takes the form

$$v_j(x) = k_j^+ \exp\left(Z_{\sigma_j}^T \mathrm{Ln}(x)\right) - k_j^- \exp\left(Z_{\pi_j}^T \mathrm{Ln}(x)\right). \tag{11.13}$$

A vector of concentrations $x^* \in \mathbb{R}_+^m$ is called a *thermodynamic equilibrium* if $v(x^*) = 0$. A chemical reaction network $\dot{x} = Sv(x)$ is called *detailed-balanced* if it admits a thermodynamic equilibrium $x^* \in \mathbb{R}_+^m$. Necessary and sufficient conditions for the existence of a thermodynamic equilibrium are usually referred to as the *Wegscheider* *conditions*, generalizing the classical results of [28]; see, e.g., [8, 25]. These imply that once a thermodynamic equilibrium x^* is given, the set of *all* thermodynamic equilibria is given by

$$\mathscr{E} := \{x^{**} \in \mathbb{R}^m_+ \mid S^T \mathrm{Ln}\,\left(x^{**}\right) = S^T \mathrm{Ln}\,\left(x^*\right)\} \tag{11.14}$$

Let now $x^* \in \mathbb{R}^m_+$ be a thermodynamic equilibrium. Define the "conductance" $\kappa_j(x^*) > 0$ of the jth reaction as the common value of the forward and reverse reaction rate at thermodynamic equilibrium x^*, i.e.,

$$\kappa_j(x^*) := k_j^+ \exp\left(Z_{\sigma_j}^T \mathrm{Ln}\,(x^*)\right) = k_j^- \exp\left(Z_{\pi_j}^T \mathrm{Ln}\,(x^*)\right), \tag{11.15}$$

for $j = 1, \ldots, r$. Then the mass action reaction rate (11.13) of the jth reaction can be rewritten as

$$v_j(x) = \kappa_j(x^*)\left[\exp\left(Z_{\sigma_j}^T \mathrm{Ln}\left(\frac{x}{x^*}\right)\right) - \exp\left(Z_{\pi_j}^T \mathrm{Ln}\left(\frac{x}{x^*}\right)\right)\right],$$

where for any vectors $x, z \in \mathbb{R}^m$ the quotient vector $\frac{x}{z} \in \mathbb{R}^m$ is defined elementwise. Defining the $r \times r$ diagonal matrix of conductances as

$$\mathscr{K} := \mathrm{diag}\big(\kappa_1(x^*), \ldots, \kappa_r(x^*)\big), \tag{11.16}$$

it follows that the mass action reaction rate vector $v(x)$ of a balanced reaction network equals

$$v(x) = -\mathscr{K} B^T \mathrm{Exp}\left(Z^T \mathrm{Ln}\left(\frac{x}{x^*}\right)\right),$$

and thus the dynamics of a balanced reaction network takes the form

$$\dot{x} = -ZB\mathscr{K} B^T \mathrm{Exp}\left(Z^T \mathrm{Ln}\left(\frac{x}{x^*}\right)\right), \quad \mathscr{K} > 0. \tag{11.17}$$

The matrix $\mathscr{L} := B\mathscr{K} B^T$ in (11.17) defines a weighted Laplacian matrix for the complex graph, with weights[3] given by the conductances $\kappa_1(x^*), \ldots, \kappa_r(x^*)$.

The system (11.17) defines a nonlinear version of the mass–damper system considered before. Indeed, define the Hamiltonian (up to a constant the Gibbs' free energy, cf. [25]) as the function

$$G(x) = x^T \mathrm{Ln}\left(\frac{x}{x^*}\right) + \left(x^* - x\right)^T \mathbb{1}_m, \tag{11.18}$$

[3]Note that $\mathscr{K}$, and therefore the Laplacian matrix $\mathscr{L} = B\mathscr{K} B^T$, is *dependent* on the choice of the thermodynamic equilibrium x^*. However, this dependence is minor: for a connected complex graph the matrix $\mathscr{K}$ is, *up to a positive multiplicative factor*, independent of the choice of x^* [25].

where $\mathbb{1}_m$ denotes a vector of dimension m with all ones. It is immediately checked that $\frac{\partial G}{\partial x}(x) = \mathrm{Ln}\left(\frac{x}{x^*}\right) = \mu(x)$, where μ is (up to a constant) the vector of *chemical potentials*. Then by considering the auxiliary port-Hamiltonian system

$$\begin{aligned} \dot{x} &= Z u_R, \\ y_R &= Z^T \frac{\partial G}{\partial x}(x), \end{aligned} \tag{11.19}$$

with inputs $u_R \in \mathbb{R}^c$ and outputs $y_R \in \mathbb{R}^c$, together with the nonlinear damping relation

$$u_R = -B\mathscr{K}B^T \mathrm{Exp}\,(y_R). \tag{11.20}$$

we recover the mass action reaction dynamics (11.17). Furthermore, by using the properties of the Laplacian matrix $\mathscr{L} = B\mathscr{K}B^T$ and the fact that the exponential function is strictly increasing, it can be shown that [25]

$$\gamma^T B\mathscr{K}B^T \mathrm{Exp}\,(\gamma) \geqslant 0 \text{ for all } \gamma, \tag{11.21}$$

with equality if and only if $B^T\gamma = 0$. Hence (11.20) defines a true *energy-dissipating* relation, that is, $y_R^T u_R \leqslant 0$ for all $y_R \in \mathbb{R}^c$ and $u_R \in \mathbb{R}^c$ satisfying (11.20). Therefore the mass action kinetics detailed-balanced chemical reaction network is a nonlinear port-Hamiltonian system, which can be regarded as a nonlinear mass–damper system, with Laplacian matrix $\mathscr{L} = B\mathscr{K}B^T$, and with non-quadratic Hamiltonian G and nonlinear dampers (11.20).

The consequences of this way of representing detailed-balanced mass action kinetics reaction networks for their analysis are explored in [25]. In particular, it follows that *all* equilibria are in fact thermodynamic equilibria, and a Lyapunov analysis using the Gibbs' free energy shows that starting from any initial state in the positive orthant the system will converge to a unique thermodynamic equilibrium (at least under the assumption of *persistence* of the reaction network: the vector of concentrations does not approach the boundary of the positive orthant $\mathbb{R}^m_+$), cf. [25] for details.[4]

11.2.5 Swing Equations for Power Grids

Consider a power grid consisting of n buses corresponding to the vertices of a graph, and transmission lines corresponding to its edges. A standard model for the dynamics of the ith bus is given by (see, e.g., [4, 12])

[4]For an extension of these results to *complex-balanced* mass action kinetics reaction networks we refer to [15].

$$\begin{aligned}\dot{\delta}_i &= \omega_i^b - \omega^r, \quad i = 1, \dots, n,\\ M_i \dot{\omega}_i &= -a_i(\omega_i^b - \omega^r) - \sum_{j \neq i} V_i V_j S_{ij} \sin(\delta_i - \delta_j) + u_i,\end{aligned}$$

where the summation in the last line is over all buses j which are adjacent to bus i; that is, all buses j that are directly linked to bus i by a transmission line. Here ω^r is the nominal (reference) frequency for the network, δ_i denotes the voltage angle, V_i the voltage amplitude, ω_i^b the frequency, $\omega_i := \omega_i^b - \omega^r$ the frequency deviation, and u_i the power generation/consumption; all at bus i. Furthermore, M_i and a_i are inertia and damping constants at bus i, and S_{ij} is the transfer susceptance of the line between bus i and j.

Define $z_k := \delta_i - \delta_j$ and $c_k := E_i E_j S_{ij}$, if the kth edge is pointing from vertex i to vertex j. Furthermore, define the momenta $p_i = M_i \omega_i$, $i = 1, \dots, n$. Then the equations can be written in the vector form

$$\begin{aligned}\dot{z} &= B^T M^{-1} p,\\ \dot{p} &= -AM^{-1}p - BC \operatorname{Sin} z + u,\end{aligned}$$

where z is the m-vector with components z_k, M is the diagonal matrix with diagonal elements M_i, A is the diagonal matrix with diagonal elements a_i, and C is the diagonal matrix with elements c_k. Furthermore, $\operatorname{Sin} : \mathbb{R}^m \to \mathbb{R}^m$ denotes the elementwise sin function. Defining the Hamiltonian $H(z, p)$ as

$$H(z, p) = \frac{1}{2} p^T M^{-1} p - \mathbb{1}^T C \operatorname{Cos} z, \tag{11.22}$$

with $\operatorname{Cos} : \mathbb{R}^m \to \mathbb{R}^m$ the elementwise cos function, the equations take the port-Hamiltonian form

$$\begin{bmatrix} \dot{z} \\ \dot{p} \end{bmatrix} = \begin{bmatrix} 0 & B^T \\ -B & -A \end{bmatrix} \begin{bmatrix} \frac{\partial H}{\partial z}(z, p) \\ \frac{\partial H}{\partial p}(z, p) \end{bmatrix} + \begin{bmatrix} 0 \\ I \end{bmatrix} u. \tag{11.23}$$

The Hamiltonian $H(z, p)$ is of the standard "kinetic energy plus potential energy" form, with potential energy $-\mathbb{1}^T C \operatorname{Cos} z = -\sum c_k \cos z_k$ (similar to the gravitational energy of a pendulum). Note that, as in the mass–spring(–damper) system example, the potential energy is associated to the edges of the graph, while the kinetic energy is associated to its vertices. A difference with the mass–spring–damper system example considered before is that in the current example the "damping torques" $A \frac{\partial H}{\partial p}(z, p)$ are associated to the vertices, instead of to the edges.

11.2.6 Electrical RLC Circuits

In the case of electrical RLC circuits the R, L and C-elements are all attached to the edges of the circuit graph; there is no energy storage or dissipation associated with the vertices. This corresponds to Kirchhoff's current laws, expressed as

$$BI = 0,$$

where B is the incidence matrix of the circuit graph, and $I \in \mathbb{R}^k$ is the vector of currents through the k edges of the circuit graph. As detailed in [24] an RLC network again determines a port-Hamiltonian system, but now with respect to a Dirac structure which takes into account the constraints imposed by Kirchhoff's current and voltage laws, and is only defined with respect to the currents and voltages through, respectively, across, the edges of the circuit graph.

On top of Kirchhoff's laws, the dynamics is defined by the energy storage relations corresponding to either capacitors or inductors, and dissipative relations corresponding to resistors. The energy-storing relations for a capacitor at edge e are given by

$$\dot{Q}_e = -I_e, \quad V_e = \frac{dH_{Ce}}{dQ_e}(Q_e), \tag{11.24}$$

with Q_e the charge, and $H_{Ce}(Q_e)$ denoting the electric energy stored in the capacitor. Alternatively, in the case of an inductor one specifies the magnetic energy $H_{Le}(\Phi_e)$, where Φ_e is the magnetic flux linkage, together with the dynamic relations

$$\dot{\Phi}_e = V_e, \quad -I_e = \frac{dH_{Le}}{d\Phi_e}(\Phi_e). \tag{11.25}$$

Finally, a resistor at edge e corresponds to a static relation between the current I_e through and the voltage V_e across this edge, such that $V_e I_e \leqslant 0$. In particular, a linear (ohmic) resistor at edge e is specified by a relation $V_e = -R_e I_e$, with $R_e \geqslant 0$. See [24] for the resulting differential-algebraic description of an RLC circuit.

11.3 Structure-Preserving Model Reduction by Clustering

The problem of *structure-preserving* model reduction of physical network systems is as follows. Given a large-scale physical network system of a certain type (e.g., a mass–spring–damper system). How to find to a physical network system of the same type (again a mass–spring–damper system), but of lesser complexity, which is offering a "suitable" approximation of the system? The approach discussed in this section is to consider reduction of the physical network system by *clustering* of the vertices, so as to obtain a graph with fewer vertices [21]. This idea is quite natural, and appears, e.g., in [10, 11, 18], and in the context of symmetric consensus dynamics in [13].

Consider a *partition* of the vertex set $\mathcal{V}$ of the graph $\mathcal{G}$ into $\hat{n}$ disjoint cells $C_1, C_2, \ldots, C_{\hat{n}}$, together with a corresponding $n \times \hat{n}$ *characteristic matrix* P. The columns of P equal the characteristic vectors of the cells; the characteristic vector of a cell C_i being defined as the vector with 1 at the place of every vertex contained in the cell C_i, and 0 elsewhere. With some abuse of notation we will denote the partition simply by its characteristic matrix P. We will first consider the case of *mass–damper systems*.

11.3.1 Mass–Damper Systems

Based on a partition P we reduce the mass–damper system (11.1) to

$$\dot{\hat{p}} = -(P^T BRB^T P)(P^T MP)^{-1}\hat{p} + P^T Eu \tag{11.26}$$

where $\hat{p} := P^T p \in \mathbb{R}^{\hat{n}}$ is the clustered state vector.

We observe that this is again a system of the form (11.1). In fact, the matrix $P^T B$ consists of column vectors containing exactly one $+1$ and one -1 together with some zero vectors (corresponding to edges which link vertices within a same cell). By leaving out the zero column vectors from $P^T B$ we thus obtain an $\hat{n} \times \hat{k}$ matrix $\hat{B}$, which is the incidence matrix of the *reduced graph* $\hat{\mathcal{G}}$ with vertices being the cells of the original graph, and with edges the union of all the edges between vertices in different cells (leaving out edges *within* cells). Correspondingly, we define $\hat{R}$ as the $\hat{k} \times \hat{k}$ diagonal matrix obtained from R by leaving out the rows and columns corresponding to edges between vertices in a same cell. Finally, we define the $\hat{n} \times \hat{n}$ diagonal matrix $\hat{M} := P^T MP$, and $\hat{E} := P^T E$. It follows that the reduced system (11.26) is given as

$$\dot{\hat{p}} = -\hat{B}\hat{R}\hat{B}^T\hat{M}^{-1}\hat{p} + \hat{E}u, \quad \hat{p} \in \mathbb{R}^{\hat{n}} \tag{11.27}$$

The ith component of $\hat{p}$ denotes the total momentum of the masses contained in cell C_i, while the ith diagonal element of $\hat{M}$ is the total mass of the mass contained in cell C_i. The velocities of the clustered masses of cells $C_1, \ldots, C_{\hat{n}}$ (all masses within a cell rigidly interconnected, and leaving out intermediate dampers) will converge to a common value. Note that the velocity v_i of cell C_i is defined as $v_i = \sum p_j / \sum m_j$, with the summation ranging over the indices of the vertices in cell C_i.

Remark 11.4 The reduced system (11.27) will easily contain multiple edges between its vertices, that is, between the cells of the original system (11.1). If desirable, multiple edges between two vertices can be combined into a single edge with weight r given by the *sum* of the weights of the multiple edges.

The main difference with the reduced model proposed in [13] for symmetric consensus dynamics is that in the latter paper the case $M = I$ ($n \times n$ identity matrix) is considered (corresponding to the standard continuous-time consensus algorithm), while moreover the reduced system is required to have $\hat{M} = I$ ($\hat{n} \times \hat{n}$ identity matrix). This is done at the expense of having a weighted Laplacian matrix which is *not* anymore symmetric, and therefore not anymore of the form $\hat{B}\hat{R}\hat{B}^T$.

Remark 11.5 A *Petrov–Galerkin* interpretation of the reduced model (11.26) can be given as follows. Recall that a Petrov–Galerkin reduction of a linear set of differential equations $\dot{x} = Ax$ is given as $\dot{\hat{x}} = W^T A V \hat{x}$, where V and W are matrices of dimension $n \times \hat{n}$ such that $W^T V = I_{\hat{n}}$. Now the reduced system (11.26) is a Petrov–Galerkin reduction of (11.1) with $W := P$, and

$$V := MP\hat{M}^{-1} = MP(P^T MP)^{-1} \tag{11.28}$$

It follows that $W^T V = I_{\hat{n}}$. Note furthermore that the matrix $V = MP\hat{M}^{-1}$ defined in (11.28) equals the *Moore–Penrose pseudo-inverse* of the clustering map $\hat{x} = P^T x$, with respect to the inner products M^{-1} on the space of full-order state vectors $x \in \mathbb{R}^n$ and $\hat{M}^{-1}$ on the space of reduced state vectors $\hat{x} \in \mathbb{R}^{\hat{n}}$. Furthermore, we note that $V^T M^{-1} V = \hat{M}^{-1}$, while also $W^T M W = \hat{M}$, implying orthogonality of V and W with respect to the inner products defined by M^{-1} and M respectively. Using the abstractions introduced earlier on we note that the linear maps V and W are actually defined as maps $V : \hat{\Lambda}_0 \to \Lambda_0$, $W : \hat{\Lambda}^0 \to \Lambda^0$, where $\hat{\Lambda}_0$ is the vertex space of the reduced graph, with dual space $\hat{\Lambda}^0$. Moreover, note that $M : \Lambda^0 \to \Lambda_0$ and $\hat{M} : \hat{\Lambda}^0 \to \hat{\Lambda}_0$.

Of course, from a model reduction point of view the most important question is how to choose the partition P in such a way that the reduced (clustered) physical network system is a good approximation of the original one, where "good approximation" very much depends on what needs to be approximated. In [14], continuing on [13], an explicit H_2 error expression was derived for a partition P with very special properties, namely a *(generalized) almost equitable* partition. From a linear algebraic point of view such partitions P are characterized (for mass–damper systems) by the property

$$BRB^T \operatorname{im} P \subset M \operatorname{im} P,$$

cf. [14] for the graph-theoretic definition. Take as output the velocity differences across all the edges, i.e.,

$$y = R^{\frac{1}{2}} B^T M^{-1} x = R^{\frac{1}{2}} B^T v \tag{11.29}$$

Then an explicit expression for the H_2 error between the transfer matrix G of the full-order system and $\hat{G}$ of the reduced-order system is

$$\|G - \hat{G}\|_2^2 = \frac{1}{2} \sum_{i \in \mathscr{V}_b} \left(\frac{1}{m_i} - \frac{1}{M_{\hat{i}}} \right), \tag{11.30}$$

where $\mathscr{V}_b$ is the set of boundary vertices, and $M_{\hat{i}}$ the sum of the masses belonging to the same cell as i. In particular, if the cells containing the boundary vertices are singletons, then $M_{\hat{i}} = m_i$, $i \in \mathscr{V}_b$, and thus the H_2 error is zero. More generally, if the non-boundary masses in the cells containing the boundary masses are small compared to the boundary masses, then the error is also small.

The method can be straightforwardly extended from linear systems (11.1) to nonlinear systems of the form

$$\dot{p} = -B\mathscr{R}\left(B^T \frac{\partial H}{\partial p}(p)\right) + Eu, \quad p \in \mathbb{R}^n, u \in \mathbb{R}^m, \tag{11.31}$$

where $H : \mathbb{R}^n \to \mathbb{R}$ is any Hamiltonian function, and $\mathscr{R} : \mathbb{R}^k \to \mathbb{R}^k$ is a nonlinear damping characteristic. Note that (11.1) is a special case of (11.31) with $H(p) = \frac{1}{2} p^T M^{-1} p$, and $\mathscr{R}$ given by the linear map R. Examples of (11.31) are mass–damper systems with nonlinear dampers and nonlinear hydraulic networks (with nonlinear pressure functions).

Given a partition P of the underlying graph $\mathscr{G}$ we reduce the system (11.31) to

$$\dot{\hat{p}} = -P^T B\mathscr{R}\left(B^T P \frac{\partial \hat{H}}{\partial \hat{p}}(\hat{p})\right) + P^T Eu \tag{11.32}$$

where $\hat{p} := P^T p \in \mathbb{R}^{\hat{n}}$ is the clustered state vector, and where the reduced Hamiltonian function $\hat{H} : \mathbb{R}^{\hat{n}} \to \mathbb{R}$ is defined as follows. Consider a Hamiltonian $H : \Lambda_0 = \mathbb{R}^n \to \mathbb{R}$ for which the Legendre transform $H^* : \Lambda^0 = \mathbb{R}^n \to \mathbb{R}$ is well defined. (In particular, if H is *convex* then H^* is the convex conjugate $H^*(p^*) := \sup_p \left[p^T p^* - H(p)\right]$.) Then define the reduced function $\hat{H}^* : \hat{\Lambda}^0 = \mathbb{R}^{\hat{n}} \to \mathbb{R}$ as

$$\hat{H}^*(\hat{p}^*) = H^*(P\hat{p}^*)$$

Finally, define $\hat{H} : \hat{\Lambda}_0 = \mathbb{R}^{\hat{n}} \to \mathbb{R}$ as the Legendre transform of $\hat{H}^*$. It is easily checked that for $H(p) = \frac{1}{2} p^T M^{-1} p$ the function $\hat{H}$ defined above is given as $\hat{H}(\hat{p}) = \frac{1}{2} \hat{p}^T (P^T M P)^{-1} \hat{p}$, thus generalizing the linear case considered before.

11.3.2 Mass–Spring–Damper Systems

Consider as before the physical network system corresponding to a mass–spring–damper system

$$\begin{bmatrix} \dot{q} \\ \dot{p} \end{bmatrix} = \begin{bmatrix} 0 & B_s^T \\ -B_s & -B_d R B_d^T \end{bmatrix} \begin{bmatrix} Kq \\ M^{-1} p \end{bmatrix}, \tag{11.33}$$

Where $p \in \mathbb{R}^n$ is the vector of momenta associated to the vertices of the graph, and $q \in \mathbb{R}^k$ is the vector of spring elongations associated to the k edges of the graph. As before, the matrix M is a positive diagonal matrix and R is a positive semi-definite diagonal matrix. Similarly, K is a positive semi-definite diagonal matrix. Furthermore, the total energy is given as $H(q, p) = \frac{1}{2} p^T M^{-1} p + \frac{1}{2} q^T K q$.

Let now P be a partition matrix. Then we define a reduced model for (11.33) as

$$\begin{bmatrix} \dot{\hat{q}} \\ \dot{\hat{p}} \end{bmatrix} = \begin{bmatrix} 0 & \hat{B}_s^T \\ -\hat{B}_s & -\hat{B}_d \hat{R} \hat{B}_d^T \end{bmatrix} \begin{bmatrix} \hat{K}\hat{q} \\ \hat{M}^{-1} p \end{bmatrix}, \tag{11.34}$$

where the reduced incidence matrices $\hat{B}_s$ and $\hat{B}_d$ are defined as $P^T B_s$, respectively, $P^T B_d$, with zero columns deleted. Furthermore, the elements of $\hat{q}$ correspond to the edges which survive in the reduced model. Obviously, (11.34) is again a mass–spring–damper system (which results from assuming that the masses in each cell move at the same velocity without dampers and springs to each other).

11.4 Structure-Preserving Model Reduction by Kron Reduction

Another approach to structure-preserving model reduction, somewhat complementary to the approach using clustering, is based on Kron reduction [7]. Kron reduction originates in electrical circuit and power network theory, and in its simplest form allows to eliminate internal vertices in a resistive electrical network, and to produce another resistive network which is equivalent to the original network from the point of view of the remaining vertices [7, 20]. Applied to physical network systems it means that the time derivatives of the storage variables at a subset of the vertices are set equal to zero, which amounts to the time-separation assumption that the state variables corresponding to these vertices will reach their steady state values much faster than at the other vertices.

11.4.1 Mass–Damper Systems

Consider a mass–damper system as before, cf. (11.1)

$$\begin{aligned} \dot{p} &= -BRB^T M^{-1} p + Eu, \quad p \in \mathbb{R}^n, u \in \mathbb{R}^m \\ y &= E^T M^{-1} p, \end{aligned} \tag{11.35}$$

Suppose one can split the vector of momenta p into a sub-vector p_s of *slow* variables and a sub-vector p_f of *fast* variables (we will comment on the interpretation of this later on), i.e.,

$$p = \begin{bmatrix} p_s \\ p_f \end{bmatrix}, \quad p_s \in \mathbb{R}^{n_s}, p_f \in \mathbb{R}^{n_f}, n_s + n_f = n$$

Denote the corresponding diagonal subblocks of M by M_s and M_f and of E by E_s and E_f, and split the incidence matrix as

$$B = \begin{bmatrix} B_s \\ B_f \end{bmatrix}$$

leading to

$$\begin{aligned} \begin{bmatrix} \dot{p}_s \\ \dot{p}_f \end{bmatrix} &= -\begin{bmatrix} B_s R B_s^T & B_s R B_f^T \\ B_f R B_s^T & B_f R B_f^T \end{bmatrix} \begin{bmatrix} M_s^{-1} & 0 \\ 0 & M_f^{-1} \end{bmatrix} \begin{bmatrix} p_s \\ p_f \end{bmatrix} + \begin{bmatrix} E_s \\ E_f \end{bmatrix} u \\ y &= \begin{bmatrix} E_s^T & E_f^T \end{bmatrix} \begin{bmatrix} M_s^{-1} & 0 \\ 0 & M_f^{-1} \end{bmatrix} \begin{bmatrix} p_s \\ p_f \end{bmatrix} \end{aligned} \tag{11.36}$$

Now assume that p_f will reach its steady state value much faster than p_s, in the sense that in the timescale of the dynamics of the "slow" variables p_s the "fast" variables p_f will always be at their steady state value corresponding to

$$0 = \dot{p}_f = -B_f R B_s^T M_s^{-1} p_s - B_f R B_f^T M_f^{-1} p_f + E_{fu}$$

Solving for $M_f^{-1} p_f$ then leads to the *reduced system in the slow variables*

$$\begin{aligned} \dot{p}_s &= -\left(B_s R B_s^T - B_s R B_f^T (B_f R B_f^T)^{-1} B_s R B_f^T \right) M_s^{-1} p_s + \hat{E} u \\ y &= \hat{E}^T M_s^{-1} p_s \end{aligned} \tag{11.37}$$

where $\hat{E} := E_s + B_s R B_f^T (B_f R B_f^T)^{-1} E_f$. (Note that we make here the assumption that the "fast" vertices do not form a connected component of the graph, implying that the sub-matrix $B_f R B_f^T$ of the Laplacian matrix is invertible; see, e.g., [20].)

The matrix $B_s R B_s^T - B_s R B_f^T (B_f R B_f^T)^{-1} B_s R B_f^T$ is a *Schur complement* of the Laplacian matrix BRB^T. As such it is again [7, 20] a weighted Laplacian matrix, that is

$$B_s R B_s^T - B_s R B_f^T (B_f R B_f^T)^{-1} B_s R B_f^T = \hat{B} \hat{R} \hat{B}^T,$$

where $\hat{B}$ is the incidence matrix of a *reduced graph* with vertex set the set of "slow" vertices, i.e., the vertices corresponding to the slow state variables x_s. The edge set of this reduced graph may be quite different from the edge set of the original graph;

in particular, new edges may arise between the slow vertices. Furthermore, $\hat{R}$ is a diagonal matrix defining the (new) weights.

Crucial questions from an approximation point of view are how to separate into slow and fast vertices, and what can be said about the approximation properties of the reduced system. From an eigenvalue point of view the light masses are candidates for fast vertices, because the convergence of their momenta to steady state will be fast compared to the heavy masses (assuming that the variation of damping constants in the network is not too large).

Another question concerns the extension of the Kron reduction approach to mass–spring–damper systems, also in connection with related work on model reduction of power networks known under the name of *slow-coherency theory* [3, 5, 17].

11.4.2 Detailed-Balanced Chemical Reaction Networks

Structure-preserving model reduction of chemical reaction networks based on Kron reduction was proposed and explored in [15, 16, 25]. Consider as before, see (11.17), a detailed-balanced chemical reaction network

$$\dot{x} = -ZB\mathscr{K}B^T \mathrm{Exp}\left(Z^T \mathrm{Ln}\left(\frac{x}{x^*}\right)\right)$$

Similar to the case of mass–damper systems reduction is performed by separating the *complexes* into "slow" and "fast" ones. Reorder the complexes in such a way that the slow complexes come first. Then partition $\mathscr{L} = B\mathscr{K}B^T$ and Z correspondingly as

$$\mathscr{L} = \begin{bmatrix} \mathscr{L}_{ss} & \mathscr{L}_{sf} \\ \mathscr{L}_{fs} & \mathscr{L}_{ff} \end{bmatrix}, \quad Z = \begin{bmatrix} Z_s \ Z_f \end{bmatrix}, \tag{11.38}$$

where "s" refers to the "slow" complexes, and "f" to the "fast" ones. Consider then the auxiliary dynamical system

$$\begin{bmatrix} \dot{y}_s \\ \dot{y}_f \end{bmatrix} = -\begin{bmatrix} \mathscr{L}_{ss} & \mathscr{L}_{sf} \\ \mathscr{L}_{fs} & \mathscr{L}_{ff} \end{bmatrix} \begin{bmatrix} w_s \\ w_f \end{bmatrix}$$

and impose the constraint $\dot{y}_f = 0$. As for mass–damper systems it follows that $w_f = -\mathscr{L}_{ff}^{-1}\mathscr{L}_{fs}w_s$, which by substitution into the equation for $\dot{y}_s$ leads to the reduced dynamics

$$\dot{y}_s = -\left(\mathscr{L}_{ss} - \mathscr{L}_{sf}\mathscr{L}_{ff}^{-1}\mathscr{L}_{fs}\right) w_s = -\hat{\mathscr{L}} w_s$$

corresponding to the Schur complement $\hat{\mathscr{L}} := \mathscr{L}_{ss} - \mathscr{L}_{sf}\mathscr{L}_{ff}^{-1}\mathscr{L}_{fs} = \hat{B}\hat{\mathscr{K}}\hat{B}^T$. Substituting $w_s = \mathrm{Exp}\,(Z_s^T \mathrm{Ln}\,\left(\frac{x}{x^*}\right))$, and making use of $\dot{x} = Z_s\dot{y}_s + Z_f\dot{y}_f = Z_s\dot{y}_s$, we then obtain the reduced network

$$\dot{x} = -\hat{Z}\hat{B}\hat{\mathscr{K}}\hat{B}^T \mathrm{Exp}\left(\hat{Z}^T \mathrm{Ln}\left(\frac{x}{x^*}\right)\right), \quad \hat{\mathscr{K}} > 0, \tag{11.39}$$

where we have denoted $\hat{Z} := Z_s$. This is again a *detailed-balanced chemical reaction network* governed by mass action kinetics, with a reduced number of complexes and stoichiometric matrix $\hat{S} := \hat{Z}\hat{B}$; see [16, 25] for further details.

11.5 Outlook

In this paper we have concentrated on physical network systems with *symmetric* Laplacian matrices, that is, Laplacian matrices L of the form $L = BRB^T$ for some incidence matrix B and diagonal matrix R of weights. This is often the case, as has been illustrated on a number of examples. However, not all physical network systems are like this. For instance, in transportation networks one will typically start with a Laplacian matrix L which is *not* symmetric, and satisfies $\mathbb{1}^T L = 0$, but *not* $L\mathbb{1} = 0$. (In fact, for chemical reaction networks this is already the case; the detailed-balanced form (11.17) arises from rewriting the equations; see [25, 26].) The case of nonsymmetric Laplacian matrices will be explored in [27].

References

1. Arcak, M.: Passivity as a design tool for group coordination. IEEE Trans. Autom. Control **52**(8), 1380–1390 (2007)
2. Bollobas, B.: Modern Graph Theory, Graduate Texts in Mathematics 184. Springer, New York (1998)
3. Biyik, E., Arcak, M.: Area aggregation and time-scale modeling for sparse nonlinear networks. Syst. Control Lett. **57**(2), 142149 (2007)
4. Bürger, M., De Persis, C., Trip, S.: An internal model approach to (optimal) frequency regulation in power grids. In: Proceedings IEEE 53rd Annual Conference on Decision and Control (CDC), pp. 223–228, 15–17 Dec 2014
5. Chow, J.H., Allemong, J.J., Kokotovic, P.V.: Singular perturbation analysis of systems with sustained high frequency oscillations. Automatica **14**(3), 271–279 (1978)
6. DePersis, C., Kallesoe, C.S.: Pressure regulation in nonlinear hydraulic networks by positive and quantized control. IEEE Trans. Control Syst. Technol. **19**(6), 1371–1383 (2011)
7. Dörfler, F., Bullo, F.: Kron reduction of graphs with applications to electrical networks. IEEE Trans. Circ. Syst. I: Regul. Pap. **60**(1), 150–163 (2013)
8. Feinberg, M.: Necessary and sufficient conditions for detailed balancing in mass action systems of arbitrary complexity. Chem. Eng. Sci. **44**(9), 1819–1827 (1989)
9. Godsil, C., Royle, G.: Algebraic graph theory. Springer-Verlag, New York (2001)
10. Imura, J.-I.: Clustered model reduction of large-scale complex networks. In: Proceedings of the 20th International Symposium on Mathematical Theory of Networks and Systems (MTNS), Melbourne (2012)
11. Ishizaki, T., Kashida, K., Imura, J.-I., Aihara, K.: Network clustering for SISO linear dynamical networks via reaction-diffusion transformation. In: Proceeding of 18th IFAC World Congress, Milan, Italy, pp. 5639–5644 (2011)

12. Machovski, J., Bialek, J., Bumby, J.: Power System Dynamics: Stability and Control, 2nd edn. Wiley, Chichester (2008)
13. Monshizadeh, N., Camlibel, M.K., Trentelman, H.L.: Projection based model reduction of multi-agent systems using graph partitions. IEEE Trans. Control Netw. Syst. **1**(2), 145–154 (2014)
14. Monshizadeh, N., van der Schaft, A.J.: Structure-preserving model reduction of physical network systems by clustering. In: Proceedings 53rd IEEE Conference on Decision and Control, Los Angeles, CA, USA, Dec (2014)
15. Rao, S., van der Schaft, A.J., Jayawardhana, B.: A graph-theoretical approach for the analysis and model reduction of complex-balanced chemical reaction networks. J. Math. Chem. **51**, 2401–2422 (2013)
16. Rao, S., van der Schaft, A.J., van Eunen, K., Bakker, B.M., Jayawardhana, B.: A model reduction method for biochemical reaction networks. BMC Syst. Biol. **8**, 52 (2014)
17. Romeres, D., Dörfler, F., Bullo, F.: Novel results on slow coherency in consensus and power networks. Eur. Control conf. **742–747** (2013)
18. Sandberg, H., Murray, R.M.: Model reduction of interconnected linear systems. Optimal Control Appl. Methods **30**(3), 225–245 (2009)
19. van der Schaft, A.J.: L_2-Gain and Passivity Techniques in Nonlinear Control, Lectures Notes in Control and Information Sciences, Vol. 218, Springer, Berlin (1996). 2nd rev. and enlarged edn. Springer, London (2000)
20. van der Schaft, A.J.: Characterization and partial synthesis of the behavior of resistive circuits at their terminals. Systems & Control Letters **59**, 423–428 (2010)
21. van der Schaft, A.J.: On model reduction of physical network systems. In: Proceeding of 21st International Symposium on Mathematical Theory of Networks and Systems (MTNS2014), pp. 1419–1425. Groningen, The Netherlands, 7–11 July 2014
22. van der Schaft, A., Jeltsema, D.: Port-Hamiltonian systems theory: an introductory overview. Found. Trends Syst. Control **1**(2/3), 173–378 (2014)
23. van der Schaft, A.J., Maschke, B.M.: The Hamiltonian formulation of energy conserving physical systems with external ports. Archiv für Elektronik und Übertragungstechnik **49**, 362–371 (1995)
24. van der Schaft, A.J., Maschke, B.M.: Port-Hamiltonian systems on graphs. SIAM J. Control Optim. **51**(2), 906–937 (2013)
25. van der Schaft, A.J., Rao, S., Jayawardhana, B.: On the mathematical structure of balanced chemical reaction networks governed by mass action kinetics. SIAM J. Appl. Math. **73**(2), 953–973 (2013)
26. van der Schaft, A.J., Rao, S, Jayawardhana, B.: Complex and detailed balancing of chemical reaction networks revisited. J. Math. Chem. **53**(6), 1445–1458 (2015)
27. van der Schaft, A.J.: Modeling of physical network systems. In: preparation (2015)
28. Wegscheider, R.: Über simultane Gleichgewichte und die Beziehungen zwischen Thermodynamik und Reaktionskinetik homogener Systeme. Zetschrift für Physikalische Chemie **39**, 257–303 (1902)

Chapter 12
Interconnections of $\mathscr{L}^2$-Behaviors: Lumped Systems

Shiva Shankar

Abstract J.C. Willems' fundamental work in electrical circuit theory spawns many questions regarding energy and its transfer across ports. This paper proposes the inverse limit of the Sobolev spaces (on $\mathbb{R}$) as the appropriate space of signals in which to address these questions. Some of the first questions are those of interconnections of circuits in this signal space, and of the elimination of latent variables. The answers to these questions in the setting of the Sobolev limit naturally lead to questions on implementability and on the regularity of implementing controllers in the sense of H.L. Trentelman and Willems.

AMS classification: 93B05

12.1 Introduction

This article is inspired by Harry Trentelman's work on the regular implementation of controllers [1, 2, 4, 10]. It addresses the preliminary questions of interconnections and of elimination in the Sobolev spaces in order to be able to extend his results to electrical circuits in the setting of Jan Willems' paper [12].

In this chapter, Willems makes a fundamental distinction between *terminals* and *ports*—'terminals are for interconnection, ports are for energy transfer'. While the trajectory of a circuit is specified by the values at various times of the voltages and currents at the terminals (which are local objects) and determined by laws described by differential equations (i.e., by laws local in time), the notions of energy and its transfer require a global description of signals (i.e., nonlocal in time), and global objects that Willems designates as ports. This chapter adopts the classical description of energy as the integral of a quadratic form, and suggests that the correct signal space

S. Shankar (✉)
Chennai Mathematical Institute, Plot No. H1, SIPCOT IT Park, Kelambakkam, Siruseri, Chennai (Madras) 603103, India
e-mail: sshankar@cmi.ac.in

M.N. Belur et al. (eds.), *Mathematical Control Theory II*, Lecture Notes in Control and Information Sciences 462, DOI 10.1007/978-3-319-21003-2_12

to study the behavior of circuits are the $\mathscr{L}^2$-Sobolev spaces. Thus, we assume that the signals, namely the voltages and currents that occur in the circuit, as well as all their derivatives, are $\mathscr{L}^2$ functions. This assumption locates these signals in the inverse limit of the Sobolev spaces.

The programme is to now study the behavior of circuits in the Sobolev limit, i.e., to determine the achievable behavior of circuits, to define interconnections, to determine whether controllers admit regular implementations, to address issues of causality and to determine whether a behavior is realizable, etc. in this signal space. The answers will depend on this choice of signal space, as did the answers when the choice was the space $\mathscr{D}$ of compactly supported smooth functions instead of $\mathscr{C}^\infty$ [4]. The case of the Sobolev limit is similar to the case of $\mathscr{D}$ in many respects, but there are also differences which are highlighted below.

This chapter confines itself to only a few of the above questions that have been influenced by the work of Trentelman. It also hopes to extend the general framework established by Willems in his study of the behavior of electrical circuits.

12.2 Preliminaries

As explained above, energy considerations require the signals in an electric circuit to be $\mathscr{L}^2$ functions. The space $\mathscr{L}^2(\mathbb{R})$ is not a module over the ring of differential operators, hence to formulate circuit theory in the algebraic language of behaviors, we need to consider a suitable subspace of it (or a suitable embedding). For the purposes of this chapter we choose the inverse limit of the Sobolev spaces [8] to locate signals; this means that the signals and all their derivatives (a priori, in the distributional sense) are in $\mathscr{L}^2$.

Let $\mathscr{A} := \mathbb{C}[\frac{\mathrm{d}}{\mathrm{dt}}]$ be the ring of ordinary differential operators. For every s in $\mathbb{R}$, the Sobolev space $\mathscr{H}^s(\mathbb{R})$ of order s is the space of tempered distributions f whose Fourier transform $\hat{f}$ is a measurable function such that

$$\|f\|_s = \Big(\frac{1}{2\pi}\int_{\mathbb{R}} |\hat{f}(\xi)|^2(1+|\xi|^2)^s d\xi\Big)^{\frac{1}{2}} < \infty$$

$\mathscr{H}^s$ is a Hilbert space with norm $\|\ \|_s$. When $s > t$, $\mathscr{H}^s \hookrightarrow \mathscr{H}^t$ is a continuous inclusion. If $p(\frac{d}{dt})$ is an element of $\mathscr{A}$ of order r, then it maps $\mathscr{H}^s$ into $\mathscr{H}^{s-r}$. Consider, the family $\{\mathscr{H}^s, s \in \mathbb{R}\}$ a decreasing family of vector spaces, its inverse limit $\overleftarrow{\mathscr{H}} = \varprojlim \mathscr{H}^s$ is isomorphic to the intersection $\bigcap_{s\in\mathbb{R}} \mathscr{H}^s$ of the Sobolev spaces, and is an $\mathscr{A}$-module. This intersection contains the Schwartz space $\mathscr{S}$ of rapidly decreasing functions but is strictly larger than it. As $\mathbb{N}$ is cofinal in $\mathbb{R}$, this limit is also the inverse limit of the countable family $\{\mathscr{H}^s,\ s = 0, 1, 2\ldots\}$. Further, if each $\mathscr{H}^s$ is given its Hilbert space topology, then the intersection $\overleftarrow{\mathscr{H}}$ with the inverse limit topology, which is the weakest topology such that each inclusion $\overleftarrow{\mathscr{H}} \hookrightarrow \mathscr{H}^s$ is continuous, is a Fréchet space.

By the Sobolev Embedding Theorem [3], the signals in $\overleftarrow{\mathscr{H}}$ are smooth functions, and hence, the derivatives of an element in it are derivatives in the classical sense.

Following Willems [11, 12], we consider the signals of interest that do occur - currents, voltages, energy etc.—to be constrained by the laws governing the circuit, such as KVL, KCL, as well as those determined by various circuit elements. These are laws local in time, given by differential operators. Thus, we study systems represented by kernels of operators defined by differential equations:

$$\begin{array}{rrcl} P(\frac{\mathsf{d}}{\mathsf{dt}}) : & \overleftarrow{\mathscr{H}}^k & \longrightarrow & \overleftarrow{\mathscr{H}}^\ell \\ & f = (f_1, \ldots, f_k) & \mapsto & P(\frac{\mathsf{d}}{\mathsf{dt}}) f \end{array} \tag{12.1}$$

where $P(\frac{\mathsf{d}}{\mathsf{dt}})$ is an $\ell \times k$ matrix whose entries $p_{ij}(\frac{\mathsf{d}}{\mathsf{dt}})$ are from the ring $\mathscr{A}$. This kernel depends only on the submodule $\mathscr{P}$ of $\mathscr{A}^k$ generated by the rows of the matrix $P(\frac{\mathsf{d}}{\mathsf{dt}})$, it being isomorphic to $\mathsf{Hom}_{\mathscr{A}}(\mathscr{A}^k/\mathscr{P}, \overleftarrow{\mathscr{H}})$. It is the behavior of the submodule $\mathscr{P}$ in $\overleftarrow{\mathscr{H}}$, and will be denoted $\mathscr{B}_{\overleftarrow{\mathscr{H}}}(\mathscr{P})$.

The Fourier transform translates the kernel of (12.1) to the kernel $\mathscr{B}_{\widehat{\overleftarrow{\mathscr{H}}}}(\mathscr{P}_\xi)$ of the map

$$\begin{array}{rrcl} P(\xi) : & \widehat{\overleftarrow{\mathscr{H}}}^k & \longrightarrow & \widehat{\overleftarrow{\mathscr{H}}}^\ell \\ & \hat{f} = (\hat{f}_1, \ldots, \hat{f}_k) & \mapsto & P(\xi) \hat{f} \end{array}$$

where $\widehat{\overleftarrow{\mathscr{H}}}$, the set of Fourier transforms of elements in $\overleftarrow{\mathscr{H}}$, has the structure of an $\mathscr{A}_\xi := \mathbb{C}[\xi]$-module given by multiplication $p_{ij}(\xi)\hat{f}_j(\xi)$, the translate of differentiation $p_{ij}(\frac{\mathsf{d}}{\mathsf{dt}}) f_j(t)$ in $\overleftarrow{\mathscr{H}}$, and where $\mathscr{P}_\xi$ is the submodule of $\mathscr{A}_\xi^k$ generated by the rows of the matrix $P(\xi)$—it equals $\widehat{\mathscr{P}}$, the polynomials corresponding to all the differential operators in $\mathscr{P}$ under Fourier transformation. Thus $\mathscr{B}_{\widehat{\overleftarrow{\mathscr{H}}}}(\mathscr{P}_\xi) = \widehat{\mathscr{B}_{\overleftarrow{\mathscr{H}}}(\mathscr{P})}$.

This is the behavioral framework in which the questions of controllability, realizability, interconnections, and of regularity of controllers are best addressed. There is ,however, a representation issue that must be first answered, namely that different submodules of $\mathscr{A}^k$ determine the same behavior in the Sobolev limit $\overleftarrow{\mathscr{H}}$. This is the Nullstellensatz question of [5], and we turn to it next.

Given, a behavior $\mathscr{B}_{\overleftarrow{\mathscr{H}}}(\mathscr{P})$ defined by $P(\frac{\mathsf{d}}{\mathsf{dt}}) : \overleftarrow{\mathscr{H}}^k \to \overleftarrow{\mathscr{H}}^\ell$ (where the rows of $P(\frac{\mathsf{d}}{\mathsf{dt}})$ generate the submodule $\mathscr{P}$ of $\mathscr{A}^k$), consider the submodule $\overline{\mathscr{P}}$ of $\mathscr{A}^k$ of all elements $p(\frac{\mathsf{d}}{\mathsf{dt}})$ such that the kernel of the map $p(\frac{\mathsf{d}}{\mathsf{dt}}) : \overleftarrow{\mathscr{H}}^k \to \overleftarrow{\mathscr{H}}$ contains $\mathscr{B}_{\overleftarrow{\mathscr{H}}}(\mathscr{P})$. This submodule $\overline{\mathscr{P}}$, called the (Willems) closure of $\mathscr{P}$ with respect to $\overleftarrow{\mathscr{H}}$, is the largest submodule of $\mathscr{A}^k$ whose behavior is $\mathscr{B}_{\overleftarrow{\mathscr{H}}}(\mathscr{P})$.

Proposition 12.2.1 *Let $\mathscr{P}$ be a submodule of $\mathscr{A}^k$. Its closure $\overline{\mathscr{P}}$ with respect to the Sobolev limit $\overleftarrow{\mathscr{H}}$ equals $\{p(\frac{\mathsf{d}}{\mathsf{dt}}) \in \mathscr{A}^k \mid a(\frac{\mathsf{d}}{\mathsf{dt}})p(\frac{\mathsf{d}}{\mathsf{dt}}) \in \mathscr{P}, 0 \neq a(\frac{\mathsf{d}}{\mathsf{dt}}) \in \mathscr{A}\}$, so that*

$\mathscr{P}$ equals its closure with respect to $\overleftarrow{\mathscr{H}}$ if and only if $\mathscr{A}^k/\mathscr{P}$ is torsion free, and hence free (as $\mathscr{A}$ is a principal ideal domain).

Proof The proof is identical to the proof for the calculation of the closure in $\mathscr{S}$ or in $\mathscr{D}$; a similar statement and proof is valid also over the ring of partial differential operators [5]. □

Thus, in the space $\overleftarrow{\mathscr{H}}$, it suffices to consider behaviors of submodules $\mathscr{P}$ such that $\mathscr{A}^k/\mathscr{P}$ is free.

The above description of the closure implies the following corollary, just as in [7].

Corollary 12.2.1 *With respect to the Sobolev limit $\overleftarrow{\mathscr{H}}$, $\overline{\mathscr{P}_1 \cap \mathscr{P}_2} = \overline{\mathscr{P}_1} \cap \overline{\mathscr{P}_2}$ (i.e. the closure of an intersection is the intersection of the closures).* □

We next consider the question of behavioral controllability [11]. It turns out that the behavior of a lumped system in the Sobolev limit always admits an image representation, and hence is always controllable. This is because $\overleftarrow{\mathscr{H}}(\mathbb{R})$ is a flat $\mathbb{C}[\frac{\mathrm{d}}{\mathrm{dt}}]$-module [8].

Remark The Sobolev limit is however not faithfully flat. In this respect $\overleftarrow{\mathscr{H}}(\mathbb{R})$ is similar to the space $\mathscr{S}(\mathbb{R})$ (but not to $\mathscr{D}(\mathbb{R})$ which is also faithfully flat). These similarities vanish when we consider these spaces on $\mathbb{R}^n, n \geqslant 2$. The significant differences now are reflected in the widely different structure of distributed behaviors in these two spaces [8].

12.3 Interconnections

We now turn to the main problem addressed in this chapter, namely that of interconnections. The behavioral analogues of the classical series and parallel interconnections of circuits are the sums and intersections of behaviors. It is always the case that the intersection of the behaviors of $\mathscr{P}_1$ and $\mathscr{P}_2$ is the behavior of $\mathscr{P}_1 + \mathscr{P}_2$, but in the Sobolev limit, the sum of these two behaviors may not be the behavior of the intersection $\mathscr{P}_1 \cap \mathscr{P}_2$, indeed it may not be a behavior at all. This is again similar to the situation in $\mathscr{S}$ [7].

Example 1 Let $\mathscr{P}_1$ and $\mathscr{P}_2$ be cyclic submodules of $\mathscr{A}^2$ generated by $(1, 0)$ and $(1, -\frac{d}{dt})$, respectively. Then $\mathscr{P}_1 \cap \mathscr{P}_2$ is the 0 submodule, so that $\mathscr{B}_{\overleftarrow{\mathscr{H}}}(\mathscr{P}_1 \cap \mathscr{P}_2)$ is all of $\overleftarrow{\mathscr{H}}^2$. On the other hand, $\mathscr{B}_{\overleftarrow{\mathscr{H}}}(\mathscr{P}_1) = \{(0, f) \mid f \in \overleftarrow{\mathscr{H}}\}$ and $\mathscr{B}_{\overleftarrow{\mathscr{H}}}(\mathscr{P}_2) = \{(\frac{dg}{dt}, g) \mid g \in \overleftarrow{\mathscr{H}}\}$. Thus an element (u, v) in $\overleftarrow{\mathscr{H}}^2$ is in $\mathscr{B}_{\overleftarrow{\mathscr{H}}}(\mathscr{P}_1) + \mathscr{B}_{\overleftarrow{\mathscr{H}}}(\mathscr{P}_2)$ only if $u = \frac{dg}{dt}, v = f + g$, where f and g are arbitrary elements in $\overleftarrow{\mathscr{H}}$. Let now u be any smooth function that decays as $\frac{1}{t}$ as $t \to \pm\infty$. Then u is in $\overleftarrow{\mathscr{H}}$, whereas $g(t) = \int_{-\infty}^{t} u \, dt$, which grows as $\log t$, is not. It follows now that $(u, 0)$ is in $\mathscr{B}_{\overleftarrow{\mathscr{H}}}(\mathscr{P}_1 \cap \mathscr{P}_2)$ but is however not in $\mathscr{B}_{\overleftarrow{\mathscr{H}}}(\mathscr{P}_1) + \mathscr{B}_{\overleftarrow{\mathscr{H}}}(\mathscr{P}_2)$. □

The question remains whether the sum of the two behaviors in the above example could be the behavior of some nonzero submodule of $\mathscr{A}^k$. That it cannot is the content of the following proposition, again similar to the case of the signal spaces considered in [7].

Proposition 12.3.1 *Let $\mathscr{P}_1$, $\mathscr{P}_2$ be submodules of $\mathscr{A}^k$. Then $\mathscr{B}_{\overleftarrow{\mathscr{H}}}(\mathscr{P}_1 \cap \mathscr{P}_2)$ is the smallest behavior containing both $\mathscr{B}_{\overleftarrow{\mathscr{H}}}(\mathscr{P}_1)$ and $\mathscr{B}_{\overleftarrow{\mathscr{H}}}(\mathscr{P}_2)$, and hence also the smallest behavior containing $\mathscr{B}_{\overleftarrow{\mathscr{H}}}(\mathscr{P}_1) + \mathscr{B}_{\overleftarrow{\mathscr{H}}}(\mathscr{P}_2)$.*

Proof Suppose $\mathscr{B}_{\overleftarrow{\mathscr{H}}}(\mathscr{P})$ contains both $\mathscr{B}_{\overleftarrow{\mathscr{H}}}(\mathscr{P}_1)$ and $\mathscr{B}_{\overleftarrow{\mathscr{H}}}(\mathscr{P}_2)$. Then $\overline{\mathscr{P}}$ is contained in both $\overline{\mathscr{P}_1}$ and $\overline{\mathscr{P}_2}$, and hence is also contained in $\overline{\mathscr{P}_1} \cap \overline{\mathscr{P}_2}$, which is by Corollary 12.2.1 equal to $\overline{\mathscr{P}_1 \cap \mathscr{P}_2}$. This implies that $\mathscr{B}_{\overleftarrow{\mathscr{H}}}(\overline{\mathscr{P}})$ contains $\mathscr{B}_{\overleftarrow{\mathscr{H}}}(\overline{\mathscr{P}_1 \cap \mathscr{P}_2})$, and hence that

$$\mathscr{B}_{\overleftarrow{\mathscr{H}}}(\mathscr{P}_1 \cap \mathscr{P}_2) = \mathscr{B}_{\overleftarrow{\mathscr{H}}}(\overline{\mathscr{P}_1 \cap \mathscr{P}_2}) \subset \mathscr{B}_{\overleftarrow{\mathscr{H}}}(\overline{\mathscr{P}}) = \mathscr{B}_{\overleftarrow{\mathscr{H}}}(\mathscr{P})$$

Thus any behavior which contains both $\mathscr{B}_{\overleftarrow{\mathscr{H}}}(\mathscr{P}_1)$ and $\mathscr{B}_{\overleftarrow{\mathscr{H}}}(\mathscr{P}_2)$ also contains $\mathscr{B}_{\overleftarrow{\mathscr{H}}}(\mathscr{P}_1 \cap \mathscr{P}_2)$. $\square$

The problem now is to locate the obstruction to the sum of two behaviors in the Sobolev limit being a behavior, and to determine when this obstruction vanishes.

Consider the following exact sequence

$$0 \to \mathscr{A}^k/(\mathscr{P}_1 \cap \mathscr{P}_2) \xrightarrow{i} \mathscr{A}^k/\mathscr{P}_1 \oplus \mathscr{A}^k/\mathscr{P}_2 \xrightarrow{\pi} \mathscr{A}^k/(\mathscr{P}_1 + \mathscr{P}_2) \to 0$$

where $i([x]) = ([x], [x])$, $\pi([x], [y]) = [x - y]$ (here [ ] indicates the class of an element of $\mathscr{A}^k$ in the various quotients). It follows that

$$0 \to \mathsf{Hom}_{\mathscr{A}}(\mathscr{A}^k/(\mathscr{P}_1 + \mathscr{P}_2), \overleftarrow{\mathscr{H}}) \to \mathsf{Hom}_{\mathscr{A}}(\mathscr{A}^k/\mathscr{P}_1, \overleftarrow{\mathscr{H}}) \oplus \mathsf{Hom}_{\mathscr{A}}(\mathscr{A}^k/\mathscr{P}_2, \overleftarrow{\mathscr{H}})$$

$$\xrightarrow{d} \mathsf{Hom}_{\mathscr{A}}(\mathscr{A}^k/(\mathscr{P}_1 \cap \mathscr{P}_2), \overleftarrow{\mathscr{H}}) \xrightarrow{\delta} \mathsf{Ext}^1_{\mathscr{A}}(\mathscr{A}^k/(\mathscr{P}_1 + \mathscr{P}_2), \overleftarrow{\mathscr{H}})$$

$$\longrightarrow \mathsf{Ext}^1_{\mathscr{A}}(\mathscr{A}^k/\mathscr{P}_1, \overleftarrow{\mathscr{H}}) \oplus \mathsf{Ext}^1_{\mathscr{A}}(\mathscr{A}^k/\mathscr{P}_2, \overleftarrow{\mathscr{H}}) \to \cdots$$

is also exact, where $d(f, g) = f + g$. By Proposition 12.2.1, $\mathscr{A}^k/\mathscr{P}_1$ and $\mathscr{A}^k/\mathscr{P}_2$ can be chosen to be free, hence the second set of Ext terms above are 0, and the sequence becomes the exact sequence

$$\cdots \to \mathsf{Hom}_{\mathscr{A}}(\mathscr{A}^k/\mathscr{P}_1, \overleftarrow{\mathscr{H}}) \oplus \mathsf{Hom}_{\mathscr{A}}(\mathscr{A}^k/\mathscr{P}_2, \overleftarrow{\mathscr{H}}) \xrightarrow{d}$$

$$\mathsf{Hom}_{\mathscr{A}}(\mathscr{A}^k/(\mathscr{P}_1 \cap \mathscr{P}_2), \overleftarrow{\mathscr{H}}) \xrightarrow{\delta} \mathsf{Ext}^1_{\mathscr{A}}(\mathscr{A}^k/(\mathscr{P}_1 + \mathscr{P}_2), \overleftarrow{\mathscr{H}}) \to 0$$

This implies that the sum of the behaviors of $\mathscr{P}_1$ and $\mathscr{P}_2$ is the behavior of $\mathscr{P}_1 \cap \mathscr{P}_2$ (which is to say that the morphism d above is surjective) if and only if the connecting morphism δ is the zero morphism, and the necessary and sufficient condition for this is that $\mathsf{Ext}^1_{\mathscr{A}}(\mathscr{A}^k/(\mathscr{P}_1 + \mathscr{P}_2), \overleftarrow{\mathscr{H}})$ vanish.

To determine when this is so, consider the following free resolution of $\mathscr{A}^k/(\mathscr{P}_1 + \mathscr{P}_2)$

$$0 \to (\mathscr{P}_1 + \mathscr{P}_2) \xrightarrow{i} \mathscr{A}^k \to \mathscr{A}^k/(\mathscr{P}_1 + \mathscr{P}_2) \to 0$$

(here $(\mathscr{P}_1 + \mathscr{P}_2)$ is finitely generated and torsion free, hence free). Applying the functor $\mathsf{Hom}_{\mathscr{A}}(-, \overleftarrow{\mathscr{H}})$ gives the sequence

$$0 \to \mathsf{Hom}_{\mathscr{A}}(\mathscr{A}^k, \overleftarrow{\mathscr{H}}) \xrightarrow{\pi} \mathsf{Hom}_{\mathscr{A}}(\mathscr{P}_1 + \mathscr{P}_2, \overleftarrow{\mathscr{H}}) \to 0$$

and thus $\mathsf{Ext}^1_{\mathscr{A}}(\mathscr{A}^k/(\mathscr{P}_1+\mathscr{P}_2), \overleftarrow{\mathscr{H}})$ vanishes exactly when the above morphism π is surjective. In other words, the necessary and sufficient condition now for the solution of the problem of the sum of behaviors is that *every* morphism $\phi : (\mathscr{P}_1+\mathscr{P}_2) \to \overleftarrow{\mathscr{H}}$ should lift to a morphism $\overline{\phi} : \mathscr{A}^k \to \overleftarrow{\mathscr{H}}$.

Let $(P_1+P_2)(\frac{\mathsf{d}}{\mathsf{dt}})$ be a matrix whose rows is a basis for the free module $(\mathscr{P}_1+\mathscr{P}_2)$. Without loss of generality assume that this matrix is given in its Smith canonical form, so that all the diagonal entries, namely its invariant factors, say $d_1(\frac{\mathsf{d}}{\mathsf{dt}}), \ldots, d_r(\frac{\mathsf{d}}{\mathsf{dt}})$, are nonzero (here r is the rank of $(\mathscr{P}_1+\mathscr{P}_2)$), and d_i divides d_j if $i \leqslant j$). A morphism $\phi : (\mathscr{P}_1 + \mathscr{P}_2) \to \overleftarrow{\mathscr{H}}$ is then given by an r-tuple of maps $\phi_i : (d_i(\frac{\mathsf{d}}{\mathsf{dt}})) \to \overleftarrow{\mathscr{H}}, i = 1, \ldots, r$ where $(d_i(\frac{\mathsf{d}}{\mathsf{dt}}))$ is the principal ideal generated by the ith invariant factor, and the ϕ_i in turn are given by mapping $d_i(\frac{\mathsf{d}}{\mathsf{dt}})$ to arbitrary f_i in $\overleftarrow{\mathscr{H}}$ (as $\mathscr{A}$ is a domain). Thus, the question now is whether the maps ϕ_i lift to maps $\overline{\phi_i} : \mathscr{A} \to \overleftarrow{\mathscr{H}}$.

By Fourier transformation (as in the previous section), this question translates to the following:

When does *every* map $\psi : (d(\xi)) \to \widehat{\overleftarrow{\mathscr{H}}}$ defined on the principal ideal generated by $d(\xi)$ extend to $\mathbb{C}[\xi]$?

This is now elementary, for if $\psi(d(\xi)) = \hat{f}(\xi)$, then necessarily the extension must map 1 to $\frac{\hat{f}}{d}(\xi)$, which is a priori in the space $\mathscr{S}'$ of tempered distributions on $\mathbb{R}$. For it to be in $\widehat{\overleftarrow{\mathscr{H}}}$ it is sufficient that $d(\xi)$ not have real zeros, as then $\frac{1}{d}(\xi)$ is bounded on $\mathbb{R}$. This condition is also necessary, for if $\hat{f}(\xi)$ is nonzero at a real zero of $d(\xi)$, then $\frac{\hat{f}}{d}(\xi)$ is clearly not in $\mathscr{L}^2(\mathbb{R})$.

The above discussion thus proves the following theorem:

Theorem 12.3.1 *Let $\mathscr{P}_1$, $\mathscr{P}_2$ be submodules of $\mathscr{A}^k$. Then $\mathscr{B}_{\overleftarrow{\mathscr{H}}}(\mathscr{P}_1)+\mathscr{B}_{\overleftarrow{\mathscr{H}}}(\mathscr{P}_2)$ is a behavior, necessarily equal to $\mathscr{B}_{\overleftarrow{\mathscr{H}}}(\mathscr{P}_1 \cap \mathscr{P}_2)$, if and only if the nonzero associated primes of $\mathscr{A}^k/(\mathscr{P}_1 + \mathscr{P}_2)$ do not have real zeros.*

Proof In the notation of the discussion above, $\mathscr{A}^k/(\mathscr{P}_1 + \mathscr{P}_2)$ is isomorphic to $\mathscr{A}^{(k-r)} \bigoplus_{i=1}^{r} \mathscr{A}/(d_i(\frac{\mathsf{d}}{\mathsf{dt}}))$; here r is the rank of the free submodule $(\mathscr{P}_1 + \mathscr{P}_2)$ of $\mathscr{A}^k$, and the $d_i(\frac{\mathsf{d}}{\mathsf{dt}})$ are its invariant factors. The linear factors of these $d_i(\frac{\mathsf{d}}{\mathsf{dt}})$ are precisely the nonzero associated primes of $\mathscr{A}^k/(\mathscr{P}_1 + \mathscr{P}_2)$. □

Remark This theorem provides an explanation to Example 1 above. There $\mathscr{P}_1$ and $\mathscr{P}_2$ are cyclic submodules of $\mathscr{A}^2$ generated by $(1, 0)$ and $(1, -\frac{d}{dt})$, so that $\mathscr{A}^2/(\mathscr{P}_1 + \mathscr{P}_2)$ is isomorphic to $\mathscr{A}/(\frac{\mathsf{d}}{\mathsf{dt}}) \simeq \mathscr{A}_\xi/(\xi)$. The ideal (ξ) is an associated prime of $\mathscr{A}_\xi/(\xi)$ which has a real zero, namely the point 0!

Remark The set of all behaviors can be topologized as in [9]. Then for behaviors in a Zariski open set, the sum of any two of them is also a behavior. This is because the condition that a polynomial with complex coefficients have real zeros is a Zariski closed condition.

12.4 Elimination

We now consider the problem of elimination of latent variables. This is patterned after [6], and we omit many of the details.

Consider the split exact sequence

$$0 \to \mathscr{A}^p \underset{\underset{\longleftarrow}{\pi_1}}{\overset{i_1}{\longrightarrow}} \mathscr{A}^{p+q} \underset{\underset{\longleftarrow}{i_2}}{\overset{\pi_2}{\longrightarrow}} \mathscr{A}^q \to 0$$

Applying the functor $\mathsf{Hom}_{\mathscr{A}}(-, \overleftarrow{\mathscr{H}})$ yields the split exact sequence

$$0 \to (\overleftarrow{\mathscr{H}})^p \underset{\underset{\longrightarrow}{i_1}}{\overset{\pi_1}{\longleftarrow}} (\overleftarrow{\mathscr{H}})^{p+q} \underset{\underset{\longrightarrow}{\pi_2}}{\overset{i_2}{\longleftarrow}} (\overleftarrow{\mathscr{H}})^q \to 0$$

Let $\mathscr{P}$ be a submodule of $\mathscr{A}^{p+q}$, so that $\mathscr{B}_{\overleftarrow{\mathscr{H}}}(\mathscr{P})$ is a behavior in $\overleftarrow{\mathscr{H}}^{p+q}$. The question elimination addresses is whether $\pi_2(\mathscr{B}_{\overleftarrow{\mathscr{H}}}(\mathscr{P}))$ is also a behavior. More generally, the following proposition relates the $\mathscr{A}$-submodules $i_2^{-1}(\mathscr{B}_{\overleftarrow{\mathscr{H}}}(\mathscr{P}))$, $\pi_2(\mathscr{B}_{\overleftarrow{\mathscr{H}}}(\mathscr{P}))$, $\mathscr{B}_{\overleftarrow{\mathscr{H}}}(i_2^{-1}(\mathscr{P}))$ and $\mathscr{B}_{\overleftarrow{\mathscr{H}}}(\pi_2(\mathscr{P}))$ of $\overleftarrow{\mathscr{H}}^q$.

Proposition 12.4.1

$$i_2^{-1}(\mathscr{B}_{\overleftarrow{\mathscr{H}}}(\mathscr{P})) = \mathscr{B}_{\overleftarrow{\mathscr{H}}}(\pi_2(\mathscr{P})) \subset \pi_2(\mathscr{B}_{\overleftarrow{\mathscr{H}}}(\mathscr{P})) \subset \mathscr{B}_{\overleftarrow{\mathscr{H}}}(i_2^{-1}(\mathscr{P}))$$

Proof The statement is true for every $\mathscr{A}$-submodule of $\mathscr{D}'$, in particular for $\overleftarrow{\mathscr{H}}$; indeed it is true also for the ring of partial differential operators [6]. □

Example 2 Let $\mathscr{P}$ be the cyclic submodule of $\mathscr{A}^2$ generated by $(\frac{d}{dt}, -1)$. Then $\mathscr{B}_{\overleftarrow{\mathscr{H}}}(\mathscr{P}) = \{(f, \frac{df}{dt}) \mid f \in \overleftarrow{\mathscr{H}}\}$ and it is easy to see that $\pi_2(\mathscr{B}_{\overleftarrow{\mathscr{H}}}(\mathscr{P})) = \{\frac{df}{dt} \mid f \in \overleftarrow{\mathscr{H}}\}$ is not a differential kernel in $\overleftarrow{\mathscr{H}}$—this also follows from the next proposition. □

Proposition 12.4.2 *In the above notation,* $\mathscr{B}_{\overleftarrow{\mathscr{H}}}(i_2^{-1}(\mathscr{P}))$ *is the smallest behavior containing* $\pi_2(\mathscr{B}_{\overleftarrow{\mathscr{H}}}(\mathscr{P}))$.

The proof is identical to the proof of Proposition 4.2 in [6]. □

The first split exact sequence above implies that the following sequence

$$0 \to \mathscr{A}^q / i_2^{-1}(\mathscr{P}) \longrightarrow \mathscr{A}^{p+q}/\mathscr{P} \longrightarrow \mathscr{A}^p/\pi_1(\mathscr{P}) \to 0$$

is exact. Thus

$$0 \to \mathsf{Hom}_{\mathscr{A}}(\mathscr{A}^p/\pi_1(\mathscr{P}), \overleftarrow{\mathscr{H}}) \xrightarrow{i_1} \mathsf{Hom}_{\mathscr{A}}(\mathscr{A}^{p+q}/\mathscr{P}, \overleftarrow{\mathscr{H}}) \xrightarrow{\pi_2}$$

$$\mathsf{Hom}_{\mathscr{A}}(\mathscr{A}^q/i_2^{-1}(\mathscr{P}), \overleftarrow{\mathscr{H}}) \xrightarrow{\delta} \mathsf{Ext}^1_{\mathscr{A}}(\mathscr{A}^p/\pi_1(\mathscr{P}), \overleftarrow{\mathscr{H}}) \xrightarrow{i_1} \mathsf{Ext}^1_{\mathscr{A}}(\mathscr{A}^{p+q}/\mathscr{P}, \overleftarrow{\mathscr{H}}) \to \cdots$$

is also exact, where the morphism π_2 is just the restriction of the π_2 in the second split exact sequence above to the $\mathscr{A}$-submodule $\mathsf{Hom}_{\mathscr{A}}(\mathscr{A}^{p+q}/\mathscr{P}, \overleftarrow{\mathscr{H}}) \simeq \mathscr{B}_{\overleftarrow{\mathscr{H}}}(\mathscr{P})$ of $\overleftarrow{\mathscr{H}}^{p+q}$. By Proposition 12.2.1, $\mathscr{A}^{p+q}/\mathscr{P}$ can be chosen to be free, hence the above exact sequence becomes the exact sequence

$$\cdots \to \mathsf{Hom}_{\mathscr{A}}(\mathscr{A}^{p+q}/\mathscr{P}, \overleftarrow{\mathscr{H}}) \xrightarrow{\pi_2} \mathsf{Hom}_{\mathscr{A}}(\mathscr{A}^q/i_2^{-1}(\mathscr{P}), \overleftarrow{\mathscr{H}})$$

$$\xrightarrow{\delta} \mathsf{Ext}^1_{\mathscr{A}}(\mathscr{A}^p/\pi_1(\mathscr{P}), \overleftarrow{\mathscr{H}}) \to 0$$

It follows now by the above proposition that $\pi_2(\mathscr{B}_{\overleftarrow{\mathscr{H}}}(\mathscr{P}))$ is a behavior if and only if the morphism δ in the above exact sequence is the zero morphism, and a necessary and sufficient condition for this is that $\mathsf{Ext}^1_{\mathscr{A}}(\mathscr{A}^p/\pi_1(\mathscr{P}), \overleftarrow{\mathscr{H}})$ be equal to 0.

To determine when this is so, consider the following free resolution of $\mathscr{A}^p/\pi_1(\mathscr{P})$

$$0 \to \pi_1(\mathscr{P}) \xrightarrow{i} \mathscr{A}^p \to \mathscr{A}^p/\pi_1(\mathscr{P}) \to 0$$

where (as in the previous section) $\pi_1(\mathscr{P})$ is finitely generated and torsion free, hence free (recollect that $\mathscr{A} = \mathbb{C}[\frac{d}{dt}]!$). Applying $\mathsf{Hom}_{\mathscr{A}}(-, \overleftarrow{\mathscr{H}})$ gives the sequence

$$0 \to \mathsf{Hom}_{\mathscr{A}}(\mathscr{A}^p, \overleftarrow{\mathscr{H}}) \xrightarrow{\pi} \mathsf{Hom}_{\mathscr{A}}(\pi_1(\mathscr{P}), \overleftarrow{\mathscr{H}}) \to 0$$

so that the necessary and sufficient condition for the projection $\pi_2(\mathscr{B}_{\overleftarrow{\mathscr{H}}}(\mathscr{P}))$ to be a behavior is that the morphism π above be surjective. Thus, in this notation, we have the following theorem.

Theorem 12.4.1 *The projection $\pi_2(\mathscr{B}_{\overleftarrow{\mathscr{H}}}(\mathscr{P}))$ is a behavior, necessarily equal to $\mathscr{B}_{\overleftarrow{\mathscr{H}}}(i_2^{-1}(\mathscr{P}))$, if and only if the nonzero associated primes of $\mathscr{A}^p/(\pi_1(\mathscr{P}))$ do not have real zeros.*

The proof is similar to the proof of Theorem 12.3.1. □

Remark This theorem provides an explanation to Example 2. There $\mathscr{P}$ is the cyclic submodule of $\mathscr{A}^2$ generated $(\frac{\mathsf{d}}{\mathsf{dt}}, -1)$, so that $\mathscr{A}/(\pi_1(\mathscr{P}))$ equals $\mathscr{A}/(\frac{\mathsf{d}}{\mathsf{dt}}) \simeq \mathscr{A}_\xi/(\xi)$. The ideal (ξ) is an associated prime of $\mathscr{A}_\xi/(\xi)$ which has a real zero, namely the point 0.

We have now established three results in the inverse limit $\overleftarrow{\mathscr{H}}(\mathbb{R})$ of the Sobolev spaces—namely the Nullstellensatz, the calculation of the obstruction to a sum of two behaviors being a behavior, as well as the calculation of the obstruction to a projection of a behavior being a behavior. Just as in [4], these are the basic results necessary to answer questions regarding the regular implementation of controllers. These questions, together with applications to electrical circuits in the framework of [12], will be pursued elsewhere.

References

1. Belur, M.N., Trentelman, H.L.: Stabilization, pole placement, and regular implementability. IEEE Trans. Autom. Control **47**, 735–744 (2002)
2. Fiaz, S., Trentelman, H.L.: Regular implementability and stabilization using controllers with pre-specified input/output partition. IEEE Transac. Autom. Control **54**, 1561–1568 (2009)
3. G.B. Folland, Real Analysis, Wiley-Interscience, 1999
4. Napp Avelli, D., Shankar, S., Trentelman, H.L.: Regular implementation in the space of compactly supported functions. Syst. Control Lett. **57**, 851–855 (2008)
5. Shankar, S.: The Nullstellensatz for systems of PDE. Adv. Appl. Math. **23**, 360–374 (1999)
6. Shankar, S.: Geometric completeness of distribution spaces. Acta Applicandae Mathematicae **77**, 163–180 (2003)
7. Shankar, S.: A Cousin problem for systems of PDE. Mathematische Nachrichten **280**, 446–450 (2007)
8. Shankar, S.: On the dual of a flat module in TOP. Linear Algebra Appl. **433**, 1077–1081 (2010)
9. Shankar, S.: The Hautus test and genericity results for controllable and uncontrollable behaviors. SIAM J. Control Optim. **52**, 32–51 (2014)
10. Trentelman, H.J., Napp Avelli, D.: On the regular implementability of nD systems. Syst. Control Lett. **56**, 265–271 (2007)
11. Willems, J.C.: The behavioral approach to open and interconnected systems. IEEE Control Syst. Mag. **27**, 46–99 (2007)
12. Willems, J.C.: Terminals and ports. IEEE Circuits Syst. Mag. **4**, 8–26 (2010)

Chapter 13
On State Observers—Take 2

Jochen Trumpf

Abstract This is the author's second attempt to provide a characterization for asymptotic functional state observers in the category of linear time-invariant finite-dimensional systems in input/state/output form in terms of a Sylvester-type matrix equation with a proof that only uses state-space and transfer function methods. The characterizing equation was already proposed in Luenberger's original work on state observers, but to prove that it is not only sufficient but also necessary when the observed system has no stable uncontrollable modes turns out to be surprisingly hard. The crux of the problem is that in a classical observer interconnection both the output and the input of the observed system enter the observer and hence also the observer error system as separate inputs. They are not independent signals, though, since they are jointly constrained by the equations of the observed system. The first attempt by the author (see the list of references) contained a subtle error in the proof of the main result. To fix this error, some new intermediate results are needed and the final proof is sufficiently different to warrant this paper. As a bonus, details on how to observe stable uncontrollable modes are also provided. The presentation is mostly self contained with only occasional references to standard results in linear system theory. It is an absolute pleasure to dedicate this paper to my friend and colleague Harry Trentelman on the occasion of his 60th birthday. Harry and I have worked together on linear system theory for the last 5 years and our behavioral internal model principle for observers (joined work with Jan Willems) provides an alternative proof for the result reported here (Trumpf et al. IEEE Trans. Autom. Control 59, 1737–1749 (2014)).

J. Trumpf (✉)
Research School of Engineering, Australian National University,
Canberra, Australia
e-mail: Jochen.Trumpf@anu.edu.au

M.N. Belur et al. (eds.), *Mathematical Control Theory II*,
Lecture Notes in Control and Information Sciences 462,
DOI 10.1007/978-3-319-21003-2_13

13.1 Problem Formulation

Consider the linear time-invariant finite-dimensional system in state-space form given by

$$\begin{aligned} \dot{x} &= Ax + Bu, \\ y &= Cx, \\ z &= Vx, \end{aligned} \tag{13.1}$$

where $A \in \mathbb{R}^{n \times n}$, $B \in \mathbb{R}^{n \times m}$, $C \in \mathbb{R}^{p \times n}$ and $V \in \mathbb{R}^{r \times n}$.

We will be interested in the characterization of asymptotic observers for z given u and y. In particular, we will be interested in observers of the following type usually considered in the geometric control literature:

$$\begin{aligned} \dot{v} &= Kv + Ly + Mu, \\ \hat{z} &= Pv + Qy, \end{aligned} \tag{13.2}$$

where $K \in \mathbb{R}^{s \times s}$, $P \in \mathbb{R}^{r \times s}$ and the other matrices are real and appropriately sized. Note that P can be rectangular (tall or wide) and/or not of full rank. The asymptotic condition for this type of observer is

$$\lim_{t \to \infty} [\hat{z}(t) - z(t)] = 0 \tag{13.3}$$

for every choice of input u and initial conditions $x(0)$ and $v(0)$. We then say that system (13.2) is an *asymptotic observer* for system (13.1).

The problem considered in this paper is to characterize when a *given* observer of the form (13.2) is an asymptotic observer for a *given* observed system (13.1). See [1], in particular Sect.3.1, for a detailed discussion of the relevant literature.

13.2 Problem Reduction

In the observer *characterization* problem, both the observed system and the observer are given and fixed, so we cannot modify them without changing the problem. We can, however, show results of the type Observer A is an asymptotic observer for System A if and only if Observer B is an asymptotic observer for System B, where both Observer A and B as well as System A and B are related by equations (one of the pairs may even be identical). This then allows to reduce the problem in the case where Observer B and/or System B are simpler than the A variety.

As a first result of this type, we show that only the observable part of the observer (13.2) is relevant for the observer characterization problem. Consider the (dual)

Kalman decomposition for the pair (P, K): There exists an invertible $S \in \mathbb{R}^{s\times s}$ such that

$$SKS^{-1} = \begin{bmatrix} K_{11} & 0 \\ K_{21} & K_{22} \end{bmatrix} \quad \text{and} \quad PS^{-1} = \begin{bmatrix} P_1 & 0 \end{bmatrix},$$

where the pair (P_1, K_{11}) is observable. Now split

$$SL = \begin{bmatrix} L_1 \\ L_2 \end{bmatrix}, \quad SM = \begin{bmatrix} M_1 \\ M_2 \end{bmatrix} \quad \text{and} \quad Sv = \begin{bmatrix} v_1 \\ v_2 \end{bmatrix},$$

and consider the reduced observer

$$\begin{aligned} \dot{v}_1 &= K_{11}v_1 + L_1 y + M_1 u, \\ \hat{z} &= P_1 v_1 + Qy. \end{aligned} \tag{13.4}$$

Note that this observer is an observable system, i.e., v_1 is observable from $((u, y), \hat{z})$ in this observer. We call such observers *observable asymptotic observers*. We now have the following result, cf. [2, Proposition 3.69].

Proposition 13.1 *System* (13.2) *is an asymptotic observer for system* (13.1) *if and only if the reduced system* (13.4) *is an (observable) asymptotic observer for system* (13.1).

Proof The proof follows from the observation that, given (u, y), system (13.4) started with $v_1(0)$ produces the same output as system (13.2) started with $v(0)$. ∎

In order to simplify the notation, we will assume in the next section that (P, K) itself is observable. We only need to replace (K, L, M, P, Q) by $(K_{11}, L_1, M_1, P_1, Q)$ in the resulting characterization to recover the general case.

In a second step, we can simplify the observed system by removing any uncontrollable stable modes. The corresponding linear functions of the state go to zero irrespective of the applied input u and hence do not need to be observed at all. We can make this discussion more precise as follows. Consider the unstable/stable Kalman decomposition for the pair (A, B): There exists an invertible $T \in \mathbb{R}^{n\times n}$ such that

$$TAT^{-1} = \begin{bmatrix} A_{11} & A_{12} & A_{13} \\ 0 & A_{22} & 0 \\ 0 & 0 & A_{33} \end{bmatrix} \quad \text{and} \quad TB = \begin{bmatrix} B_1 \\ 0 \\ 0 \end{bmatrix},$$

where (A_{11}, B_1) is controllable, A_{22} is anti-Hurwitz and A_{33} is Hurwitz. Now split

$$CT^{-1} = \begin{bmatrix} C_1 & C_2 & C_3 \end{bmatrix}, \quad VT^{-1} = \begin{bmatrix} V_1 & V_2 & V_3 \end{bmatrix}, \quad \text{and} \quad Tx = \begin{bmatrix} x_1 \\ x_2 \\ x_3 \end{bmatrix}.$$

We now have the following result.

Proposition 13.2 *Let* $\dim(x_3) < \dim(x)$. *Then system* (13.2) *is an asymptotic observer for system* (13.1) *if and only if it is an asymptotic observer for the reduced system*

$$\begin{aligned}\begin{bmatrix}\dot{x}_{1r}\\ \dot{x}_{2r}\end{bmatrix} &= \begin{bmatrix}A_{11} & A_{12}\\ 0 & A_{22}\end{bmatrix}\begin{bmatrix}x_{1r}\\ x_{2r}\end{bmatrix} + \begin{bmatrix}B_1\\ 0\end{bmatrix}u_r,\\ y_r &= \begin{bmatrix}C_1 & C_2\end{bmatrix}\begin{bmatrix}x_{1r}\\ x_{2r}\end{bmatrix},\\ z_r &= \begin{bmatrix}V_1 & V_2\end{bmatrix}\begin{bmatrix}x_{1r}\\ x_{2r}\end{bmatrix}.\end{aligned} \tag{13.5}$$

The latter is in the sense that $u := u_r$ *and* $y := y_r$ *in the observer yields* $\lim_{t\to\infty}[\hat{z}(t) - z_r(t)] = 0$ *for all choices of* $x_{1r}(0)$, $x_{2r}(0)$, $v(0)$ *and* u_r.

The proof will use the following characterization of output stability, cf. [2, Proposition 3.50]. We prove a slightly extended version.

Lemma 13.3 *Consider the linear time-invariant finite-dimensional system in state-space form given by*

$$\begin{aligned}\dot{x} &= Ax + Bu,\\ y &= Cx,\end{aligned}$$

where $A \in \mathbb{R}^{n\times n}$, $B \in \mathbb{R}^{n\times m}$ *and* $C \in \mathbb{R}^{p\times n}$. *Then* $\lim_{t\to\infty} y(t) = 0$ *for all choices of* $x(0)$ *and* u *if and only if* $CR(A, B) = 0$ *and the corestriction of* A *to the quotient space* $\mathbb{R}^n/\mathfrak{N}(C, A)$ *is Hurwitz. Here,* $R(A, B) = \begin{bmatrix}B & AB & \dots & A^{n-1}B\end{bmatrix}$ *is the reachability matrix of the pair* (A, B) *and* $\mathfrak{N}(C, A) \subset \mathbb{R}^n$ *is the unobservable subspace of the pair* (C, A). *If in addition* (C, A) *is observable then* A *is Hurwitz and* $B = 0$.

Proof Assume there exists $x_0 \in \mathfrak{R}(A, B) := \operatorname{Im} R(A, B)$, the reachable subspace of the pair (A, B), with $Cx_0 \neq 0$. Since $x_0 \in \mathfrak{R}(A, B)$, there exists u and a corresponding trajectory x that oscillates between 0 and x_0, contradicting $\lim_{t\to\infty} y(t) = \lim_{t\to\infty} Cx(t) = 0$. Hence $CR(A, B) = 0$. If $\mathfrak{N}(C, A) = \mathbb{R}^n$ there is nothing to prove for the corestriction of A. Assume $\mathfrak{N}(C, A) \neq \mathbb{R}^n$ and assume that A is not stable on $\mathbb{R}^n/\mathfrak{N}(C, A)$. Then there exists $0 \neq x_0 \in \mathbb{R}^n$ with $x_0 \notin \mathfrak{N}(C, A)$ and $Ax_0 = \lambda x_0$ for a $\lambda \in \mathbb{C}$ with $\mathrm{Re}\lambda \geq 0$. It is $x_0 \notin \operatorname{Ker} C$ since the span of x_0 is A-invariant but $\mathcal{N}(C, A)$ is the largest A-invariant subspace of $\operatorname{Ker} C$. Choosing $x(0) = x_0$ and $u = 0$ yields a trajectory with $\lim_{t\to\infty} y(t) = \lim_{t\to\infty} Cx(t) \neq 0$, a contradiction. Hence A is stable on $\mathbb{R}^n/\mathfrak{N}(C, A)$.

Conversely, let $CR(A, B) = 0$ and A be stable on $\mathbb{R}^n/\mathfrak{N}(C, A)$. Let $x(0) = x_0 \in \mathbb{R}^n$ and u be arbitrary. Then

$$y(t) = Cx(t) = Ce^{At}x_0 + C\int_0^t e^{A(t-\tau)}Bu(\tau)d\tau,$$

where the integral is an element of $\mathfrak{R}(A, B)$ and hence of Ker C. Let $\mathbb{R}^n = \mathfrak{N}(C, A) \oplus \mathfrak{W}$ and decompose $x_0 = n_0 + w_0$ with $n_0 \in \mathfrak{N}(C, A)$ and $w_0 \in \mathfrak{W}$. Since $\mathfrak{N}(C, A)$ is A-invariant and contained in Ker C it follows that $y(t) = Ce^{At}w_0$. But A is stable on $\mathfrak{W}$ and hence $\lim_{t\to\infty} y(t) = 0$.

If in addition (C, A) is observable then $\mathfrak{N}(C, A) = \{0\}$ and A is Hurwitz. Furthermore, $C\begin{bmatrix} B & AB & \dots & A^{n-1}B \end{bmatrix} = 0$ implies

$$\begin{pmatrix} C \\ CA \\ \vdots \\ CA^{n-1} \end{pmatrix} B = 0$$

and hence $B = 0$ by observability. ■

Proof of Proposition 13.2 Note that system (13.1) and the Kalman decomposed system

$$\begin{bmatrix} \dot{x}_1 \\ \dot{x}_2 \\ \dot{x}_3 \end{bmatrix} = \begin{bmatrix} A_{11} & A_{12} & A_{13} \\ 0 & A_{22} & 0 \\ 0 & 0 & A_{33} \end{bmatrix} \begin{bmatrix} x_1 \\ x_2 \\ x_3 \end{bmatrix} + \begin{bmatrix} B_1 \\ 0 \\ 0 \end{bmatrix} u,$$

$$y = \begin{bmatrix} C_1 & C_2 & C_3 \end{bmatrix} \begin{bmatrix} x_1 \\ x_2 \\ x_3 \end{bmatrix},$$

$$z = \begin{bmatrix} V_1 & V_2 & V_3 \end{bmatrix} \begin{bmatrix} x_1 \\ x_2 \\ x_3 \end{bmatrix}$$

have the same external behavior in terms of the variables (u, y, z). In order to enable a direct comparison of the trajectories of system (13.1) and the reduced system (13.5), whenever we are given an initial condition $x(0)$ for system (13.1), we will use $x_{1r}(0) := x_1(0)$, $x_{2r}(0) := x_2(0)$ in the reduced system (13.5). Then, system (13.1) and the reduced system (13.5) have the same output, i.e., $y = y_r$ and $z = z_r$, if $x_3(0) = 0$ and $u_r = u$.

Let system (13.2) be an asymptotic observer for system (13.1), then $\lim_{t\to\infty}[\hat{z}(t) - z(t)] = 0$ for every choice of $x(0)$, $v(0)$ and u, in particular for those $x(0)$ with $x_3(0) = 0$. But in that case $x_{1r}(0) := x_1(0)$, $x_{2r}(0) := x_2(0)$ and $u_r := u$ in system (13.5) implies $y = y_r$, and hence the observer output is the same for both observed systems. Moreover, $z = z_r$ in this case and hence $\lim_{t\to\infty}[\hat{z}(t) - z_r(t)] = 0$ in the observer interconnection with the reduced system. It follows that system (13.2) is an asymptotic observer for the reduced system (13.5).

Conversely, let system (13.2) be an asymptotic observer for the reduced system (13.5). Fix $x_{1r}(0)$, $x_{2r}(0)$ and $v(0)$ and let

$$x(0) := T^{-1} \begin{bmatrix} x_{1r}(0) \\ x_{2r}(0) \\ x_3(0) \end{bmatrix}$$

in system (13.1) where $x_3(0)$ is arbitrary but fixed. Then $x_2 = x_{2r}$ and $\lim_{t\to\infty} x_3(t) = 0$ for all choices of u_r and u. Define $\delta := x_1 - x_{1r}$ then

$$\dot{\delta} = A_{11}\delta + B_1(u - u_r) + A_{13}x_3, \qquad \delta(0) = 0.$$

Since (A_{11}, B_1) is controllable, there exists a feedback matrix F such that $A_{11}+B_1F$ is Hurwitz. Consider the auxiliary system

$$\dot{\delta}_F = (A_{11} + B_1F)\delta_F + A_{13}x_3, \qquad \delta_F(0) = 0,$$

then $\lim_{t\to\infty} \delta_F(t) = 0$ by [3, Corollary 3.22]. Now the choice $u_r := u - F\delta_F$ yields $\delta = \delta_F$ and hence $\lim_{t\to\infty} \delta(t) = \lim_{t\to\infty}[x_1(t)-x_{1r}(t)] = 0$. It follows that $\lim_{t\to\infty}[y(t)-y_r(t)] = 0$, $\lim_{t\to\infty}[z(t)-z_r(t)] = 0$ and $\lim_{t\to\infty}[u(t)-u_r(t)] = 0$, i.e., the external behaviors of system (13.1) and system (13.5) are asymptotically equal under the above correspondence of trajectories.

According to Proposition 13.1, the reduced observer (13.4) is an asymptotic observer for the reduced system (13.5), and by Lemma 13.3, K_{11} is Hurwitz: Choose $x(0) = 0$ and $u = 0$ to obtain $y = 0$ and $\lim_{t\to\infty} \hat{z}(t) = 0$ for all choices of $v_1(0)$. We can also connect this reduced observer to system (13.1) instead of to the reduced system (13.5), and since the external behaviors of these two systems are asymptotically equal, another application of [3, Corollary 3.22] shows that the two resulting observer outputs are asymptotically equal. This implies that the reduced order observer (13.4) is also an asymptotic observer for system (13.1) (since $\lim_{t\to\infty}[z(t) - z_r(t)] = 0$), and by Proposition 13.1, so is system (13.2). ∎

Again, we will simplify the notation in the next section by assuming that system (13.1) has no stable uncontrollable modes and can recover the general case by replacing the system matrices (A, B, C, V) in the resulting characterization with the reduced system matrices of system (13.5).

We finish this section by treating the remaining case not covered by Proposition 13.2, namely the case where system (13.1) is completely uncontrollable and stable, i.e., where A is Hurwitz and $B = 0$.

Proposition 13.4 *Let A be Hurwitz and let $B = 0$ in system* (13.1). *Then system* (13.2) *is an asymptotic observer for system* (13.1) *if and only if $PR(K, M) = 0$ and the corestriction of K to the quotient space $\mathbb{R}^s/\mathfrak{N}(P, K)$ is Hurwitz. Here, $R(K, M) = \begin{bmatrix} M & KM & \dots & K^{s-1}M \end{bmatrix}$ is the reachability matrix of the pair (K, M) and $\mathfrak{N}(P, K) \subset \mathbb{R}^s$ is the unobservable subspace of the pair (P, K).*

Proof Let system (13.2) be an asymptotic observer for system (13.1) with A Hurwitz and $B = 0$. Then $\lim_{t\to\infty} z(t) = 0$ and hence $\lim_{t\to\infty} \hat{z}(t) = 0$ for all choices of $x(0)$, $v(0)$ and u, in particular for $x(0) = 0$ (hence $y = 0$), and all choices of $v(0)$ and

u. By Lemma 13.3 then $PR(K, M) = 0$ and the corestriction of K to the quotient space $\mathbb{R}^s/\mathfrak{N}(P, K)$ is Hurwitz.

Conversely, assume that $PR(K, M) = 0$ and that the corestriction of K to the quotient space $\mathbb{R}^s/\mathfrak{N}(P, K)$ is Hurwitz. Then $P_1 R(K_{11}, M_1) = 0$ and K_{11} is Hurwitz in the reduced observer (13.4). By Lemma 13.3, $\lim_{t\to\infty} \hat{z}(t) = 0$ for all choices of $v_1(0)$ and u and $y = 0$. Since $\lim_{t\to\infty} y(t) = 0$ for all choices of $x(0)$ and u, an application of [3, Corollary 3.22] yields $\lim_{t\to\infty} \hat{z}(t) = 0$ for all choices of $v_1(0)$, $x(0)$ and u, and hence the reduced observer (13.4) is an asymptotic observer for system (13.1) (since $\lim_{t\to\infty} z(t) = 0$). By Proposition 13.1, so is system (13.2). ■

13.3 The Main Characterization Result

We are now in a position to state the main characterization result for asymptotic state observers. We will briefly discuss the error in the previous proof attempt [1, Theorem 9] after we have given the new proof. The proof references the following two technical lemmas proved in [1] that we restate here for convenience but without proof. Consider the linear time-invariant finite-dimensional system in state-space form given by

$$\begin{aligned} \dot{x} &= Ax + Bu, \\ y &= Cx + Du, \end{aligned} \tag{13.6}$$

where $A \in \mathbb{R}^{n\times n}$, $B \in \mathbb{R}^{n\times m}$, $C \in \mathbb{R}^{p\times n}$ and $D \in \mathbb{R}^{p\times m}$.

Lemma 13.5 ([1, Lemma 3]) *Let all uncontrollable modes of system* (13.6) *be unstable. Then, for every $Q \in \mathbb{R}^{q\times n}$ with $Q \neq 0$ there exists an initial condition x_0 and an input u such that* $\lim_{t\to\infty} Qx(t) \neq 0$.

Lemma 13.6 ([1, Proposition 5]) *If* $\lim_{t\to\infty} y(t) = 0$ *for all choices of x_0 and u in system* (13.6) *then its transfer function* $G(s) = C(sI - A)^{-1}B + D \equiv 0$ *and in particular $D = 0$. If, moreover, all uncontrollable modes of system* (13.6) *are unstable then $C = 0$.*

Theorem 13.7 *Let all uncontrollable modes of system* (13.1) *be unstable. Then system* (13.2) *is an observable asymptotic observer for z given u and y if and only if there exists a matrix $U \in \mathbb{R}^{s\times n}$ such that*

$$\begin{aligned} UA - KU - LC &= 0, \\ M - UB &= 0, \\ V - PU - QC &= 0, \end{aligned} \tag{13.7}$$

K is Hurwitz and (P, K) is observable.

Proof Let system (13.2) be an observable asymptotic observer for z given u and y and define $e := \hat{z} - z$. Then

$$e = Pv + Qy - Vx = Pv - (V - QC)x.$$

Assume, to arrive at a contradiction, that $\operatorname{Im}(V - QC) \not\subset \operatorname{Im}(P)$. Then there exists an invertible $S \in \mathbb{R}^{r\times r}$ such that

$$SP = \begin{bmatrix} P_1 \\ 0 \end{bmatrix} \text{ and } S(V - QC) = \begin{bmatrix} V_1 \\ V_2 \end{bmatrix}$$

with $V_2 \neq 0$. Now $\lim_{t\to\infty} Se(t) = 0$ implies $\lim_{t\to\infty} V_2 x(t) = 0$ for all initial conditions $x(0)$ and all inputs u, a contradiction to Lemma 13.5. We conclude that $\operatorname{Im}(V - QC) \subset \operatorname{Im}(P)$, and hence there exists a matrix $U \in \mathbb{R}^{s\times n}$ such that $V - QC = PU$. This implies the third equation in (13.7).

Define $d := v - Ux$ then

$$\begin{aligned} \dot{d} &= \dot{v} - U\dot{x} \\ &= Kv + Ly + Mu - UAx - UBu \\ &= Kv - KUx + KUx + LCx + Mu - UAx - UBu, \end{aligned}$$

and hence the observation error $e = Pv - (V - QC)x = Pv - PUx$ is governed by the error system

$$\begin{aligned} \dot{d} &= Kd - (UA - KU - LC)x + (M - UB)u, \\ e &= Pd. \end{aligned} \tag{13.8}$$

The first two equations in (13.7) now follow immediately from an application of Proposition 13.8 stated below. Apply Lemma 13.3 to the resulting error system

$$\begin{aligned} \dot{d} &= Kd, \\ e &= Pd \end{aligned}$$

to see that K must be Hurwitz.

Conversely, assume that the system matrices of systems (13.1) and (13.2) fulfill Equation (13.7) with K Hurwitz, then $\lim_{t\to\infty} e(t) = 0$ follows immediately from the form of the error system (13.8). In its derivation, we have only used the third equation in (13.7). ∎

Before we state and prove the missing Proposition 13.8 below, let us briefly discuss what is wrong in the proof of [1, Theorem 9]. In that proof, the conclusion after the derivation of the error system (13.8) uses the argument *[...] follows immediately from [...] the fact that* (P, K) *is observable (hence* $e(t) \to 0$ *implies* $d(t) \to 0$). While this assertion refers only to the error system (13.8), and is hence actually true *a posteriori*, it is not generally true as an *a priori* assertion about observable systems, which is how it is being used in the logic of the proof given in [1].

By definition, observability of a linear state-space system means that zero output *and* input imply zero state, but the property makes no (direct) statement about limits or about the case where the input is nonzero.

One could think that the argument can be saved by the fact that it is only used as an assertion on the totality of *all* solutions of the error system, as in $e(t) \to 0$ for *all* solutions implies $d(t) \to 0$ for *all* solutions. Indeed, Lemma 13.3 at first seems to support this. Note, however, that Lemma 13.3 cannot be applied to the error system (13.8), since the two inputs $v = (x, u)$ are not independent signals. In fact, this is the reason why Theorem 13.7 is difficult to prove, cf. [1, Remark 8].

Fixing the above error requires the following generalization to [1, Proposition 7].

Proposition 13.8 *Consider the composite system*

$$\begin{aligned} \dot{x} &= Ax + Bu, \\ \dot{d} &= Kd + Rx + Su, \\ e &= Pd, \end{aligned} \tag{13.9}$$

and assume that $\lim_{t\to\infty} e(t) = 0$ *for all choices of* $x(0)$, $d(0)$ *and* u. *If all uncontrollable modes of* $\dot{x} = Ax + Bu$ *are unstable and* (P, K) *is observable then* $R = 0$ *and* $S = 0$.

The proof of this proposition will be given with the help of the following technical lemma.

Lemma 13.9 *Let* $(P, K) \in \mathbb{R}^{r\times s} \times \mathbb{R}^{s\times s}$ *be observable and let* $A \in \mathbb{R}^{n\times n}$ *and* $R \in \mathbb{R}^{s\times n}$. *Then*

$$\begin{bmatrix} 0 & P \end{bmatrix} \begin{bmatrix} A & 0 \\ R & K \end{bmatrix}^i \begin{bmatrix} x \\ 0 \end{bmatrix} = 0 \tag{13.10}$$

for all $x \in \mathbb{R}^n$ *and all* $i \in \mathbb{N}$ *implies* $R = 0$.

Proof We have

$$\begin{bmatrix} A & 0 \\ R & K \end{bmatrix} \begin{bmatrix} x \\ 0 \end{bmatrix} = \begin{bmatrix} A \\ R \end{bmatrix} x$$

and, using (13.10) with $i = 1$, also $PRx = 0$ for all $x \in \mathbb{R}^n$.

Assume that

$$\begin{bmatrix} A & 0 \\ R & K \end{bmatrix}^i \begin{bmatrix} x \\ 0 \end{bmatrix} = \begin{bmatrix} A^i \\ \sum_{l=1}^{i} K^{l-1} R A^{i-l} \end{bmatrix} x \tag{13.11}$$

for all $x \in \mathbb{R}^n$ and some $i \in \mathbb{N}$ and

$$PK^{l-1}Rx = 0 \tag{13.12}$$

for all $x \in \mathbb{R}^n$ and all $l = 1, \ldots, i$. Then

$$\begin{bmatrix} A & 0 \\ R & K \end{bmatrix}^{i+1} \begin{bmatrix} x \\ 0 \end{bmatrix} = \begin{bmatrix} A & 0 \\ R & K \end{bmatrix} \begin{bmatrix} A^i \\ \sum_{l=1}^{i} K^{l-1} R A^{i-l} \end{bmatrix} x$$
$$= \begin{bmatrix} A^{i+1} \\ RA^i + K \sum_{l=2}^{i+1} K^{(l-1)-1} R A^{(i-(l-1))} \end{bmatrix} x$$
$$= \begin{bmatrix} A^{i+1} \\ \sum_{l=1}^{i+1} K^{l-1} R A^{(i+1)-l} \end{bmatrix} x$$

for all $x \in \mathbb{R}^n$, where we have used hypothesis (13.11) in the first line. But then (13.10) implies that

$$0 = P \left(\sum_{l=1}^{i+1} K^{l-1} R A^{(i+1)-l} \right) x = P K^i R x$$

for all $x \in \mathbb{R}^n$, where we have used hypothesis (13.12) in the final conclusion.

By induction, it follows that $PK^{i-1}(Rx) = 0$ for all $i \in \mathbb{N}$ and all $x \in \mathbb{R}^n$. By observability of (P, K) this implies $Rx = 0$ for all $x \in \mathbb{R}^n$ and hence $R = 0$. ■

Proof of Proposition 13.8 Apply Lemma 13.6 to system (13.9) to obtain

$$P(sI - K)^{-1} \left[-R(sI - A)^{-1} B + S \right] \equiv 0$$

and hence $-R(sI - A)^{-1}B + S \equiv 0$ since (P, K) is observable. Since S is constant and $R(sI - A)^{-1}B$ is strictly proper, it follows that $S = 0$ and $R(sI - A)^{-1}B \equiv 0$. If $\dot{x} = Ax + Bu$ was controllable, we would be done at this point, since then $R(sI - A)^{-1}B \equiv 0$ would imply $R = 0$. With the help of Lemma 13.3 and Lemma 13.9 above we can, however, treat the more general case of this proposition.

Given that all uncontrollable modes of $\dot{x} = Ax + Bu$ are unstable, there exists an invertible $S \in \mathbb{R}^{n \times n}$ such that

$$SAS^{-1} = \begin{bmatrix} A_{11} & A_{12} \\ 0 & A_{22} \end{bmatrix} \text{ and } SB = \begin{bmatrix} B_1 \\ 0 \end{bmatrix},$$

where $A_{11} \in \mathbb{R}^{n_1 \times n_1}$, the pair (A_{11}, B_1) is controllable and all eigenvalues of $A_{22} \in \mathbb{R}^{n_2 \times n_2}$ have nonnegative real parts (Kalman decomposition). Define

$$\begin{bmatrix} R_1 & R_2 \end{bmatrix} := RS^{-1} \quad \text{and} \quad \begin{bmatrix} x_1 \\ x_2 \end{bmatrix} := Sx,$$

where the block sizes are as for the matrix SAS^{-1} above. Then $R(sI - A)^{-1}B \equiv R_1(sI - A_{11})^{-1}B_1 \equiv 0$ and hence $R_1 = 0$ because (A_{11}, B_1) is controllable. It follows that

$$\begin{aligned} \dot{x}_2 &= A_{22}x_2, & x_2(0) &= x_{02}, \\ \dot{d} &= Kd + R_2x_2, & d(0) &= d_0, \\ e &= Pd. \end{aligned} \tag{13.13}$$

By Lemma 13.3, $\lim_{t\to\infty} e(t) = 0$ for all choices of x_{02} and d_0 in (13.13) implies that the corestriction of the linear map

$$\begin{bmatrix} A_{22} & 0 \\ R_2 & K \end{bmatrix} : \mathbb{R}^{n_2+s} \to \mathbb{R}^{n_2+s}$$

to the quotient space $\mathbb{R}^{n_2+s}/\mathfrak{N}$ is stable, where

$$\mathfrak{N} := \mathfrak{N}\left(\begin{bmatrix} 0 & P \end{bmatrix}, \begin{bmatrix} A_{22} & 0 \\ R_2 & K \end{bmatrix}\right) = \bigcap_{i\in\mathbb{N}} \operatorname{Ker}\left(\begin{bmatrix} 0 & P \end{bmatrix}\begin{bmatrix} A_{22} & 0 \\ R_2 & K \end{bmatrix}^{i-1}\right)$$

denotes the unobservable subspace. A straightforward computation shows that

$$\begin{bmatrix} x_2 \\ d \end{bmatrix} \in \mathfrak{N} \quad \text{implies} \quad d \in \mathfrak{N}(P, K),$$

i.e., $d = 0$ since (P, K) is observable. This shows $\mathfrak{N} \subset \mathbb{R}^{n_2} \times \{0\}$. On the other hand, since all eigenvalues of A_{22} have nonnegative real parts, the corestriction of the above linear map to any quotient space of the form $\mathbb{R}^{n_2+s}/\mathfrak{S}$ with $\mathfrak{S} \subsetneq \mathbb{R}^{n_2} \times \{0\}$ can not be Hurwitz. It follows that $\mathfrak{N} = \mathbb{R}^{n_2} \times \{0\}$. By Lemma 13.9 above this implies $R = 0$. ■

Note that the use of Proposition 13.8 eliminates the need for [1, Lemma 6] and with it the use of the theory of pole-zero cancelations, making the final proof of Theorem 13.7 slightly more elementary.

13.4 Conclusion

The overall picture now is as follows. In the case where A is Hurwitz and $B = 0$ in system (13.1), the full characterization of asymptotic state observers (13.2) is given by $PR(K, M) = 0$ and the corestriction of K to the quotient space $\mathbb{R}^s/\mathfrak{N}(P, K)$ being Hurwitz (Proposition 13.4). Note that in this case, the characterization is independent of the system matrices (A, B, C, V). Otherwise, the characterization is given by Eq. (13.7) and K Hurwitz (Theorem 13.7), where we may have to replace the system matrices (A, B, C, V) with the reduced system matrices of system (13.5) if system (13.1) has stable uncontrollable modes (Proposition 13.2), and the observer matrices (K, L, M, P, Q) by the observer matrices $(K_{11}, L_1, M_1, P_1, Q)$ of the reduced observer (13.4) if the observer (13.2) is not observable (Proposition 13.1).

Acknowledgments The author wishes to thank Uwe Helmke for pointing out the error in the proof of [1, Theorem 9].

References

1. Trumpf, J.: On state observers. In: Hüper, K., Trumpf, J. (eds.) Mathematical System Theory—Festschrift in Honor of Uwe Helmke on the Occasion of his Sixtieth Birthday, pp. 421–435. CreateSpace (2013)
2. Trumpf, J.: On the geometry and parametrization of almost invariant subspaces and observer theory. Ph.D. thesis, Universität Würzburg, Germany (2002)
3. Trentelman, H., Stoorvogel, A., Hautus, M.: Control Theory for Linear Systems. Springer, London (2001)
4. Trumpf, J., Trentelman, H., Willems, J.: Internal model principles for observers. IEEE Trans. Autom. Control **59**, 1737–1749 (2014)

Chapter 14
When Is a Linear Complementarity System Disturbance Decoupled?

A.R.F. Everts and M.K. Camlibel

Abstract In this chapter, we study the disturbance decoupling problem for linear complementarity systems that form a class of piecewise affine systems. Direct application of the existing results for piecewise affine systems to linear complementarity systems leads to somewhat bulky conditions. By exploiting the compact description of linear complementarity systems, this chapter provides crisp conditions that are far more insightful than those for general piecewise affine systems.

14.1 Introduction

The disturbance decoupling problem amounts to finding a feedback law that eliminates the effect of disturbances on the output of a given input/state/output dynamical system. The investigation of this problem has been the starting point for the development of geometric control theory [1–3]. For both linear and (smooth) nonlinear systems, geometric control theory has been proven to be very efficient in solving various control problems, including the disturbance decoupling problem (see, e.g., [4–8]).

Dedicated to Harry Trentelman on the occasion of his sixtieth birthday

A.R.F. Everts · M.K. Camlibel (✉)
Johann Bernoulli Institute for Mathematics and Computer Science,
University of Groningen, Groningen, 9700 AV Groningen, The Netherlands
e-mail: m.k.camlibel@rug.nl

A.R.F. Everts
e-mail: a.r.f.everts@rug.nl

M.N. Belur et al. (eds.), *Mathematical Control Theory II*,
Lecture Notes in Control and Information Sciences 462,
DOI 10.1007/978-3-319-21003-2_14

In the context of hybrid dynamical systems, the results on disturbance decoupling are limited to switched linear systems [9, 10] and piecewise affine systems [11]. The major difference between switched linear systems and piecewise affine systems is the nature of the switching behavior. For piecewise affine systems the switching behavior is state-dependent whereas it is state-independent for switched linear systems.

For state-independent switching case, the solution of the disturbance decoupling problem can be obtained by following mainly the footsteps of the (non-switching) linear case. Indeed, a noteworthy consequence of the state-independent switching is that the set of reachable states under the influence of disturbances is a subspace. This allows one to generalize the so-called controlled invariant subspaces of linear systems to switched linear systems. Such a generalization leads to elegant necessary and sufficient conditions [9, 10] for a switched linear system to be disturbance decoupled. In the same papers, disturbance decoupling problems by different feedback schemes have also been solved based on these necessary and sufficient conditions.

However, a similar approach breaks down in the case of state-dependent switching as the set of reachable states under the influence of disturbances is not anymore a subspace, not even a convex set in general. As such, neither the results nor the approach adopted for the state-independent case can be applied to state-dependent switching case. By taking a novel approach that takes into account the state-dependent switching behavior of piecewise affine systems, the paper [11] provided a set of necessary conditions and a set of sufficient conditions under which a continuous piecewise affine dynamical system is disturbance decoupled. Although these conditions do not coincide in general, some special cases in which they do coincide were pointed out in [11]. Furthermore, [11] presented conditions for the existence of mode-independent static feedback controllers that render the closed-loop system disturbance decoupled. All conditions presented in [11] are geometric in nature and can be easily verified by utilizing extensions of the well-known subspace algorithms. Yet the conditions obtained in [11] are somewhat bulky due to the very general description of piecewise affine systems that was taken as the starting point. In this chapter, we focus on a particular class of piecewise affine systems, namely linear complementarity systems. It turns out that the compact description of linear complementarity systems leads to conditions for disturbance decoupling that are not only easily checkable but also very crisp.

The structure of the chapter is as follows. In Sect. 14.2 we introduce the notational conventions as well as the basic concepts of the geometric approach to linear systems. This will be followed by the formulation of the linear complementarity problem in Sect. 14.3 where also linear complementarity systems are introduced. In Sect. 14.4 we first define what we mean by a linear complementarity system to be disturbance decoupled. After providing some technical auxiliary results that are, in a way, of interest themselves, we present necessary and sufficient conditions that are the main results of this chapter. Finally, the chapter closes with conclusions in Sect. 14.5.

14.2 Preliminaries

Consider, the linear system $\Sigma = \Sigma(A, B, C, D)$ given by

$$\dot{x}(t) = Ax(t) + Bu(t) \tag{14.1a}$$

$$y(t) = Cx(t) + Du(t) \tag{14.1b}$$

where the input u, state x, and output y have dimensions m, n, and p, respectively. In what follows, we quickly introduce some of the fundamental notions of geometric control theory of linear systems for the sake of completeness. We refer to [8] for more details.

The *controllable subspace* of Σ is the smallest A-invariant subspace containing $\operatorname{im} B$. We will denote it by

$$\langle A \mid \operatorname{im} B\rangle := \operatorname{im} B + A\operatorname{im} B + \cdots + A^{n-1}\operatorname{im} B.$$

Note that

$$\langle A + BK \mid \operatorname{im} B\rangle = \langle A \mid \operatorname{im} B\rangle \tag{14.2}$$

for any matrix $K \in \mathbb{R}^{m\times n}$.

A subspace $\mathscr{T} \subseteq \mathbb{R}^n$ is called an *input containing conditioned invariant subspace* of Σ if

$$\begin{bmatrix} A & B \end{bmatrix}\left((\mathscr{T} \times \mathbb{R}^m) \cap \ker \begin{bmatrix} C & D \end{bmatrix}\right) \subseteq \mathscr{T}.$$

It is well known that a subspace $\mathscr{T}$ is an input containing conditioned invariant subspace if and only if there exists a matrix $L \in \mathbb{R}^{n\times p}$ such that

$$(A + LC)\mathscr{T} \subseteq \mathscr{T} \quad \text{and} \quad \operatorname{im}(B + LD) \subseteq \mathscr{T}. \tag{14.3}$$

The *strongly reachable subspace* of Σ is the smallest (with respect to the subspace inclusion) input containing conditioned invariant subspace and will be denoted by $\mathscr{T}^*(\Sigma)$.

It follows from (14.3) with the choice of $L = 0$ that the controllable subspace is an input containing conditioned invariant subspace. Hence, we have

$$\mathscr{T}^*(\Sigma) \subseteq \langle A \mid \operatorname{im} B\rangle. \tag{14.4}$$

Let K and L be $m \times n$ and $n \times p$ matrices, respectively. Also let $\Sigma_{K,L}$ denote the system $\Sigma(A + BK + LC + LDK, B + LD, C + DK, D)$. It can easily be verified that

$$\mathscr{T}^*(\Sigma_{K,L}) = \mathscr{T}^*(\Sigma). \tag{14.5}$$

A subspace $\mathscr{V} \subseteq \mathbb{R}^n$ is called an *output nulling controlled invariant subspace* of Σ if

$$\begin{bmatrix} A \\ C \end{bmatrix} \mathscr{V} \subseteq (\mathscr{V} \times \{0\}) + \operatorname{im} \begin{bmatrix} B \\ D \end{bmatrix}.$$

The *weakly unobservable subspace* of Σ is the largest (with respect to the subspace inclusion) output nulling controlled invariant subspace and will be denoted by $\mathscr{V}^*(\Sigma)$.

It is well known that the transfer matrix $D + C(sI - A)^{-1}B$ is right-invertible as a rational matrix if and only if

$$\mathscr{V}^*(\Sigma) + \mathscr{T}^*(\Sigma) = \mathbb{R}^n \text{ and } \begin{bmatrix} C & D \end{bmatrix} \text{ is of full row rank.}$$

Straightforward linear algebra arguments show that these conditions are equivalent to

$$\operatorname{im} D + C\mathscr{T}^*(\Sigma) = \mathbb{R}^p. \tag{14.6}$$

14.3 Linear Complementarity System

The problem of finding a vector $z \in \mathbb{R}^m$ such that

$$z \geqslant 0 \tag{14.7a}$$

$$q + Mz \geqslant 0 \tag{14.7b}$$

$$z^T(q + Mz) = 0 \tag{14.7c}$$

for a given vector $q \in \mathbb{R}^m$ and a matrix $M \in \mathbb{R}^{m \times m}$ is known as the linear complementarity problem. We denote (14.7) by LCP(q, M). It is well-known [12, Thm. 3.3.7] that the LCP(q, M) admits a unique solution for each q if and only if all principal minors of M are positive. Such matrices are called P-matrices in the literature of mathematical programming.

When M is a P-matrix, the unique solution of the LCP(q, M), say $z(q)$, depends on q in a Lipschitz continuous way. In particular, for each q there exists an index set $\alpha \subseteq \{1, 2, \ldots, m\}$ such that the solution $z = z(q)$ is determined by

$$z_\alpha = -(M_{\alpha\alpha})^{-1} q_\alpha \qquad z_{\alpha^c} = 0 \tag{14.8a}$$

and the following inequalities hold

$$-(M_{\alpha\alpha})^{-1} q_\alpha \geqslant 0 \qquad q_{\alpha^c} - M_{\alpha^c\alpha}(M_{\alpha\alpha})^{-1} q_\alpha \geqslant 0 \tag{14.8b}$$

where α^c denotes the set $\{1, 2, \ldots, m\} \setminus \alpha$.

Linear complementarity systems (LCSs) consist of nonsmooth dynamical systems that are obtained in the following way. Take a standard linear input/output system. Select a number of input/output pairs (z_i, w_i), and impose for each of these pairs a complementarity relation of the type (14.7) at each time instant. A wealth of examples, from various areas of engineering as well as operations research, of LCSs can be found in [13–16]. For the work on the analysis of LCSs, we refer to [17–23].

In this chapter, we will focus on the LCSs of the following form:

$$\dot{x}(t) = Ax(t) + Bz(t) + Ed(t) \tag{14.9a}$$

$$w(t) = Cx(t) + Dz(t) + Fd(t) \tag{14.9b}$$

$$0 \leqslant z(t) \perp w(t) \geqslant 0 \tag{14.9c}$$

$$y(t) = Jx(t). \tag{14.9d}$$

Here $x \in \mathbb{R}^n$ is the state, $(z, w) \in \mathbb{R}^{m+m}$ are the complementarity variables, $d \in \mathbb{R}^q$ is the disturbance, $y \in \mathbb{R}^p$ is the output, and all the matrices are of appropriate sizes.

In the sequel, we will work under the following blanket assumptions:

1. The matrix D is a P-matrix.
2. The transfer matrix $F + C(sI - A)^{-1}E$ is right-invertible as a rational matrix.

In order not to blur the main message of the chapter, we focus on LCSs that satisfy these assumptions that are technical in nature. Most of the subsequent results can be generalized to cases for which these assumptions do not hold.

Since D is a P-matrix, $z(t)$ is a piecewise linear function of $Cx(t) + Fd(t)$ (see, e.g., [12]). This means that for each initial state x_0 and locally integrable disturbance d there exist unique absolutely continuous trajectories $(x^{x_0,d}, y^{x_0,d})$ and locally integrable trajectories $(z^{x_0,d}, w^{x_0,d})$ such that $x^{x_0,d}(0) = x_0$ and the quadruple $(x^{x_0,d}, z^{x_0,d}, w^{x_0,d}, y^{x_0,d})$ satisfies the relations (14.9) for almost all $t \geqslant 0$.

Even though LCSs are nonsmooth and nonlinear dynamical systems, their local linear behavior enables elegant characterizations of certain system-theoretic properties. To give a flavor of such a result, we quote the following theorem from [21] that states necessary and sufficient conditions for controllability of an LCS.

Theorem 14.1 *An LCS of the form* (14.9a)–(14.9c) *for which d is treated as input is controllable if and only if the following conditions hold:*

1. $\langle A \mid \operatorname{im} \begin{bmatrix} B & E \end{bmatrix} \rangle = \mathbb{R}^n$.
2. *the system of inequalities*

$$\eta \geqslant 0, \quad \begin{bmatrix} \xi \\ \eta \end{bmatrix}^T \begin{bmatrix} A - \lambda I & E \\ C & F \end{bmatrix} = 0, \quad \begin{bmatrix} \xi \\ \eta \end{bmatrix}^T \begin{bmatrix} B \\ D \end{bmatrix} \leqslant 0$$

admits no solution $\lambda \in \mathbb{R}$ and $0 \neq (\xi, \eta) \in \mathbb{R}^{n+m}$.

In this chapter, we will investigate yet another system-theoretic property for LCSs, namely disturbance decoupling.

14.4 Disturbance Decoupled LCSs

We say that an LCS (14.9) is *disturbance decoupled* if for all initial states x_0, locally integrable disturbances d_1 and d_2, and $t \geqslant 0$ we have

$$y^{x_0,d_1}(t) = y^{x_0,d_2}(t).$$

In this section, we will investigate necessary and sufficient conditions for an LCS (14.9) to be disturbance decoupled. To do so, we first derive an alternative representation of LCSs that makes the underlying switching behavior more transparent.

Since D is a P-matrix, we can solve the LCP given by (14.9c) by employing (14.8). More precisely, if the quadruple (x, d, z, w) satisfies (14.9a)–(14.9c) for almost all $t \geqslant 0$ then for almost all $t \geqslant 0$ there exists an index set $\alpha_t \subseteq \{1, 2, \ldots, m\}$ such that

$$\dot{x}(t) = \big(A - B_{\bullet\alpha_t}(D_{\alpha_t\alpha_t})^{-1}C_{\alpha_t\bullet}\big)x(t) + \big(E - B_{\bullet\alpha_t}(D_{\alpha_t\alpha_t})^{-1}F_{\alpha_t\bullet}\big)d(t) \quad (14.10a)$$

whenever

$$\begin{bmatrix} -(D_{\alpha_t\alpha_t})^{-1}C_{\alpha_t\bullet} & -(D_{\alpha_t\alpha_t})^{-1}F_{\alpha_t\bullet} \\ C_{\alpha_t^c\bullet} - D_{\alpha_t^c\alpha_t}(D_{\alpha_t\alpha_t})^{-1}C_{\alpha_t\bullet} & F_{\alpha_t^c\bullet} - D_{\alpha_t^c\alpha_t}(D_{\alpha_t\alpha_t})^{-1}F_{\alpha_t\bullet} \end{bmatrix} \begin{bmatrix} x(t) \\ d(t) \end{bmatrix} \geqslant 0. \quad (14.10b)$$

For an index set $\alpha \subseteq \{1, 2, \ldots, m\}$, define

$$A_\alpha = A - B_{\bullet\alpha}(D_{\alpha\alpha})^{-1}C_{\alpha\bullet} \quad (14.11)$$

$$E_\alpha = E - B_{\bullet\alpha}(D_{\alpha\alpha})^{-1}F_{\alpha\bullet} \quad (14.12)$$

$$G_\alpha = \begin{bmatrix} -(D_{\alpha\alpha})^{-1}C_{\alpha\bullet} \\ C_{\alpha^c\bullet} - D_{\alpha^c\alpha}(D_{\alpha\alpha})^{-1}C_{\alpha\bullet} \end{bmatrix} \quad (14.13)$$

$$H_\alpha = \begin{bmatrix} -(D_{\alpha\alpha})^{-1}F_{\alpha\bullet} \\ F_{\alpha^c\bullet} - D_{\alpha^c\alpha}(D_{\alpha\alpha})^{-1}F_{\alpha\bullet} \end{bmatrix}. \quad (14.14)$$

With these definitions (14.10) can be rewritten as follows:

$$\dot{x}(t) = A_{\alpha_t}x(t) + E_{\alpha_t}d(t) \text{ whenever } \begin{bmatrix} G_{\alpha_t} & H_{\alpha_t} \end{bmatrix} \begin{bmatrix} x(t) \\ d(t) \end{bmatrix} \geqslant 0. \quad (14.15)$$

Before turning our attention to conditions for this system to be disturbance decoupled, we prove a technical auxiliary result that we will employ later.

Lemma 14.2 *For each index set $\alpha \subseteq \{1, 2, \ldots, m\}$, let $N_\alpha \in \mathbb{R}^{n\times p}$ be such that*

$$A_\alpha = A + N_\alpha C \quad \textit{and} \quad E_\alpha = E + N_\alpha F.$$

Let

$$\mathscr{S} = \sum_{\gamma \subseteq \{1,2,\ldots,m\}} \langle A_\gamma \mid \text{im } E_\gamma \rangle.$$

The following statements hold:

1. $\text{im } (N_\alpha - N_\beta) \subseteq \mathscr{S}$ *for any* $\alpha, \beta \subseteq \{1, 2, \ldots, m\}$.
2. $\mathscr{S}$ *is invariant under* A_α *for any* $\alpha \subseteq \{1, 2, \ldots, m\}$.
3. $\mathscr{S} = \langle A \mid \text{im } \begin{bmatrix} B & E \end{bmatrix} \rangle$.

Proof To prove the first statement, let Σ_γ denote the linear system $\Sigma(A_\gamma, E_\gamma, C, F)$ for $\gamma \subseteq \{1, 2, \ldots, m\}$. It follows from (14.4) that

$$\mathscr{T}^*(\Sigma_\gamma) \subseteq \langle A_\gamma \mid \text{im } E_\gamma \rangle \subseteq \mathscr{S}.$$

Then, we have

$$(A + N_\gamma C)\mathscr{T}^*(\Sigma_\gamma) \subseteq (A + N_\gamma C)\langle A_\gamma \mid \text{im } E_\gamma \rangle \subseteq \langle A_\gamma \mid \text{im } E_\gamma \rangle \subseteq \mathscr{S}.$$

Let $\tilde{\Sigma}$ denote the linear system $\Sigma(A, E, C, F)$. It follows from (14.5) that $\mathscr{T}^*(\Sigma_\gamma) = \mathscr{T}^*(\tilde{\Sigma})$ and hence that

$$(A + N_\gamma C)\mathscr{T}^*(\tilde{\Sigma}) \subseteq \mathscr{S}$$

for any $\gamma \subseteq \{1, 2, \ldots, m\}$. This yields

$$(N_\alpha - N_\beta)C\mathscr{T}^*(\tilde{\Sigma}) \subseteq \mathscr{S} \tag{14.16}$$

for any $\alpha, \beta \subseteq \{1, 2, \ldots, m\}$. Also we have

$$\text{im } (E + N_\gamma F) \subseteq \langle A_\gamma \mid \text{im } E_\gamma \rangle \subseteq \mathscr{S}$$

for any $\gamma \subseteq \{1, 2, \ldots, m\}$. Thus, we get

$$(N_\alpha - N_\beta)\text{im } F \subseteq \mathscr{S}$$

for any $\alpha, \beta \subseteq \{1, 2, \ldots, m\}$. By combining the last relation with (14.16), we obtain

$$(N_\alpha - N_\beta)\big(\text{im } F + C\mathscr{T}^*(\tilde{\Sigma})\big) \subseteq \mathscr{S}.$$

Since the transfer matrix $F + C(sI - A)^{-1}E$ is right-invertible as a rational matrix, it follows from (14.6) that $\text{im } F + C\mathscr{T}^*(\tilde{\Sigma}) = \mathbb{R}^m$. Therefore, we have

$$\text{im } (N_\alpha - N_\beta) \subseteq \mathscr{S}.$$

To prove the second statement, let $\alpha, \gamma \subseteq \{1, 2, \ldots, m\}$. Note that

$$\begin{aligned} A_\alpha \langle A_\gamma \mid \text{im } E_\gamma \rangle &\subseteq A_\gamma \langle A_\gamma \mid \text{im } E_\gamma \rangle + \text{im } (A_\alpha - A_\gamma) \\ &\subseteq \langle A_\gamma \mid \text{im } E_\gamma \rangle + \text{im } (N_\alpha - N_\gamma). \end{aligned}$$

It follows from the definition of $\mathscr{S}$ and the first statement that

$$A_\alpha \langle A_\gamma \mid \text{im } E_\gamma \rangle \subseteq \mathscr{S}.$$

Hence, we have

$$A_\alpha \mathscr{S} \subseteq A_\alpha \Big(\sum_{\gamma \subseteq \{1,2,\ldots,m\}} \langle A_\gamma \mid \text{im } E_\gamma \rangle \Big) \subseteq \sum_{\gamma \subseteq \{1,2,\ldots,m\}} A_\alpha \langle A_\gamma \mid \text{im } E_\gamma \rangle \subseteq \mathscr{S}.$$

To prove the third statement, note first that im $N_\gamma \subseteq$ im B for any $\gamma \subseteq \{1, 2, \ldots, m\}$. Hence, we have

$$\text{im } E_\gamma = \text{im } (E + N_\gamma C) \subseteq \text{im } \begin{bmatrix} B & E \end{bmatrix}.$$

This results in

$$\langle A_\gamma \mid \text{im } E_\gamma \rangle \subseteq \langle A_\gamma \mid \text{im } \begin{bmatrix} B & E \end{bmatrix} \rangle \tag{14.17}$$

for any $\gamma \subseteq \{1, 2, \ldots, m\}$. Since $A_\gamma = A + N_\gamma C$ and im $N_\gamma \subseteq$ im B, it follows from (14.2) that

$$\langle A_\gamma \mid \text{im } \begin{bmatrix} B & E \end{bmatrix} \rangle = \langle A \mid \text{im } \begin{bmatrix} B & E \end{bmatrix} \rangle.$$

In view of (14.17), this means that

$$\langle A_\gamma \mid \text{im } E_\gamma \rangle \subseteq \langle A \mid \text{im } \begin{bmatrix} B & E \end{bmatrix} \rangle$$

for any $\gamma \subseteq \{1, 2, \ldots, m\}$. Consequently, we obtain

$$\mathscr{S} = \sum_{\gamma \subseteq \{1,2,\ldots,m\}} \langle A_\gamma \mid \text{im } E_\gamma \rangle \subseteq \langle A \mid \text{im } \begin{bmatrix} B & E \end{bmatrix} \rangle. \tag{14.18}$$

Since $\langle A_\gamma \mid \text{im } E_\gamma \rangle \subseteq \mathscr{S}$ for all $\gamma \subseteq \{1, 2, \ldots, m\}$, we have

$$\begin{gathered} \langle A \mid \text{im } E \rangle \subseteq \mathscr{S} \\ \langle A - BD^{-1}C \mid \text{im } (E - BD^{-1}F) \rangle \subseteq \mathscr{S} \end{gathered}$$

for the particular choices $\gamma = \emptyset$ and $\gamma = \{1, 2, \ldots, m\}$, respectively. We know from (14.5) that the strongly reachable subspaces of the systems $\Sigma(A, E, C, F)$ and $\Sigma(A - BD^{-1}C, E - BD^{-1}F, C, F)$ coincide. Let $\mathscr{T}^*$ denote this common strongly reachable subspace. It follows from (14.4) that

$$\mathscr{T}^* \subseteq \langle A \mid \text{im } E\rangle \subseteq \mathscr{S}$$
$$\mathscr{T}^* \subseteq \langle A - BD^{-1}C \mid \text{im } (E - BD^{-1}F)\rangle \subseteq \mathscr{S}.$$

These inclusions yield

$$\begin{aligned} A\mathscr{T}^* &\subseteq A\langle A \mid \text{im } E\rangle \\ &\subseteq \langle A \mid \text{im } E\rangle \subseteq \mathscr{S} \\ (A - BD^{-1}C)\mathscr{T}^* &\subseteq (A - BD^{-1}C)\langle A - BD^{-1}C \mid \text{im } (E - BD^{-1}F)\rangle \\ &\subseteq \langle A - BD^{-1}C \mid \text{im } (E - BD^{-1}F)\rangle \subseteq \mathscr{S}, \end{aligned}$$

from which we obtain

$$BD^{-1}C\mathscr{T}^* \subseteq \mathscr{S}. \tag{14.19}$$

On the other hand, we readily have

$$\text{im } E \subseteq \langle A \mid \text{im } E\rangle \subseteq \mathscr{S}$$
$$\text{im } (E - BD^{-1}F) \subseteq \langle A - BD^{-1}C \mid \text{im } (E - BD^{-1}F)\rangle \subseteq \mathscr{S}.$$

Combining these two inclusions results in

$$BD^{-1}\text{im } F \subseteq \mathscr{S}.$$

Together with (14.19), this implies that

$$BD^{-1}(\text{im } F + C\mathscr{T}^*) \subseteq \mathscr{S}.$$

It follows from the blanket assumption and (14.6) that

$$\text{im } F + C\mathscr{T}^* = \mathbb{R}^m.$$

Thus, we get

$$\text{im } B \subseteq \mathscr{S}.$$

From the second statement of Lemma 14.2, we know that the subspace $\mathscr{S}$ is A_α-invariant for any $\alpha \subseteq \{1, 2, \ldots, m\}$. In particular, the choice of $\alpha = \emptyset$ implies that $\mathscr{S}$ is A-invariant. Since $\langle A \mid \text{im } B\rangle$ is the smallest A-invariant subspace that contains im B, we have

$$\langle A \mid \text{im } B\rangle \subseteq \mathscr{S}. \tag{14.20}$$

As we readily have $\langle A \mid \operatorname{im} E\rangle \subseteq \mathscr{S}$, the inclusion (14.20) implies that

$$\langle A \mid \operatorname{im} B\rangle + \langle A \mid \operatorname{im} E\rangle = \langle A \mid \operatorname{im} \begin{bmatrix} B & E \end{bmatrix}\rangle \subseteq \mathscr{S}.$$

Together with (14.18), this proves that $\mathscr{S} = \langle A \mid \operatorname{im} \begin{bmatrix} B & E \end{bmatrix}\rangle$. ■

Now we are ready to present necessary and sufficient conditions for an LCS to be disturbance decoupled.

Theorem 14.3 *An LCS of the form* (14.9) *is disturbance decoupled if and only if*

$$\langle A \mid \operatorname{im} \begin{bmatrix} B & E \end{bmatrix}\rangle \subseteq \ker J.$$

Proof To prove the 'only if' part, let $\gamma \subseteq \{1, 2, \ldots, m\}$. Note that

$$\begin{bmatrix} G_\gamma & H_\gamma \end{bmatrix} = \begin{bmatrix} -(D_{\gamma\gamma})^{-1} & 0 \\ -D_{\gamma^c\gamma}(D_{\gamma\gamma})^{-1} & I \end{bmatrix} \begin{bmatrix} C_{\gamma\bullet} & F_{\gamma\bullet} \\ C_{\gamma^c\bullet} & F_{\gamma^c\bullet} \end{bmatrix}. \tag{14.21}$$

Since $F + C(sI - A)^{-1}E$ is right-invertible as a rational matrix by the blanket assumption, $\begin{bmatrix} C & F \end{bmatrix}$ is of full row rank. So must be the matrix $\begin{bmatrix} G_\gamma & H_\gamma \end{bmatrix}$ due to (14.21). Then, one can find x_0 and d such that

$$\begin{bmatrix} G_\gamma & H_\gamma \end{bmatrix} \begin{bmatrix} x_0 \\ d \end{bmatrix} > 0.$$

Let $e \in \mathbb{R}^q$. Clearly, there exists a sufficiently small $\mu > 0$ such that

$$\begin{bmatrix} G_\gamma & H_\gamma \end{bmatrix} \begin{bmatrix} x_0 \\ d + \mu e \end{bmatrix} > 0.$$

Now define

$$d_1(t) = d \quad \text{and} \quad d_2(t) = d + \mu e$$

for all $t \geq 0$. Let $x_i(t)$ denote the trajectory $x^{x_0, d_i}(t)$ for $i = 1, 2$. Since x_i and d_i are continuous, there exists an $\varepsilon > 0$ such that

$$\begin{bmatrix} G_\gamma & H_\gamma \end{bmatrix} \begin{bmatrix} x_i(t) \\ d_i(t) \end{bmatrix} > 0$$

holds for all $t \in [0, \varepsilon)$. Thus, the trajectories x_1 and x_2 satisfy

$$\dot{x}_i(t) = A_\gamma x_i(t) + E_\gamma d_i(t)$$

for all $t \in [0, \varepsilon)$ and $i = 1, 2$. As the system is disturbance decoupled, we have that

$$Jx_1(t) = Jx_2(t)$$

for all $t \in [0, \varepsilon)$. Since d_1 and d_2 are constant, we obtain

$$J(A_\gamma x_0 + E_\gamma d) = J\big(A_\gamma x_0 + E_\gamma (d + \mu e)\big)$$

by differentiating and evaluating at $t = 0$. This results in

$$J E_\gamma e = 0.$$

By repeating the differentiation and evaluation at $t = 0$, we get

$$J A_\gamma^k E_\gamma e = 0$$

for all $k \geqslant 0$. Since e is arbitrary, we have

$$J A_\gamma^k E_\gamma = 0$$

for all $k \geqslant 0$. Consequently, one gets

$$\langle A_\gamma \mid \text{im } E_\gamma \rangle \subseteq \ker J.$$

Thus, we have

$$\sum_{\gamma \subseteq \{1,2,\ldots,m\}} \langle A_\gamma \mid \text{im } E_\gamma \rangle \subseteq \ker J.$$

It follows from the third statement of Lemma 14.2 that

$$\langle A \mid \text{im } \begin{bmatrix} B & E \end{bmatrix} \rangle \subseteq \ker J.$$

To prove the 'if' part, it is enough to show that

$$x^{x_0,d_1}(t) - x^{x_0,d_2}(t) \in \langle A \mid \text{im } \begin{bmatrix} B & E \end{bmatrix} \rangle$$

for any initial state $x_0 \in \mathbb{R}^n$, locally-integrable disturbances d_1 and d_2, and $t \geqslant 0$. To do so, let $\mathscr{V} := \langle A \mid \text{im } \begin{bmatrix} B & E \end{bmatrix} \rangle$ and let $v \in \mathscr{V}^\perp$. From (14.9a), we have

$$v^T \big(\dot{x}^{x_0,d_1}(t) - \dot{x}^{x_0,d_2}(t)\big) = v^T A \big(x^{x_0,d_1}(t) - x^{x_0,d_2}(t)\big) \tag{14.22}$$

for almost all $t \geq 0$. Define

$$\zeta(t) := v^T \big(x^{x_0,d_1}(t) - x^{x_0,d_2}(t)\big).$$

From (14.22) and A^T-invariance of $\mathscr{V}^\perp$, we get

$$\frac{d^k \zeta}{dt^k}(t) = v^T A^k \big(x^{x_0,d_1}(t) - x^{x_0,d_2}(t)\big)$$

for $k \geq 0$. The Cayley-Hamilton theorem implies that there exist real numbers c_i with $i = 0, 1, \ldots, n-1$ such that

$$\frac{d^n \zeta}{dt^n}(t) + c_{n-1}\frac{d^{n-1}\zeta}{dt^{n-1}}(t) + \cdots + c_1\frac{d\zeta}{dt}(t) + c_0\zeta(t) = 0.$$

Since

$$\frac{d^k \zeta}{dt^k}(0) = 0$$

for $k \geqslant 0$, we get $\zeta(t) = 0$ for all $t \geqslant 0$. Consequently, we have

$$x^{x_0,d_1}(t) - x^{x_0,d_2}(t) \in (\mathscr{V}^{\perp})^{\perp} = \mathscr{V} = \langle A \mid \operatorname{im} \begin{bmatrix} B & E \end{bmatrix} \rangle$$

which completes the proof. ■

14.5 Conclusions

This chapter studied a class of non-smooth and nonlinear dynamical systems, namely linear complementarity systems. These systems belong to the larger family of piecewise affine dynamical systems for which the disturbance decoupling problem has already been solved. In this chapter we have shown that linear subsystems of a linear complementarity system share certain geometric structure. By exploiting this geometric structure, we provided necessary and sufficient conditions for a linear complementarity system to be disturbance decoupled. Compared to already existing conditions for general piecewise affine systems, the new conditions are much crisper and more insightful.

Future research possibilities are weakening the technical blanket assumptions and studying the disturbance decoupling problem under different feedback schemes.

Acknowledgments This research was funded by NWO (The Netherlands Organisation for Scientific Research) project CODAVI (613.001.108).

References

1. Basile, G., Marro, G.: Controlled and conditioned invariant subspaces in linear system theory. J. Optim. Theory Appl. **3**, 306–315 (1969)
2. Basile, G., Marro, G.: On the observability of linear, time-invariant systems with unknown inputs. J. Optim. Th. & Appl. **3**, 410–415 (1969)
3. Wonham, W.M., Morse, A.S.: Decoupling and pole assignment in linear multivariable systems: a geometric approach. SIAM J. Control Optim. **8**, 1–18 (1970)
4. Wonham, W.M.: Linear Multivariable Control: a geometric approach. Applications of Mathematics, Springer, New york (1985)

5. Nijmeijer, H., van der Schaft, A.J.: Nonlinear Dynamical Control Systems. Springer, Berlin (1990)
6. Basile, G., Marro, G.: Controlled and Conditioned Invariants in Linear System Theory. Prentice Hall, Englewood Cliffs (1992)
7. Isidori, A.: Nonlinear Control Systems. Communications and Control Engineering, Springer, Berlin (1995)
8. Trentelman, H.L., Stoorvogel, A.A., Hautus, M.L.J.: Control Theory for Linear Systems. Springer, London (2001)
9. Otsuka, N.: Disturbance decoupling with quadratic stability for switched linear systems. Syst. Control Lett. **59**(6), 349–352 (2010)
10. Yurtseven, E., Heemels, W.P.M.H., Camlibel, M.K.: Disturbance decoupling of switched linear systems. Syst. Control Lett. **61**(1), 69–78 (2012)
11. Everts, A.R.F., Camlibel, M.K.: The disturbance decoupling problem for continuous piecewise affine systems. In: Proceeding of the 53rd IEEE Conference on Decision and Control, Los Angeles (USA) (2014)
12. Cottle, R.W., Pang, J.-S., Stone, R.E.: The Linear Complementarity Problem. Academic Press, Boston (1992)
13. Camlibel, M.K., Iannelli, L., Vasca, F.: Modelling switching power converters as complementarity systems. In: Proceeding of the 43th IEEE Conference on Decision and Control, Paradise Islands (Bahamas) (2004)
14. Schumacher, J.M.: Complementarity systems in optimization. Math. Program. Ser. B **101**, 263–295 (2004)
15. van der Schaft, A.J., Schumacher, J.: An Introduction to Hybrid Dynamical Systems. Springer, London (2000)
16. Heemels, W.P.M.H., Brogliato, B.: The complementarity class of hybrid dynamical systems. Eur. J. Control **26**(4), 651–677 (2003)
17. Camlibel, M.K., Heemels, W.P.M.H., van der Schaft, A.J., Schumacher, J.M.: Switched networks and complementarity. IEEE Transactions on Circuits and Systems I **50**(8), 1036–1046 (2003)
18. Heemels, W.P.M.H., Camlibel, M.K., Schumacher, J.M.: On the dynamic analysis of piecewise-linear networks. IEEE Trans. Circ. Syst. I Fundam. Theory Appl. **49**(3), 315–327 (2002)
19. Camlibel, M.K., Heemels, W.P.M.H., Schumacher, J.M.: On linear passive complementarity systems. Eur. J. Control, **8**(3), 220–237 (2002)
20. van der Schaft, A.J., Schumacher, J.M.: The complementary-slackness class of hybrid systems. Math. Control Signals Syst. **9**, 266–301 (1996)
21. Camlibel, M.K.: Popov-Belevitch-Hautus type controllability tests for linear complementarity systems. Syst. Control Lett. **56**, 381–387 (2007)
22. van der Schaft, A.J., Schumacher, J.M.: Complementarity modelling of hybrid systems. IEEE Trans. Autom. Control **43**(4), 483–490 (1998)
23. Heemels, W.P.M.H., Schumacher, J.M., Weiland, S.: Linear complementarity systems. SIAM J. Appl. Math. **60**(4), 1234–1269 (2000)

MIX
Papier aus verantwortungsvollen Quellen
Paper from responsible sources
FSC® C105338

If you have any concerns about our products,
you can contact us on
ProductSafety@springernature.com

In case Publisher is established outside the EU,
the EU authorized representative is:
Springer Nature Customer Service Center GmbH
Europaplatz 3, 69115 Heidelberg, Germany

Printed by Libri Plureos GmbH
in Hamburg, Germany